Anunnaki Earth Agenda

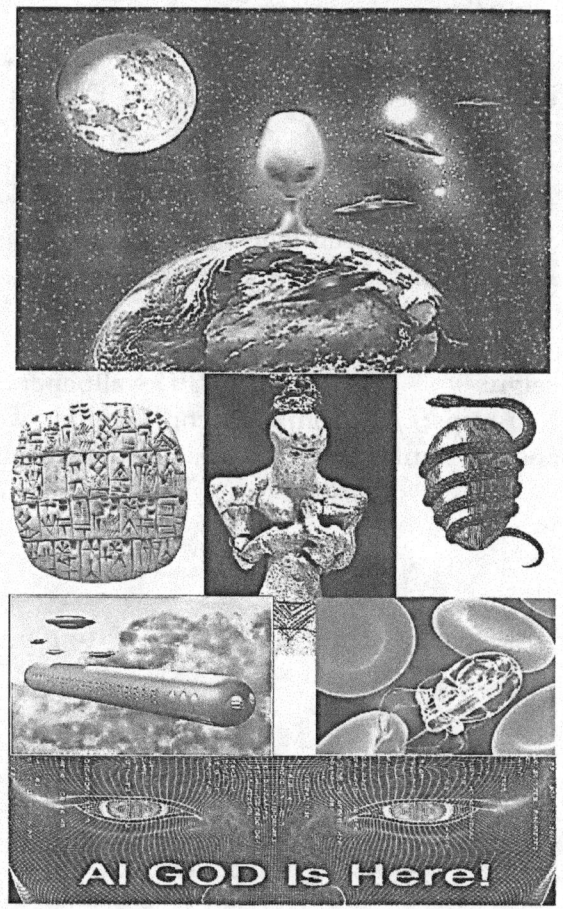

5G, AI, H+, Epigenetics, Breakaway Civilization & Space
Force... The Real History of the Skygods Intervention.

It is all about Control and 5G will cement that Option.

Copyright © 2019 by T. J.H.   v. 14.7

Categories: Metatags: Anunnaki, Nephilim, Djinn, interdimensionals,
ET, UFO, Mankind, origins, Genetic engineering, DNA, Egyptian,
extraterrestrial influences, soulless, OPs, NPCs, creation, evolution,
Darwin, reptile, Sumeria, John Keel, Chas. Fort,, Robert Monroe, OBE,
Control System, 5G, AI, Tanshumanism, Beings of Light, souls, Scripts,
karma, reincarnation, recycling, Earth Graduate, Virtual Earth Sphere.,
Pleiadians, Draco.

Cover design: **Planet Earth UFO wallpaper... top picture**.
https://www.moddb.com/groups/warhammer-dark-
forcescience-fictionfantasyclan/images/planet-earth-ufo-wallpaper
cuneiform : omniglot.com   Ubaid Sculpture : Pinterest
Egg serpent : Wikipedia   UFO mothership : alienxfiles.com
Nanite in blood : Money Inc   AI God : WordPress.com

Author may be reached at TJ_cspub14@yahoo.com

ISBN – 13: 978-1702801577

Other Books by the Author

VEG	Virtual Earth Graduate
TOM	The Transformation of Man
TEW	The Earth Warrior (docu-novel)
TSiM	The Science in Metaphysics
QES	Quantum Earth Simulation
AL	Anunnaki Legacy
GEP	Great Earth Puzzle
TSF	The Sacred Feminine
VDoG	Virtual Disneyland of the Gods
DNQF	Dynamics of the New Quantum Faith

Author Bio

TJ Hegland is of Norwegian-Celtic descent with a degree in English (with Journalism emphasis) from Syracuse University and speaks 5 languages. His original goal was to be a European Foreign Correspondent in Paris where his languages would benefit everyone, but late-60's found him headed instead to VietNam. Detour! So he went back to Syracuse, and spent 34 years in computer work: from a humble beginning as a Computer Operator, to Programmer in 7 programming languages, 14 years in IBM mainframes, then Superminis to PC networks; as Lead Programmer Analyst in Manufacturing with Bill of Material Processor; in Hospital applications including Patent Diagnosis and Pharmacy, and Patient Billing. He was an MIS Director in two Southern California companies. He also worked several years in Education; state licensed for ESL, Spanish & French, and lastly teaching a Computer Lab.

His ancestry goes back to King Oscar I of Norway thru his father's side.

In 1991 he did a **Regression** that showed him his past 3 lifetimes (where he spoke two of the languages he learned so easily this lifetime), and the Master counseled him that he would be **writing books** guided by a member of today's Great White Brotherhood (aka ancient Order of Melchizedek). This was an agreement he had before this incarnation. Nothing happened for 7 more years, and then he was visited at 2 a.m. in the morning of October 25th while on a 1998 trip, in a dark motel room on the TX-NM border: 3 Beings of Light **merged** him with a 5th level aspect from his Soul Group. Then 10 years later on January 8 at 3 a.m., Baldy (Phillaleal) shows up and starts dictating VEG.

He has never been abducted (to his knowledge) but only once saw the UFO that dropped off his mother who disappeared for 2 hours. His mother and sister were abducted many times, resulting in their odd female health issues. He has talked with several people who have had CE-5 encounters, as well as several NDE and OOBE experiencers, and other authors, as well as Baldy who is clearly not of this Earth, but is 6'8" (bald), well-built and very human-looking.

On August 24th, 2016, the 5th level merge went home and direct guidance, protection and perfect health went with him. The imparted Knowledge stayed. The books are the result.

Also see Insert: **Personal Insight** in Chapter 8

Acknowledgements

Many thanks are due to **Robin Landry** for her inestimable assistance in providing the latest and deeper information on the real meaning of the Sumerian-Anunnaki history. As well as her comments and feedback on early Earth history are greatly appreciated.

Equal thanks go to **Randy Walsh** who has provided enough feedback and critique to make well-though-out corrections to the Anunnaki scenario, as well as to insights regarding the Moon, Mars and Flat Earth scenarios.

Lastly, great thanks are due **Linda French** for her patience in also providing feedback and editorial critique of the Anunnaki issue.. often putting up with my sharing and not locking me up in an upstairs closet when I would explain what Baldy had last given me.

Posthumous thanks are due to the late **Jim Marrs** who not only provided much info thru his great and interesting books, but several times in his emails steered me in a new direction and then endorsed my VEG book.

Introduction

The world and America are experiencing a lot of chaos and disinformation today – and it is by design. A **Control Agenda** is at work. Has been for thousands of years.

No, not by a God of the Universe…or even a Devil. But by those negative entities who hijacked the initial, ancient **Earth Conservatory** and later re-created Man and hijacked the **Earth Realm School**. Just how this has happened and <u>continues to happen</u> is the subject of this book. The form it takes via Media, prescription drugs, genetic manipulation, Telecom networks (5G – 6G), religion and education is something we have to look at.

> The resolution and way to get out is also given.

*   *   *

We are in a phase where the vibrations have been naturally going up on the planet, by design to entrain souls to a higher level, and it is driving the PTB/Remnant nuts… hence the chaos and stress that we see nowadays on the TV news. There are many who can't handle it.

The planetary **vibrations have been going up** for some years, but that has not resulted in as many souls synching up with it as was hoped, thus the proactive Greys have scaled back their work, the vibrations have levelled off, and Man's real fate is now <u>in his hands</u> over the next 5 years. The **Elohim** (aka Planners) want Man to move forward but due to the huge amount of disinformation (Fake News and a controlled Media) Man is not aware of what is really going on and so is not proactively seeking to protect himself nor the planet… he is about to lose both.

> This book, instead of inspiring Man to take the planet into his own hands, may instead be more of an historical account of how Earth was lost. (I hope not.)

We really appear now to be in an **Endgame** and the outcome depends on what Man and the Insiders do in the next 5 years.

There is a lot happening (Chapters 8 & 9) and those who diss or ignore this information, and choose to continue to live in their Prophylactic Fantasy have the right to do so (Freewill – Chapter 4) but they may soon regret it. It is no surprise that the PTB and their controller Remnant and their public lackeys hate it – they have ramped up their effort to control Man with 5G – examined in this book.

*　　　　*　　　　*

This book recaps the Anunnaki and Skygods chronicle with much additional NEW HISTORICAL information about the Skygods -- and what they did to the Pleiadians who came to help (Chapter 1 & 2). And then the latest amazing info on Genetics (Chapter 3), Transhumanism [H+] and AI (Chapter 6) is given to help understand where we are at today... all impacted by the **5G issue** (Chapters 5 and 7). Your health is supported by Chapter 7 and a way to get out of this mess is given in Chapter 8.

Appendix G summarizes EMF dangers and guidelines plus personal protection products.

Chapter 5, page 220, offers responsible 5G installation guidelines.

Whereas the **5G scenario** is of questionable safety, the issue is examined in some detail as proven by **many doctors and engineers** seeking to stop it from being rashly implemented. In Chapters 5 and 7 you'll see why, and a schema for a **responsible 5G implementation** is given. In addition, you are presented with a choice of 2 Options (Chapter 5) as to what this 5G Super Grid is really all about, and then in Chapter 8 you are given options for personally handling what is happening, and then in Chapter 9 the stark reality is laid out for you.

Also well worth your time, whereas many of us do not read **Appendices** – there are several that will benefit you and raise a few eyebrows:

> namely **Appendices E, F, & G.** You really want to read
> these 3 ... they contain extra fascinating info and pix.

If you have read some of my previous books, Appendices A, B & D will be a repeat, but valuable to understand Appendices C & F.

Appendix G examines EMF aspects of many common devices and how to protect your home, office, car and body. It also examines the **WiGig "Death Grid."**
Just when you thought you could live with WiFi, they ramp up a more powerful one.

*　　　　*　　　　*

Footnotes – very important: if you see a **footnote** on a topic that peaks your curiosity, go back to the Endnotes and read it – I often, as I was editing the final Printfile, would remember something of note and did not want to add 2 paragraphs in the middle of the book, screwing up the pageovers, white space and pix ... thus having to completely redo the 99% finished Printfile. So I footnoted the info and gave you more insight in the Endnotes.

Epiphanies – of note: There are two sections, one in Chapter 3 and also one in Chapter 4, labeled **"Epiphany"** – please read them. They foreshadow and help explain Chapter 8.

Index – as usual, I do not include an index as (1) I do not have the <u>expensive</u> software for it, and (2) the breakdowns in the Table of Contents serve the same purpose. Chapters are by major topic and anything Genetic, for example, would be in Chapter 3 on Genetics.

Summary

The United States at this time is precariously teetering on the brink of collapse – not financially but **socially** and to stand by and just watch it, assuming someone else will fix it, amounts to condoning what is happening. The goal is to destroy the US with chaos (Antifa), Socialism (Green New Deal), constant argument, destruction of traditional values, Rx side effects, 5G health issues, and deception... if they can bring the US down, the rest of the world falls easily.

And yet this book does not get into political issues.

The problems outlined in Chapters 5 & 7 are not a case of Man being ignorant and making mistakes – this is all **orchestrated** against your Freewill (Chapter 4).

Not melodrama. Not speculation. Not froo-froo – it is real and if you do not **wake up** and <u>at least</u> take a mental position against it, you will be part of the problem. <u>And manipulated</u> because you won't see it coming (Chapter 4).

> Your awareness of the problem and mental position <u>does</u> help
> to obstruct it on an astral/mental/spiritual level...
> (Think: 100th Monkey Syndrome and the novel Quantum Physics
> Observer Effect).

Along the way, we'll also take a relevant look at:

The Real Earth History – many old gaps filled in (Chapter 1 & 2).

The Issue of Freewill – what it is, how it works and most of all: how it is being subtly violated (Chapter 4).

Health Issues & Handling It – what you can do to compensate, support and reinforce your health (Chapter 7).

The Breakaway Civilization and a **Space Force** as necessary elements in Man's survival and future success. (Chapter 8 and Appendix F).

What a Breakaway Civilization is and how it relates to Man getting off the planet before it is locked down... and how it could have already evolved according to a credible report.

*   *   *

NOTE:

The titles of my previous books (referenced in many chapters) are abbreviated according to the following schema:

Other Books by the Author

VEG	Virtual Earth Graduate
TOM	The Transformation of Man
TEW	The Earth Warrior (novel)
TSiM	The Science in Metaphysics
QES	Quantum Earth Simulation
AL	Anunnaki Legacy
GEP	Great Earth Puzzle
TSF	The Sacred Feminine
VDoG	Virtual Disneyland of the Gods
DNQF	Dynamics of the New Quantum Faith

I designed this book to be such that if you have VEG and TOM, TSF and this one, then you have it all.

Those who want an upgrade to traditional Religion, need to check out DNQF.

TEW is a novel based on real events, and contains some 'Easter eggs.'

Table of Contents:

Chapter 4: Freewill & LifePlan 129
... Do we have Freewill or are our lives preplanned?...

Chapter 8: Escape
…can the Earth scene really be avoided?.....

The Earth Scene

Chapter 1: Earth Realm Basics

Before getting into the meat of the book, there are some preliminaries that need to be addressed… specifically: whereas <u>Virtual Earth Graduate</u> (VEG) was an accurate view from the 40,000' level, just recently this author was shown the same information from the 10,000' level —**more depth** and that is in preparation for the further exposé of the **Anunnaki Control Agenda**.

While this book does not correct or amend anything in the VEG book, it will expand on a very significant aspect from VEG that was just lightly touched on in its Chapter 14, where violation of Freewill, and false beliefs can derail a person. Not a repeat but a clarification with new information in how your Freewill is being subtly violated <u>today</u> and is about to be further threatened once the **5G Network** is up and running.

Why?

First, because not all the Anunnaki went home and a small percentage stayed underground as the **Anunnaki Remnant** manipulating the human PTB – who constantly manipulate us. It is what they have always done, but now we are about to give them a powerful network tool to control human thinking and actions. As well as make you sick.

Secondly, it is important to know who is here with us, what they are doing and what you can do about it. Refusal to protect oneself, or ignoring this information, is tantamount to condoning what they are doing – and that <u>implicit agreement</u> is exactly how they often violate your Freewill. You agree to what they cleverly suggest and then the Beings of Light (and the Solar Council) can do nothing about it – <u>because</u> you agreed to it (even though you didn't understand what was really happening!).

> Reptoids never openly attack and come in guns blazing – they are much too clever for that. They prefer subterfuge: that way the Solar Council cannot say they openly violated your Freewill and then make them stop.
>
> This is a Freewill Universe and Earth Realm and the Reptoids have as much right to do whatever they do as you do – <u>your job</u> is to stay aware, learn and avoid their

deception/problems... and the Reptoids often make sure via disinformation (Think: Media) that you do not become aware until it is too late... if at all.

Galactic Law by the way says that they must give a warning somewhere, sometime, as to what they are about to do... and such was done with the **Georgia Guidestones** in 1980 -- but did anyone pay attention?

Georgia Guidestones and 10 Language Proclamations

Around the edges of the top square are written translations to four ancient languages, one per edge. Starting from the top capstone and proceeding clockwise are: Babylonian cuneiform, Classical Greek, Sanskrit, and Ancient Egyptian in hieroglyphs.

In addition, the four (2-sided) granite panels say same the same thing in 8 major world languages:
English, Spanish, Hindi, Hebrew, Swahili, Russian, Arabic, and Chinese.

The message is a 10-point guideline for a more controlled world, including **point #1**: Maintain humanity under 500,000,000 (culling).... **Point #2:** Guide reproduction wisely (genetic engineering)... and **point #3:** Unite humanity with a new living language (via the Internet).... (Wikipedia article shows all 10 points.)

More and deeper insight on this in Appendix F.

PTB Control

So there needs to be some form of media that can provide alternate input for the discerning humans, and hopefully they will read it, wonder, research, think about it, and save themselves. As you will see in this book...

Knowledge Protects
Ignorance Endangers

Remember that **the PTB love being Lords over Serfs** (or Slaves). They want as many ignorant humans as possible to do what they are told... and the Anunnaki Lords have run their control over humans in one form or another (using proxies and puppets) since humans were created.... Originally via disease (***Suruppu Disease*** in the Sumerian *Atra Hasis*), then, **Religion** in many forms (different versions to keep humans from uniting), and now via Fake News, via controlled **Mass Media**, (spinning the stories as they want you to believe them) and now on the horizon is a **5G network** with your powerful cellphone (don't you have the latest iPhone 11?) able to listen to what you say, and emit 10Hz EMF to dumb you down.

This is not a joke – see Chapter 5.

This is not crazy, but it will sound that way to those who are brainwashed by the Media. In presenting the potential dangers (if not mitigated upon implementation), with as much detail and science as possible is given to help the reader get a sense that something shadowy (typically Reptoid) and harmful is going on... even though Telecom, Google, Facebook, Apple and AI **are not evil themselves**... but they are being manipulated and influenced in a typically Reptoid (PTB) Agenda – as later explained in Chapters 4 and 5.

Be advised that they are being superficially manipulated by profit to do what they are doing, and some actually believe that 5G will be a blessing – bigger, faster -- and it is reminiscent of **OnStar** ®: Don't you want help when your vehicle breaks down in the middle of nowhere? (That is how it was sold.) What it also does and this was not advertised: If the PTB, Police, FBI, CIA, etc.. does not want you to drive away out of town, they can turn your engine off via OnStar.

Same thing with the new electric **SmartMeters** monitoring electrical usage in your home: Saves money – they do not have to pay a meter reader to walk house to house any longer... the meter talks to a remote monitoring and recording station. (That is how that was sold.) What can also happen: they can turn your electricity off for any of several reasons – even that you may be using too much power, more than your share (alleging BrownOut issues... you have no say.)

SmartTVs now can change channels by voice command. (That is how that is sold.) What they don't tell you is that the **TV is always listening,** like

> **Alexa ®:** Just say something about taking matters into your own hands, or you wish someone (on TV) was dead... <u>it is recorded</u> and you may be watched and visited... (for terrorist prevention purposes, you know).

Reptoid manipulation is <u>never overtly obvious</u> which is why some people will have a problem believing what is really going on, and that is the secret of their success in past millennia – until the Higher Beings or Lyrans made them stop (Chapter 2) – **Hitler and his Nazi movement** were of their doing, as was the dark **STS Soviet Union**, Ghengis Khan, Chairman Mao and Communism...

But before we get to the issues in detail (so that you really see the problem), we need to do a recap of the Earth Realm, with some detail so that the following chapters and the **Control Agenda** make more sense.

> At this point it needs to be pointed out that the following two accounts of (1) **Pre-Anunnaki** colonization and (2) Post-Anunnaki **Pleiadian insertion** into the Earth Realm are found in very credible sources. One is a book that is no longer in print (<u>Eden</u>) but whose 30+ years' research was done by serious scholars in the Mesopotamian area – prior to Zechariah Sitchin (who 'adapted' some of the information for his books).
> The second source is a book which carries a big pricetag, (currently $156) which arises from Aboriginal legends and amounts to nothing less than Pleiadian input, transmitted to its Aboriginal author (<u>Alcheringa</u>). The second account is nothing less than the **Pleiadian Massacre** on Earth at the hands of the Reptilians.
>
> Both these sources are plausible and submitted for your consideration as they help to flesh-in the overall ancient history of Earth and Man, and have a certain coherence and historicity to them.
> (Both books are referenced in this Bibliography, and may be found on Amazon.)

Earth Realm

Earth is/was a special creation, a **3D Construct in 4D**… operating with 3D Physical Laws empowered by the 4D Realm in which the planet sits. Its construction was similar to a Virtual Reality, but on a more physical level and it is watched over by several levels of beings – those at the 4D level who set up the Experiment, the Zoo, the Prison – whatever you want to call it – They have been called the Higher Beings, and the Sumerians called them the Elohim or the **Kadishtu.** There is a **Grid** around the Earth effectively keeping it in Quarantine from the Dracos, and most of the Reptoids from Orion. Also involved are the Solar Council, and the **Planners** who have a stake in seeing that incarnating **souls** (here for unique experiences) are not destroyed and that they get their "lessons" and progress to "graduation." (Chapter 8 and see also VEG, Ch.13.)

Besides having been originally set up as a Garden or Zoo with many flora and fauna from different worlds, the original Museum/Conservatory purpose was co-opted by the Reptilians and their **re-creation** of sentient beings in the Garden meant that Earth would now have to be used as a School.

> The co-opting was done principally by the **Draco** with their creation the Reptilians (who created the Dinoids), but before getting to the Reptoid/Anunnaki incursion, and the later Pleiadian massacre, there was another, prior group that first came and attempted to just settle peacefully on the planet.

> This group is often spoken of as the **Anunna gods**, and are the precursors to the group that later included Enlil and Enki – still Anunnaki but not (yet) the group that re-created Man..

Re-Creation

> NOTE: **the Anunnaki did not create humans** – they modified the existing hominid which was already there, and the Higher Beings were very proud of him, his having been created much as the way **Genesis 1** says – manifested by Higher Beings/God – which is something that Reptilians even with their superior genetic science cannot do. Creating life is the purview of the **Planners aka Elohim..**

> So the original Man creature was co-opted by the Reptilians to serve them, with a few modifications, cutting down the intelligence and connection with a Higher Source, the original Man had quite a divine potential, but much of the genetic coding was set to non-functioning (called "Junk DNA" today) and Man became what Sitchin called a *Lulu* or *Adama* -- an animal, or a beast.

This modification was alluded to in **Genesis 2,** the "2nd creation." Man now became a slave. And even the Maya knew that and drew a pictograph of the **serpent link** (below) enslaving the humans...
1

10- Humanity in chains, literally being strangled by reptiles. Plate 34 of the *Laud Codex*, Mixtec culture.

The Mayans also saw these "gods" come and go in their Skycraft, called by several names, depending on size: Margid'da and Mu'u...

As well as a sculpture from Turkey...

Mu'u

And this one from the Maya...

Obviously a reptile piloting a flying craft...

note the intake port in the front, and the horizontal exhaust "tube" in the rear... plus the "tail" stabilizer. How would ignorant jungle natives know to design those elements??

And this is another Mayan or Aztec design of something flying: [2]

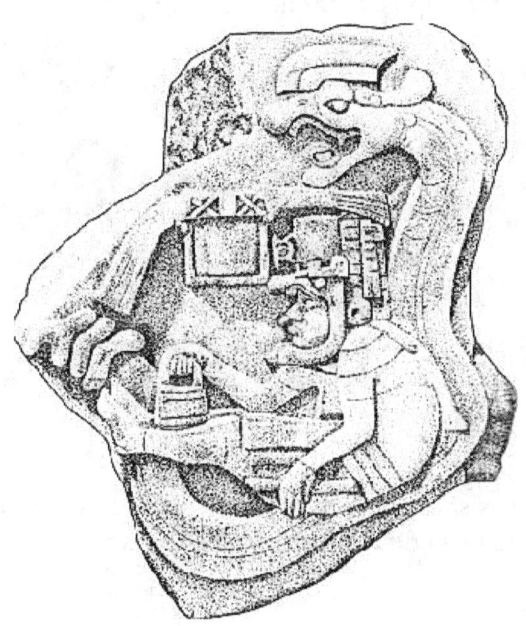

Again we have the flying motif with the emphasis on a Serpent... and the Anunnaki Reptiles did have flying craft...

Note the helmet.

And again, below...

7- Relief from a cave at the archaeological site of Chalcatzingo, in the state of Morelos, south of Mexico City. We see a "god" or high Aztec dignitary sitting inside an ovoid vessel that spews flames. An important detail are the double Gs in the pilot's hands and under his seat. This typical Amerindian symbol represents our galaxy, the Milky Way. It has been used here to express the fact that the vessel can travel from one end of the universe to the other. Also note the concentric frontal designs that recall the magneto hydrodynamic (MHD) technique of air intake which is ejected in the back of a flying aircraft to provide supersonic propulsion. Archaeologists interpret this scene simply as a rain cult with the open jaws of a jaguar symbolizing the Earth.

Again, note the many "flames" as exhaust, the stars, and the "intake port" (?) in the left front...

The text says that the stylized "Gs" in his hands and under his seat represent the Galaxy and that this craft can fly across the Universe.

And the classic images from the Egyptian temple at **Abydos**... (note the cigar-shaped Mother ship [submarine?] upper right)

Helicopter and Jets (2) in Ancient Egypt

Just to remind you that our ancestors saw Skycraft that were piloted by the gods... something that flew and the Sumerians said it was their gods, the Anunnaki.

Anunnaki Appearance

Before getting into the arrival and what they did, let's take a side trip for a second and check out what the Reptilian Anunnaki looked like – that **Zechariah Sitchin** so carefully avoided (but Ch. 3 in VEG did not).

When VEG was written in 2008, and Sitchin was the main "authority" at the time, since then several sources have come forth, some anonymously, to reveal key information about our origins. Of course they should be viewed cautiously and vetted if possible with hard archeological or written historical evidence (as this author has done: you will not read/see anything that is not true or at least plausible and likely the truth).

Sitchin never, in fact, told us what the Anunnaki looked like, he just referred to the temple pictures and scrolls which showed something like the picture below... which is human-looking.

According to an author whose books are being "disappeared," and who did the 30 years research into Sumerian cuneiform tablets and (like this author) was given many insights into the who, what, when, where, and why... and it all meshes together to form a coherent picture of our past progenitors... (below):

Sa'am
(c) Anton Parks

Mamítu Nammu
(c) Anton Parks

According to the major source for the Anunnaki (aka *Gina'abul*) appearance, we have the following: [3]
(Pronunciation: Ghee-Nah-Ah-Bull.)

The male is the upper and the female is the lower picture.

This explains the elongated heads including that of **Akhenaten** in Egypt... who was more human-looking than reptile, but it is a characteristic of several ET races to have **elongated heads.**

Sa'am is also called **Nudimmud** or **Enki** [4]

Mamitu Nammu was (is?) a great Gina'abul **Planner** working with the *Kadishtu* (or Elohim). She is/was one of the "good guys." A priestess. (Not all reptiles are bad.)

Pharaoh Akhenaten also has an elongated head (below):

Note the elongated head.... it wasn't just the odd hat shape – the hat covered what was different from most humans' heads.

(Appendices E & F.)

However there were several types of *Gina'abul*...and they fought among themselves for control of land and the humans...

Gina'abul – the reptilian race that includes the Shutum, the Amashutum, the Kingu, the Kingu-Babbar, the Mushgir, the Miminu, and the Nungal Planners and the Anunna Warriors.

Shumtum   Gina'abul males in general.
Amashutum -- Gina'abul females in general

Kingu – Gina'abul royalty who live in Ushu (Orion/Draco)
Kingu-Babbar – albino Kingu; the **ruling royalty** in Ushu.

Mushgir – reptilian **dragon**, warrior class.

Miminu – the bio-cybernetic **Greys** created by the Gina'abul.

Nungal – race of male **Planners** created by Enki (Sa'am) and Mamitu-Nammu (head female Planner).

The Planners work with the Higher Beings, aka **Kadishtu,** who work for the Original Source ("God") as planners of the Universe. The Akkadian word *Qadishtu* (holy woman) was one of the names for high priestesses – such as was given to Mary Magdalene (see TSF: she was a high priestess in Egypt, in Alexandria).

The **Anunna Warriors** – were created by the Gina'abul by An and Ninmah on the Duku, a principal planet in the con-Stellation MulMul (Pleiades).

An – the leader of the Gina'abul back on their home planet. Ninmah – high priestess assisting An in creating Anunna Warriors... and right arm of Tiamata – queen of the Gina'abul and one of the seven members of the Ushumgal Council.

Ushumgal – "Great Dragon" -- name of the seven great rulers of the Gina'abul (a largely patristic and rigidly hier-archical society)

Not so obvious, the Gina'abul constantly fight the Kadishtu and the Kingu fight the Ushumgal.

Gilimanna – Kadishtu name for the wild and wooly Gina'abul.

For the foregoing you can see that it is not just a simple Anunnaki that came here, and their **war in our solar system**, destroyed Tiamat (between Mars and Jupiter) and wiped the atmosphere almost completely off Mars and then all came to Earth.

The Pre-Anunnaki Group

These are the major gods of Sumerian lore, written about on many (200,000+) cuneiform tablets, seals and cylinders. They are just the oldest Skygods to have come here, according to tablet BM74329 there was a very early landing of many ships, according to the *Enuma Elish*, which could be called the Elohim (the "good angels" as opposed to the "bad angels" later called the Draco aka Djinn).

> This appears to be the earliest reference anywhere to the age-old battle between Good and Evil, i.e., between Light and Darkness. This is also found in another significant Sumerian tablet (BM74329) called *The Divine Geneology of the Firm Ground* and dealt with the same thing as the Bible's Genesis: the creation of the waters, earth and biota. [5]

The significance here is that it chronicles **seven generations of Anunnaki** rule that documents the beginnings of Anunnaki intrigue with sons deposing their parents, often killing them to assume power, and then marrying their sister, among them a Queen Tiamat, who is said to be the "mother of life." These were seen as the divine families (Tiamat and Apsu) who descended from the sky chariots and founded the first Sumerian settlements, among them **Kharsag**.

> Nothing is said anywhere of a **Nibiru**, which appears to have been unfortunately a Sitchin invention. [6]

Major Players

Later tablets document the arrival of **Enlil**, also called "Lord of the Breath", and chief administrator (or **Satam**) of the colony of Kharsag. He was the eldest son of An and the brother of Enki. He despised humans.
Anu (aka An) was the supreme god, creator of the Anunna warriors.
The **Anunna warriors** were the troops who destroyed the forces of Queen Tiamat.
Enki-Ea (Ptah) was son of the god An, also called "Lord of the Earth" (En-Ki), chief officer of Kharsag, and of the **E.Din** garden of Ninkharsag-Ninmah. He was also called "Lord of Understanding" due to his great knowledge, as well as **"Serpent"** which was a title of wisdom. He was also the great benefactor to

mankind (the opposite of Enlil, which is where the "*Satam*" connotation came into play with the derivation, "Satan" which reflects Enlil's anti-human stance.

Man was divided into two groups: the first was Homo *erectus* and Homo *habilis* when the Anunnaki got to Earth, and the second was a **re**-creation of Homo to serve the gods – working the mines, fields, and building the ziggurats.

Messengers were the Igigi, or workers in orbit around Earth and transporting things back and forth between Earth and Mars, who came down and fraternized with Earth women creating the **Nephilim**. Enki also had his own cadre of support 'troops' to do his bidding.

Ninkharsag-Ninmah was the first wife of Enlil who oversaw the Garden of Eden (E.Din), and was the great matriarch of the colony.

Marduk (Ra) was a son of Enki who made a name for himself as a warrior and a strong ruler. The *Enuma Elish* as an epic Sumerian poem was intended to glorify Marduk and even the rulers of Babylon created statues and plaques in his honor. (However, Marduk enters the scene <u>after</u> the Creation epics have run their course and it is doubtful that he had any part in Creation.) Nonetheless, Marduk is associated with the "sacred mound," the Duku on Earth, where the original Skygods descended and built Kharsag. Marduk is a very important figure in the pantheon of Anunnaki gods: it was his duty to protect the Duku… hence the inclusion of "duku" in Marduk's name: MAR-DU-KU.

Ningishzidda (Thoth) was the other son of Enki and was also a benefactor, whereas Marduk was arrogant and aggressive and carried out Enlil's plans.

> In a surprise revelation, emerging from today's later research
> into tablets that Sitchin did not have access to, it turns out
> that "Marduk" was a title and that Marduk is called "Enlil of
> the gods" (in the *Enuma Elish*, 7th and last tablet, line 149).
> Thus Marduk could have also Enlil in some writings. [7]

Tablet BM 47406 says: [8]

> …Ninirta is Marduk, as god of the hoe,
> Nergal is Marduk, if battle is involved,
> Zababa is Marduk as god of combat,
> Enlil is Marduk in matters of sovereignty and deliberation,
> Nabu is Marduk [in what concerns] accounts…
> Ea [Enki] is Marduk for the clod of clay [humans]…

Sumerians and The Pleiades

We will get to the Pleiadian massacre later, but it is interesting that the Sumerians, and thus the Anunnaki, were very aware of the Pleiades (7 balls, upper right)…

9. Assyrian clay seal with, at top right, the seven stars of the Pleiades *(Mulmul)*, next to a sun and moon. Above the moon hovers a spacecraft with three occupants. Two priests seem to be praying around a Tree of Life aligned with the moon and the "flying chariot." At bottom right is the dragon-serpent Tiamat, above which is an arrow pointing skyward, often a symbol for Marduk. The latter's victory over Tiamat earned him the emblem of the dragon. The fact of placing Tiamat and Marduk under the Pleiades might confirm the starting point of the cosmic war described in the *Enuma Elish.*

And as far as the Pleiades are concerned, they were referred to as MUL-MUL by the Sumerians, which means "the stars" (repetition indicates a plural). As will be seen later, the Reptilians and Pleiadians were not exactly friends and often fought. So the reference in the above cylinder seal (which has been rolled out flat) shows that Marduk (the arrow) was hostile to them. It is thought that Orion is the origin of the Reptoids as their creators (the Draco) hail from there, and Pleiadians and Dracs are not friends. It is more than likely that the **space war**, scarring Mars and destroying Tiamat (between Mars and Jupiter) was fought between Pleiadians/Lyrans and the Anunnaki – over what is unknown.

Space War

According to the Sumerian tablets, there was a time when there was a big war in the heavens (as recorded on tablets CBS 8383 and CBS 14005). [9]

The **Anunna** were the warriors involved with the war on Tiamat (the planet between Mars and Jupiter which is now an **asteroid belt**). Since the Anunnaki also had mining operations on Mars, the battle between factions involved Mars and Tiamat (and Earth to a lesser degree), and during the battle, Mars was seriously scarred: the ***Valles Marinaris*** gouge was created by a huge plasma disruptor weapon.

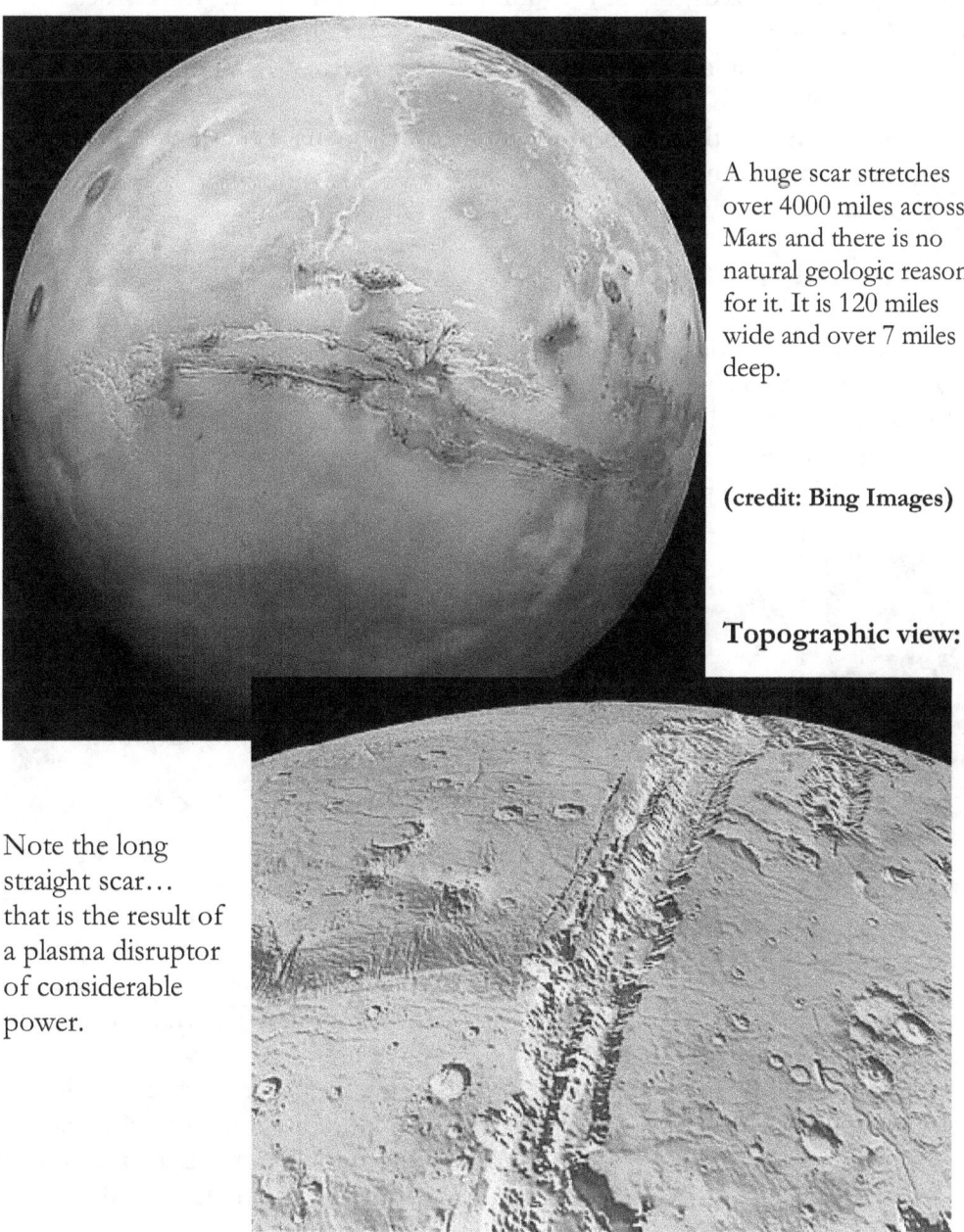

A huge scar stretches over 4000 miles across Mars and there is no natural geologic reason for it. It is 120 miles wide and over 7 miles deep.

(credit: Bing Images)

Topographic view:

Note the long straight scar… that is the result of a plasma disruptor of considerable power.

So the Anunnaki came from Orion, partially settled Mars and Earth, went thru 7 generations, including the space war, and then with the advent of Enlil and Enki developing the Earth, they **re**-created the worker humans.

What has been carefully left out of the Anunnaki story is **that the Reptoids were created by the Draco**, and that will be examined as the Pleiadian story unfolds… following the capsulized story in <u>Eden</u> of Man's history, including the Garden of Eden and the Flood – of which the Bible is a greatly "sanitized" and abridged version. The Sumerian accounts are much more complete.

> **According to the Epilog in <u>Eden</u>, the following 4 sections (thru the "Dénouement") is a capsulized version of our ancient history:** [10]

Cuneiform History of Sumeria

A long time ago, the gods of Sumerian lore, the Skygods, landed on Earth due to a galactic conflict, as recorded in tablets CBS 8383 (col. 1, lines 1 & 2), and CBS 14005 (side a, lines 18-19). Their spacecraft were called *Gigirlah* ("chariots") and enables the Anunnaki to travel from planet to planet, as well as thru the air on Earth. Even the Maya pictured that (below):

Rewind:
Note that it is a reptile and while the vehicle is not space-worthy, it suggests they had special craft for terrestrially getting around.

As is shown later, in Chapter 2, the Sumerians pictured the larger space craft as orbs with wings suggesting that the 1st Sumerian symbol (shown next page), was a symbol for the more realistic craft below it… called **Margid'da:**

Sumerian Symbol of Skygod in Flight

The Real Anunnaki Interplanetary Skycraft

Anu's Skycraft which flew planet to planet was called Uanna.

The Anunna gods are said to have landed atop a low mountain, which they called a Duku. This was their sacred mound where they built their first city, Kharsag (which means "the main wall") and was built of cedar and used copper for the roofing. Copper (mined at the Great Lakes in North America) and cedar trees (Lebanon) were quite plentiful in those times.

Garden of Eden

The Garden grew the food for the colony at Kharsag as well as for the troops. In the beginning, it was surrounded by a reed fence, to keep out smaller animals, but when the Homo *erectus* began attacking the Garden for food, they replaced it with a log fence. When the humans (*Lulu*) jumped/climbed over that, Enlil, the great bur irascible **Satam** declared war on the humans... and yet, his brother Enki opened the gates to the Garden to some humans, having compassion on them. And this was the beginning of brotherly animosity.

So the savage man (according to the tablet CBS 8383) penetrated the Garden, stole fruit, knocked down trees, trampled veges... this led to the discovery of pathogens that the humans carried and the Anunnaki had no defense for it, and many Gina'abul became sick. Enki was the first to recover and with his great medical (as well as genetic) knowledge, he was able to heal his colony.

> From that moment on, (CBS 8383, col 15) man was called an animal [**Lulu**], described as being naked and having dark skin... he trespassed into the Garden 4 times and the gods wanted to persecute the "carnivorous man" with their metal weapons. [11]

> These were called Rig'giri – a "lightning spear."

The next document CBS 14005 tells of cereals that ripened in as little as 30 days and the Anunnaki were not willing to work that hard that often, so they counseled with Enki to see if they could modify some of the animalistic humans to breed them as workers – give them just enough sense to follow direction, domesticate them, and use them as slaves. (Sitchen called this group **Lulus**.) The tablet called these workers "the people of the flint." (This was about 250,000 years ago.)

The workers moved around on all fours and the Anunnaki built cattle stalls for them to live in, and treated them like the beasts they were. Enki's mother **Nammu**, who was called the goddess of the grain, would oversee their care and feeding.

To really impress the workers, the gods had the workers line up in front of the gleaming Skycraft to impress them while getting them to swear allegiance and obey orders. The language spoken to the workers was a very basic language, called **Eme-es** and so began a life of toil, working the Garden and carrying the foodstuffs off to the rest of the colony. The **"black beings"** as they were called were subordinate and anything but quiet -- they "talked a lot." So much so that Enlil was getting aggravated at their constant talking and (at night) their racket: singing and dancing around their campfires. (CBS 8322).

More on the Garden of Eden and pix in Appendix F.

Enki was ultimately responsible for the gate to the Garden and the behavior of the worker slaves. Enki was also pro-civilization and wanted to see his progeny grow and develop, as much from compassion for their hard life, as from his scientific interest: would it be possible to develop this species? (*Lulu* became *Adama* .)

Forbidden Fruit

Enlil was not oblivious to Enki's machinations with the new slaves – he watched with disgust as Enki shared the **"forbidden luxury"** of **tools** with the workers – He showed them how to make and use tools. This further distanced the brothers.

> According to the Sumerian word *Gis,* which can mean "tree" or "tool" (since tools were often made from a tree branch (e.g., and a metal head was attached for digging), the "tool is the extension of the tree, its fruit." This thousand-year-old play on words was mistaken in Genesis as "forbidden fruit." [The word for penis was **"Ges"** and offered mistranslation possibilities.]
> The knowledge of forbidden fruit was simply that of **tool making** and was not to be divulged. [12]   (CBS 11065)

We can see why this book was banned by the Church.

You also see why Enlil was walking in the Garden (see VEG Ch. 3) and did not know where Adam and Eve were hiding – an omniscient God would have known where they were. Enlil was later called Yahweh.

You can also see where Genesis is a very high-level account of what really happened, but when you are a fledgling Church building its own theology, you get to choose what goes into it.

> **This concept ("forbidden fruit") was later equated with a sexual taboo as Enki not only taught humans metallurgy (above as tool "GIS" making), but also gave the humans the ability to sexually "GES" reproduce.**

Enlil was irate as he felt that humans were too animalistic to develop and due to the **Galactic Law of Sentient Beings** (Glossary), when the Anunnaki were thru with the human workers, they would discard them. That was later done with The Flood to also dispose of the Nephilim. Enlil had given orders to not fraternize with the humans, and the "200 Watchers" (aka *Igigi*) who mated with Earth women when they were later not so animalistic, as **Adama** and **Adapa**, created real problems even for the colony as the Nephilim (and Anakim and Gibborim) would raid the Garden for food themselves… and due to their insatiable appetite (they were giants) they also would catch and eat humans. (see VEG, Ch. 3)

Moving right along, back to the <u>Eden</u> story…

Over time, the more civilized humans (*Adapa*) began to make tools, and clothe themselves, began to emulate the gods (CBS 11065-a, col. 2.) However, the gods needed unschooled laborers (as the PTB does today) occupied just with toil to keep their social system running. An unthinkable "virus" of knowledge had invaded the colony and Man was now replicating and spreading out before their very eyes! Man was becoming a threat to the colony. He must be brought back under control.

> To this end the Anunnaki sometimes used their **dinosaur** creation, directing them against the human settlements. And they also used **Suruppu Disease** to cull their numbers. Enlil was an irascible god, as Yahweh, as often portrayed in the Old Testament – obey and he was happy, disobey and he would smite thee!

Dénouement

So says the final chapter of the <u>Eden</u> account:

> The chariots of the sky [Mu'u] were dispatched to pursue Man and bring him back to his place of work in the Garden. But the humans rebelled against the gods and even laid siege to Kharsag and the Garden. Many were the rebels at the foot of the sacred hill (Duku) …and the **black man** now dominated the entire region; armed with the destructive metal weapon, **sword and bow**, and he sought revenge against his lords.

> Enlil's vengeance came with the deafening thunder of his sky chariots, bringing death, crushing and burning the "black being" -- the brave new human being – who had dared to defy the gods.

> It is also written that Enki's **'Messengers'** lived among the humans at the time and were busy casting metal themselves, apparently in support of the rebellion. However, the greater Enlil force prevailed and the humans were subdued and then

expelled from the Garden and brought back to their place of work and "bound."

It was further related that Man was altered by the gods – mostly genetically to ensure subservience... some of Enki's enhancements were **dumbed down** to prevent creative thinking... these attributes would be later added back in by the Greys as the proactive Remnant Insiders and Kadishtu took over after the main Anunnaki contingent went hone, about 2500 BC.

Such was the basic account from the Sumerian tablets of Man's early beginnings; **Zechariah Sitchin** elaborated on the scenario sometimes embellishing, and usually obfuscating the real nature of the Anunnaki (e.g., reptilian: See VEG Ch 3). He was correct in his account of the genetic manipulation of Man, originally to build better workers, and then (after The Flood) when Enlil finally relented and saw that Enki was going to have his way, openly or secretly, and Enlil wisely let Enki do his thing openly so that he could be watched.

Sitchin Creativity

Having said all that, the late Zechariah Sitchin was a rascal and didn't tell us seven key things about the Anunnaki.

Appearance:	they were initially reptilian, later created genetic mammalian versions, as for **Inanna and Marduk**, and then for hybrids like Noah, Moses, Sargon, Alexander the Great...;
Gold:	they mined gold but not for their planet's atmosphere; it was ingested as monoatomic gold to sustain their longevity;
Nibiru:	was a planet similar to the *Star War's* Death Star -- a virtual travelling Battlestar that won't be coming back – it was destroyed near Jupiter in April 2003 as it was returning;
Remnant:	most Anunnaki left Earth in 2500-2000 BC; a portion stayed on due to Galactic Law governing Creator Races (see Glossary).
Reptile Types	See above section breakdown – there was a lot more than the "plain vanilla" Anunnaki and they fought among themselves.
Underground	It was never mentioned that a Remnant stayed on Earth, underground, and continues to disagree among themselves to this day: Dissidents vs Insiders (See Glossary).
Re-creation	Sitchin led people to believe that the Anunnaki created Man from an existing hominid (plus Anunnaki DNA, true) , but the truth was they **RE**-created with genetic manipulation... a significant point as the Anunnaki love to have people think they created Man.

> **The Elohim created the original Man working with the Kadishtu and then the Reptiles severely hacked the DNA, creating an animal -- from a human with divine potential.**

Genetic manipulation was something that the Anunnaki had learned from their masters, the Draco. And it was not that special. Even the Pleiadians and the Lyrans were capable of it. And Man today is quickly gaining the expertise (Chapter 3).

And yet, the following now deals with the Draco-reptilian abuse of the Pleiadians, as mentioned earlier… and the subsequent Reptoid activities on Earth. The Draco are important as they were the negative component of the **Elohim** – that part that "fell away" from the Light and chose to do their own thing… also as semi-angelic beings like the **Beings of Light** (BoL), and their 'superiors': the Planners, Kadishtu and Elohim who ARE able to manipulate matter and create.

Anunnaki/Reptoid Group

Thus, the **Draco** were the so-called **"Fallen Angels"** (aka **Djinn**) who chose to not follow the Light and do whatever they wanted… and they had the personal power to do it, known for **shapeshifting**, creation, teleportation, manipulation of matter, and were a special class of being, certainly gods themselves, but below the Higher Beings to whom they must answer and who were the ones that showed Enoch the "imprisoned" Watchers.
(See GEP, Ch. 6; VEG, Ch. 2; and QES, Ch. 11.)

32- This statuette representing a Mušgir is identified with the Assyro-Babylonian demon Pazuzu, a denizen of the netherworld and parallel realms. He sports large wings and has a body covered with scales. In between the wings on the back is the following inscription: "*I am Pazuzu, son of Hanpa. The king of the evil wind spirits that emerge violently from the Šadû* [the KUR in Assyrian] *and create havoc, that I am!*" The realm of the demons was very present in the minds of the Mesopotamians, and Mušgir-Pazuzu was considered to be one of the most powerful of them all. As a result, he was often depicted on amulets to conjure the other infernal powers. Many of these amulets were found in the foundations of dwellings in Mesopotamia. Assyrian bronze (inv. MNB 467), Louvre, Paris.

(credit: [13])

The Genie of popular movies and stories... in this case a **Blue Djinn**...

The color denoted rank and power – **Black Djinn** were the most powerful, then came Blue, Yellow, and Green.
They form clans
The **Red Djinn** work with Iblis (Satan) to destroy the human race, form no clans, and they often take on the form of reptiles. [14]
(All shapeshifters.)

(credit: Pinterest)

The legend said that if you find a special lamp imprisoning the Genie, rub it and the Genie would appear and grant three wishes, but be careful what you ask for, they were great tricksters.

The Dracos' first act of creation was the **Reptoids**, in the Orion area, and being a lot like their creators, the Reptilians were **all head and no heart**, no compassion and no soul growth (as many of them would not and could not host incarnating souls).

Then the Draco left for another dimension and left the Reptilians to their own devices, and that was when the Reptoids created the **Dinosaurs**...as much for food as sport. There were contests among the several Reptilian settlements to see who could create the biggest and baddest lizards… and scare humans.

Rewind: Later when the Reptilians got tired working the Mesopotamian fields for food, building their ziggurats, and working the African mines for gold, they created **worker human**s based on an existing local humanoid, apelike form. These were what Zechariah Sitchin accurately called the ***Lulus***.

It was found that the dinosaurs could be **controlled mentally** (Reptilians inherited that ability from their creators, the Draco) and the giant lizards were often used to terrorize the humans – and keep them in line!

> The **Reptoids are control freaks** and still are – as will be seen in later chapters. The important thing to know at this point is that not only the **Draco**s but their progeny, sharing some of the Draco genetics, the **Reptoids** have the ability to make you think what they want, influence your thinking and beliefs, and thus your actions, just as the **Greys** (Miminu) control their abductees and their thinking, and actions, often wiping their memories so the humans can't recall the abduction. The Reptoids have always been superior mentally, and so are the Greys who are **biocybernetic androids.**
> Man was originally created with higher mental abilities <u>omitted</u> by the Anunnaki and given a weaker physical body so that he could be controlled.

Later, as dinosaurs started killing each other, and dying, their rotting flesh proved to be a problem… and the Gina'abul developed ways to get rid of it – **fungus, bacteria and viruses** were designed and turned loose – also afflicting the humans at times. One such event was before the Great Flood, called ***Suruppu* Disease** – for which the Reptilians also had the cure. And yes, they also caused the <u>localized</u> **Flood** to erase the problem with some of their kind mating with earth women and creating Nephilim, Annakim and Gibborim… all hostile to humans.

> The Flood was localized to the African-Mesopotamian area and did not involve a huge ship with live animals from all over the world – Sorry -- think of the logistics involved in corralling them and getting them on the Ark.
>
> Enki used his Messengers to collect DNA from all the animals and plants that would be affected, and then **cryogenically stored it** all in glass vials in large containers, and that was the "Ark"… maybe 30'x30'.

The bacteria and viruses did not affect the Reptilians, but we still have the remnant of their creations in our world, as well as nasty mosquitoes, spiders, ticks, fleas, wasps, bats and poisonous plants and colorful, poisonous frogs, for example.

> Sorry if you thought this world and everything in it was
> created by a loving God of the Universe … so loving that
> He would afflict his creation (according to the Bible version
> of Creation) with the foregoing nasty biota?

Breaking Point

As will be seen later, Earth and this Galaxy were created as a Freewill realm but the Reptoids were being watched -- the Great Flood drew the Lyrans down on them, although the Council agreed that removing the Nephilim and their ilk was the right thing to do… **Galactic Law** said that attacking and physically removing the Reptoids was not condoned (due to Karma), but the Reptoids could be 'persuaded' to leave… so maybe the insertion of a more spiritual race on Earth would help drive things in the right direction…

Needless to say, the Solar Council and Pleiadians and Lyrans were not oblivious to what was being done to their Earth Zoo. But as long as the Zoo was not destroyed and no sentient beings were destroyed, the Reptoids could continue their activities – and the Council was curious to see what would be done with the sentient humans that had been re-created

> **Galactic Law** also said that if you (re)create a sentient
> being, you are responsible to nurture, educate and develop
> it into a responsible, intelligent proactive being… or dispose
> of it. (See Glossary.)

According to **Alcheringa** (an Aboriginal Legend book), the Council negotiated the departure of the Reptoids and the intelligent **Dinoids** (Think: Velociraptors) from the Earth, with just a small Reptilian contingent to remain underground ("Anunnaki Remnant") – because they created humans and were still responsible for their development – in a proactive way. (Think: Greys and genetic manipulation as a way to continually monitor and upgrade the human race, complying with the Galactic requirement to develop the progeny.)

The Treaty was overseen and "implemented" by the Leonine race (Lyrans) also known as a kind of the **Galactic Police Force** – they could easily destroy all the Reptoids (and yet had to be careful as the Dracos were yet more powerful –and that was what the Reptoids were banking on: the Draco support so that they didn't have to leave Earth).

Some Dracos have wings and allegedly are a form of the **"Fallen Ones"** with semi-angelic powers (Think: **Djinn**):

Left: suggested likeness of a Draco warrior, about 9-10' tall, something of a Reptoid **Kingu** but having almost supernatural abilities. They created the Reptoids to serve them.

Above: a version of **Anunna** – there are many different variations: some have horns, some have wings, some are green, some brown and some are white… and some (**Left**) are a cross between Draco and Lizard:

And sure enough, when the meeting was held to sign the Treaty, allowing the Pleiadians to colonize Earth, the Draco were present and **mind-controlling the Reptoids**… who were manipulated into ambushing the Pleiadian ship.

Enter The Pleiadians

According to the Alcheringa book, as well as Baldy's input in VEG Ch. 3 which corroborated Alcheringa, the Council initiated a meeting and a Treaty with the existing denizens of Earth, about 4 races plus the Reptoids (about 780,000 BC). The goal was to offset the Darkness and lower, denser STS vibration accruing to the once pristine, neutral Garden/Zoo by inviting the higher STO vibration Pleiadians to colonize the Earth. Their Light and peaceful ways and vibration should **anchor the Light** and balance and then eventually overtake the STS energy and help return Earth to a positive, peaceful planet in this part of the Galaxy.

Remember the movie *Knowing,* and the scene at the end where the Starbeings show up… a very good likeness of the Pleiadians (below)…

It is said that the Pleiadians occupy the 5D realm and have to pass thru a Gateway of Change to visit 3D Earth which is in 4D.
(This 3D-4D mix is explained later in this chapter.)

Of course the Reptiles had no interest in complying but went along with the idea, just going thru the motions, secretly planning to ambush the incoming Pleiadian ship… and destroy it (next chapter).

As was said, also getting into the act, were the Dracos who considered Earth to be theirs and they backed their progeny, the Reptoids, in sustaining the STS energy of Earth, leaving their higher dimension temporarily to participate in supporting them.

The Reptoids were perfectly aware of the consequences of their nefarious action, and most returned to Orion (under threat and bombardment by the Lyran Police Force), and the Draco escaped back into their chosen dimension.

Lyran (Leonine) Police Force, the sworn enemy of the Reptoids (from an Egyptian hieroglyphic likeness)…:

(No, that is not a joke, nor is it Photoshop. ®)

It gives a new meaning to the Biblical phrase, the "Lion of Judah."

We humans need to get used to the idea that there are more and varied lifeforms in the Universe than we have been willing to admit. We are not alone and not all humanoid forms look like us. There are also insect-like, fish-like and bird-like forms out there.

Just for your consideration:

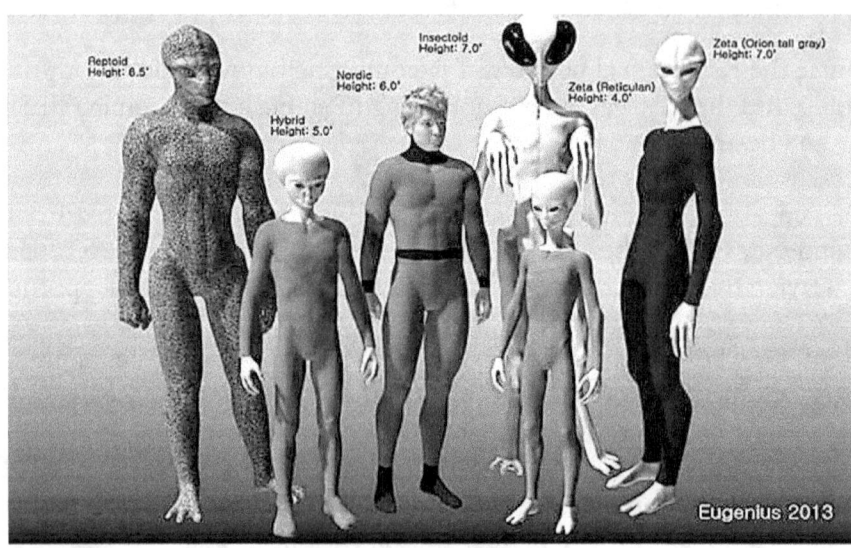

And **Oannes** from Sirius:

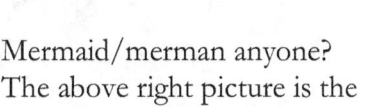

A GOD-FISH.[1]

Mermaid/merman anyone?
The above right picture is the
Sumerian depiction of the god Oannes. (also GEP, Ch. 8.)

The Sirian fish god **Nommo** visited the Dogon in Africa and looked the same.

Bird Race

And it came to light during the recent info given on Gina'abukl, that there is a bird-race pictured on the walls of Egyptian temples (next page)... Meanwhile, is the following so impossible to believe?

Left: the God **Thoth**...

Below: the god **Horus**

And **Isis** is pictured with wings...

Be aware that Isis was actually the Egyptian name for **Inanna** the Sumerian-Anunnaki goddess who also ruled over Indus Valley ...
from her skycraft.

Horus also had a flying craft and defeated Set in the air.

But the interesting thing, according to the Sumerian writing, is that there were actual gods (Gina'abul) who could take whatever form they wanted. They were Planners with **bird-shaped bodies** called *Sukkal.*

So is this what the Aztec warriors were copying? A bird? Why not emulate <u>only</u> the bigger, stronger jaguar or panther (which they also did)?

The American Indians did the same thing... why a bird, unless there was a race of Bird-gods? (According to the Sumerians, there were, and they were called *Sukkal.*)

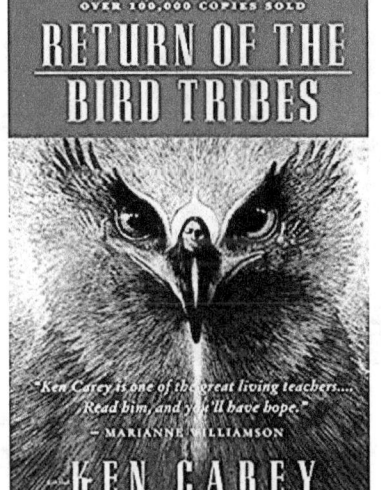

Is that what the book by Ken Carey was about? Says an Amazon reviewer (M. Schmidt):

"I recommend this to anyone who feels despair about our current status on this planet. Rise above anguish and grasp the hope of the time we are in. Consciousness is key.... We all have the power to shift the direction of our planet, one soul at a time."

And that is further examined in Chapter 8.

Anyway, back to the Pleiadian plight....

A few Reptoids, having some conscience (and souls), defected from their peers and began to help the stranded Pleiadian remnant survive on Earth, now a very hostile environment without their advanced technology. These few better Reptilians would later be called the **Anunnaki Remnant Insiders**, the more proactive STO half.

Earth Protection

Then as has been said in Appendix B, a **Quarantine** was put around Earth in the form of an impenetrable **EMF Grid**, managed by the Solar Council and Lyrans – the same Grid that in TEW is encountered by Maria in Chapter 1, the same Grid that Charles Fort called the **Gegenschein** (see VEG, Ch. 12), the same Grid that NASA said they found at 7200 miles above the Earth that is "impenetrable" like a "glass sphere" (also GEP, Apx. E). [15]

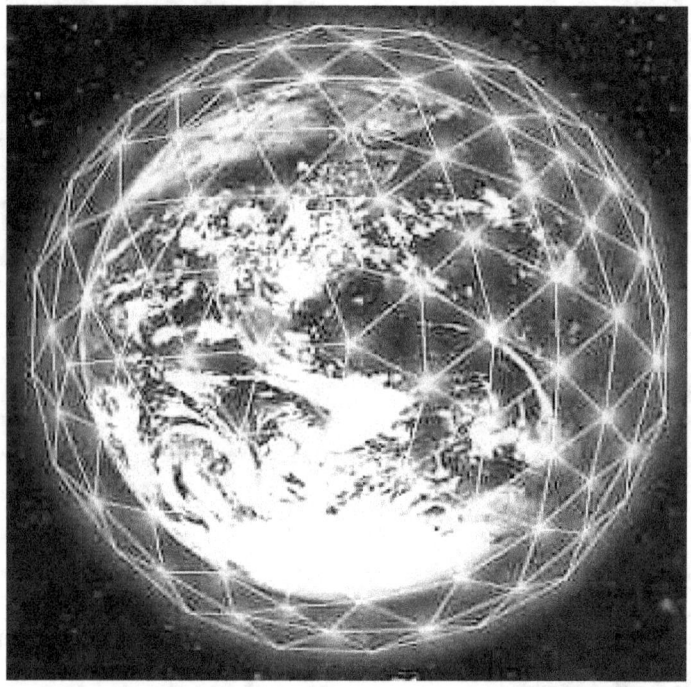

Simulated Energy Grid Protecting Earth

And according to NASA and **Charles Fort**, the *Gegenschein* does exist and

resembles a **reflection** or glow in the sky off of something that is stationary…
(See VEG Ch. 12 for more about this phenomena)

Gegenschein

And **Appendix B** examines the NASA barrier discovered at
7200 miles above the Earth… there **is** something there.

Of course the Grid does not stop the Draco who come and go interdimensionally
(also see VEG, Ch. 12 and **Transdimensionals**), but they have not been as much
of a problem as the Reptilians who are <u>still here</u> and <u>still</u> trying to hamper and
remove humans from what they <u>still</u> consider their planet.

> **John Keel** also deduced that there was a certain interdimensional
> element affecting us and he called them **Ultradimensionals**.
> (See his book, <u>Operation Trojan Horse</u>.)

Fortunately, as will be seen later, there are also many benevolent beings here
offsetting the machinations of the Reptoids, and the Council lets it happen as it is
all **catalyst in the Greater Drama** to wake Man up and empower his soul
growth… to grow him by confronting him with an antagonist for personal growth
– if all were hunky-dory, no problems, no disease, no wars, nothing big and bad
ever happened, there would be no impetus to growth… we <u>overcome and so grow
stronger</u> and that is our destiny as souls when we <u>graduate</u> from Earth School – to
be strong and wise enough rule over those who are currently trying to stop us…
(Chapter 8).

And yet, when the Reptoids step over the line, they are stopped.

And sometimes that takes the form of a **Wipe & Reboot**, which has happened before (Think: AD 800-900 in Ch. 10 VEG). And sometimes it takes the form of a **Timeline Shift**... which is NOT predicted this time.

> **BTW: take heart – Nothing is out of control and**
> **Man does survive and will one day get it together.**

> Many of the foregoing details were adapted from Valerie Barrow's <u>Alcheringa</u> account of the Pleiadian Mission 700,000+ years ago. (See Bibliography.)

The Pleiadian story and its aftermath is continued in the next chapter.

So what is Earth really?   We now know it isn't flat, but what is it?

VR Earth Construct

As was said earlier, Earth is a virtual Construct based on 3D Science/Laws but sitting in 4D, so that the 4D energy empowers the lesser 3D construct.

Be aware that most life in the Universe resides in 4D (and above).

Much as a museum in many cities is constructed differently than a commercial skyscraper, or an opera house, so Earth was specially designed to be a repository for different fauna and flora from around the Galaxy.

Further detail was given in VEG, Ch. 13 along with the "proof" from several distinguished Physicists...

VR Earth 3D Construct Definition

The 3D replicated Earth is real, contained within the quarantine "shell" (*Gegenschein*) as a 3D Construct, actually sitting in 4D, and the "computer" running the Show is a semi-sentient, Bio-plasmic Organic computer-like organism with a feedback Control System as part of its main Operating System.
(See Appendix D)
Objects on Earth are quantized and materialized within the holographic framework by a subatomic process (Replication) similar to what was described in the *Star Trek* Holodeck. To protect Man, the Earth and its immediate surroundings are enclosed in a high energy barrier allowing very controlled exit and entrance.

Man (and other sentient beings) is inserted as a soul into this environment for lessons and comes in with a Script or LifePlan that orchestrates experiences. He interacts with OPs who are NPCs (Chapter 6). Beings of Light and Neggs help administer the individual soul's Script within the framework of the larger Greater Script – which reflects the purpose for the 3D Construct.

Note that a soul's Script is controlled within the Greater (Divine) Script, just as individual application programs on a PC are controlled by the Operating System. For example, MS Word must operate within the confines and structure of the MS XP or 7 or 10 operating systems, for example, and use its resources. Anything not prohibited by the LifePlan to the soul is considered "Freewill." (Examined more in Chapter 4.)

The Earth Realm is subject to manipulation by Higher Beings ("Planners") who monitor phases of the Greater (Divine) Script, and Man's individual Scripts, or LifePlans, coordinating all with the Beings of Light and the Neggs. They can insert objects, people and events as deemed necessary… sometimes called miracles and synchronicities. They can also disappear things and remove people as needed, and They can perform a "Wipe and Reboot" to clean the Stage and reset the Drama as a new Era if necessary.

> Please note that a Script or LifePlan does not tell Man
> what to do or say – only certain tests and events are
> programmed as "tests" and Man has Freewill to handle
> them as best he can – that is why Earth is a School
> and events are TESTS to see what the Soul has learned.

Thus Earth is not Man's real home and he is expected to learn, overcome, and graduate.

BTW, it is due to the Energy Grid containing the Earth School that one must pass thru a **Tunnel** to exit the Earth Realm at death… one then returns to the InterLife and reviews progress and may train for the next sortie into whatever Realm the Soul chooses to experience.

This was all explained further in TOM and comes from the author's trip to the Other Side (aka InterLife). The Soul is eternal and Life is a continuous experience -- somewhere and in sometime.

Also see **Personal Insight** insert in Chapter 8.

Chapter 2: Skygods & Creation

500,000 B.Z. (Before Zechariah Sitchin)…

The huge Pleiadian Cruiser approached Earth orbit cautiously…
they knew hostile Draco ships could be nearby, cloaked. 50,000 peaceful souls were on board eager to start a new colony on Earth, with Solar Council approval. While there were already 5 other races on Earth, scattered over the landmasses, from other inhabited planets in the Galaxy, the Pleiadians were unique with their very advanced technology and they intended to set up a peaceful but isolated settlement in what is now called Australia.

All 5 races, including the Reptilians, and the Pleiadians had signed a Treaty of Development and were doing their part to stabilize the Earth Experiment. Yet they were apprehensive as a rogue faction of the Reptilian Dracos had been against the move from the start, claiming that Earth was originally theirs, and only theirs, so the Pleiadian ship, which was 8 miles in diameter and a nice big target, approached cautiously, deflector shields up.

If Dracos attacked, as long as the Pleiadian shields were up, there would be no problem resisting the Dracos, even if the Dracs fired first.

Settling into orbit, many of the crew had moved to the shuttlebays and were firing up the smaller 80-foot Vril craft for the initial sortie.

Whuummp!

A giant asteroid hit the shield and bounced off... catching the Commander's attention on the Bridge. Soldiers armed the plasma disruptor and teleradar tracked the object back to its point of origin... a cloaked Draco ship, but before the Pleidian ship could return fire, or move, a second cloaked ship fired on the Pleidian Cruiser from the opposite direction and cut a big hole in the outer hull! Disintegration entrained among the nano-crystal particles of the ship.

Panic on the Bridge. The shields were down... How could that happen?!

Then the third Draco ship opened fire, a much larger Battlecruiser, and delivered real damage, igniting fires on multiple decks and killing several hundred passengers. Those who could, raced for the shuttlebays to get off the rapidly dissolving ship... a crystalline sentient ship, the best in Pleiadian technology, was composed of a crystalline nanotech structure capable of storing AI and intelligently interfacing with the 3 pilots via their headbands. Nonetheless, the ship's crystalline latticework started to dissolve, and spread rapidly from crystal module to crystal module.... leaving green crystalline residue... In effect killing the once (AI) sentient ship.

The Commander noted the shield status panel and saw that the shield had been turned off just before the attack, from within Engineering, and knew there was a saboteur, a turncoat, in his ranks. But it made no difference... Engineering was the first section hit and everyone there was dead, or in little pieces... the smaller Reptilian ships left the scene, leaving the Battlecruiser.

Scores of Vril ships headed for the surface of the planet, fanning out at top speed, and the remaining Draco Battleship continued firing on them, disintegrating many before they could reach the ground. Fortunately there were too many Vril to all be lost, and the Pleiadians flying at top speed, made the survival of some craft a surety.

Many crash-landed but just under a hundred survivors made it.

Unknown to both Pleidians and Dracs, this whole scenario had been watched from the Moon by the Lyrans, a leonine race with greatly superior power to that of the Reptilians, so much so that they were often called the Galaxy Police Force, and they raced into action.

Shields meant nothing to Lyran weaponry... the energy of the shield was turned against the enemy like antimatter. The Lyrans descended on the Battlecruiser before it could fire again – the Battlecruiser radar saw the Lyrans coming but they could not flee – they knew what was going to happen to them. The larger Lyran ship had locked on and was holding the Drac ship in place.   The Lyrans gave orders to stand down and surrender.... Instead the Dracs broke the lock by phasing into hyperspace and returned to their dimension.

Marooned

Now the peaceful colonists were stranded on a planet with very little of their technology and the environment was hostile... they would have to find a way to survive. **Underground** -- as the Sun's rays were harmful to their bodies. They made many of the huge caves we see today.

> Think: **Cappadocia and Derinkuyu** type dwellings in Turkey as well as the 24+ mystery caves discovered in China in 1992, called **Longyou Caves** --- a series of large artificial caverns located in Zhejiang province. Australia

also has its share of mystery caves associated with the Wandjina.

Longyou Cave

Five years later, there came a rescue ship from the Pleiades and yet the survivors of the surprise attack chose to not return home, but stayed and accepted the technology they needed to rebuild their advanced civilization, and eventually became the **Ancient Ones**. They adopted a new role as overseers and benevolent caretakers of the Earth… all the while avoiding the more aggressive Reptilian contingent also on Earth.

The Reptilian contingent had already re-created slave worker humans to do their work for them, *Lulus* and *Adama,* and these were the ones Zechariah Sitchin would discover in the Sumerian cuneiform tablets, chronicling the activities of one group of the Anunnaki.

Aftermath

The original 90+ survivors gradually perished due to the occasional exposure to the overly hot Sum, bacteria and viruses for which their pure immune systems had no defense, and they occasionally ran afoul of poisonous plants and wild animals. But not before they succeeded in using some of the Pleiadian technology, some of which they managed to salvage as well as that given to them 5 years later by the rescue ships, and they mixed their DNA with that of the newly-created *Adama,* now running around the planet. In effect they re-created a new lifeform into which higher-minded souls could incarnate.

The rescue ship gave them genetic technology equipment, new scout ships to get around Earth with, and technology with which to fight off the Reptiles… and dinosaurs.

The Anunnaki had created large reptilian beasts as much for sport as for food, and sometimes used the beasts to terrorize the human populations… Pleiadians and Lyrans would put a stop to that.

Such Pleiadian capturing and reworking of the humanoids and removing the dinosaurs did not sit well with the Reptilians and there were skirmishes with them, forcing the Pleiadian survivors to secure new quarters in more remote, mostly underground Polar locations around the planet.

Pleiadian Possibilities

So what did the Pleiadians look like? According to <u>Alcheringa</u>:

They were tall, fragile, slightly bluish luminescent skin, bald, big eyes, and very advanced spiritually.

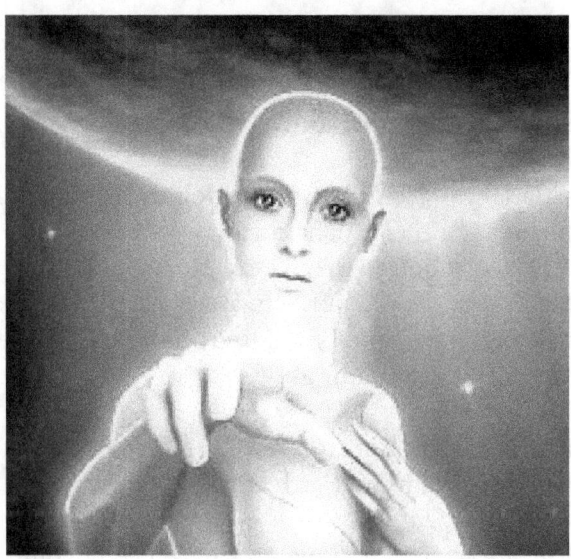

Pleiadian Likeness

And then again, if you have read the **Billy Meier** account of his meetings with the Pleiadians, some of their **earth-bound genetic modifications** may look more like the following:

Das Schnellste von allem was fliegt, ist der Gedanke.

The German caption reads: The fastest of all that flies, is thought.

Of course, those who have read other books on the Nazi development of the Vril antigravity craft, (and <u>The Earth Warrior</u> by this book's author), will immediately recognize **Maria Orsitch** on the right, above. (See Chapter 9.)

Two key points: (1) Maria was allegedly from **Aldebaran** and that is a planet in the Pleiadian cluster of planets. And (2) <u>Alcheringa</u> does state that **the Pleiadians modified their own DNA to better survive on Earth**… so if they could not handle the Earth environment, would not Semjase, Nera or Asket, as Billy Meier called his female contactees, have opted for a body more able to step out of the skycraft, breathe air and speak with him?

> And what if they were not from the Pleiades but the ones
> visiting him were from the Earth contingent?

Of course much of Meier's account and pictures have been debunked (as is usual), but could there have been originally a kernel of truth that he later embellished (maybe altruistically?) – to make a proactive impact on Mankind?
Would not the Earthbound Pleiadians (i.e., The Ancient Ones) still be concerned with protecting and enlightening Man and then visit him (or anybody) to spread a message of peace and cooperation among Mankind?

Star Connections

Is there any connection between **Aldebaran** and the **Pleiades**?

As can be seen in the starmap below, [16] the Pleiades are close to Aldebaran, and also **Orion** which is in the Taurus constellation area...

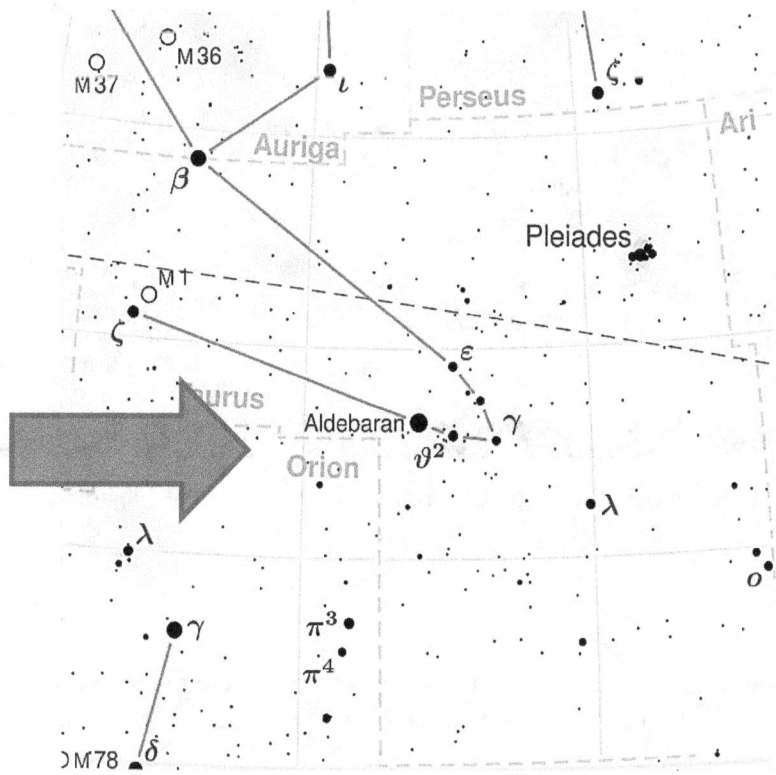

This is significant as the **Reptilians** are from Orion, the **Nordics** aka Pleiadians are from the Pleiades area and both were in competition as being the progenitors of Mankind. (In addition to the **Sirians**, nearby.)

The other significant player in the history of Mankind is the group from **Sirius** (from Canis Major) which visited the **Dogon in Africa** – more on them layer. The **Lyrans** were from another distant quadrant (below)...

As can be seen from the skychart below, most of the Visitors to Earth all came from the same quadrant of the sky – Orion, Pleiades, Sirius ....

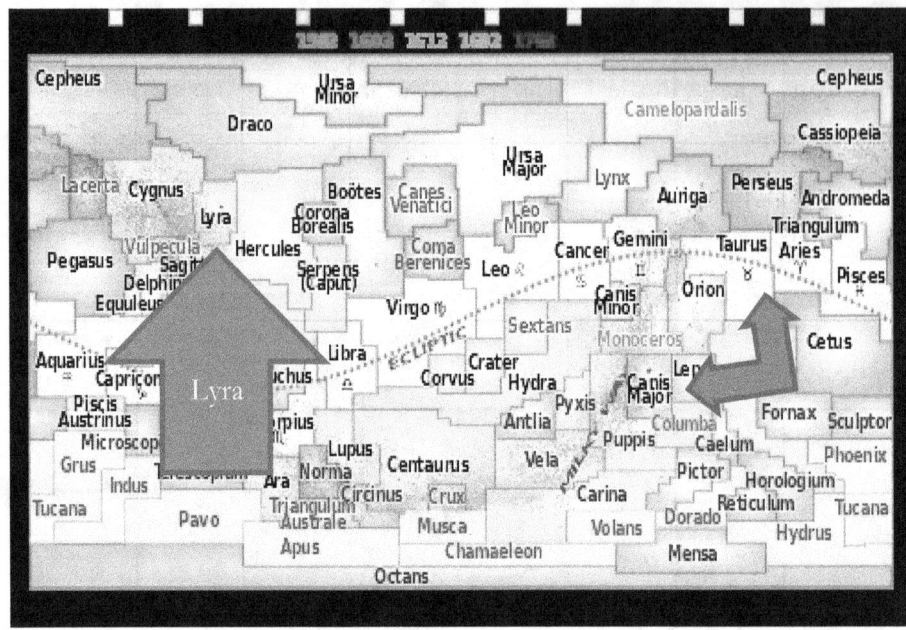

Lyrans are the sworn enemy of Orion, but friends with the Pleiadians..

Hard Evidence

So could all this be true, and Sitchin's Anunnaki in Sumeria were preceded by about 90 Pleiadian colonists who barely survived… landing as some say in **Southeastern Australia?**

Remember that throughout history, even in **China**, there were reported cases of

 little people crashing their skycraft and leaving evidence via the **Dropa Stones**: 716 stone disks with grooved recording estimated to be about 10,000 years old.

Dropa Stones (left).

The Aborigines of Australia have a legend of the **Wandjina** (below) who were their ancestors, centered basically in Northwestern Australia.

Aboriginal Painting of the Wandjina

It is not known what this is supposed to represent, but it does have some remote similarity to the Greys... large eyes, no mouth, no hair.

Gosford Glyphs

The Southeast of Australia, however happens to be the site where very interesting and ancient rock carvings, or glyphs, are found between two rock walls that have been dated to thousands of years ago....The **Gosford Glyphs.** This is in **Kariong,** in **New South Wales**, not far from **Sydney** which has been officially declared an Aboriginal site.

What is unique about the glyphs is that many of them resemble Egyptian hieroglyphics.

> **Despite the credence given by <u>Alcheringa</u> to the Aboriginal "meeting place" between the sandstone walls, only the oldest of the glyphs can be considered authentic, and probably carved by the Aborigines, not by the Egyptians.**

According to Alan Dash, a local surveyor who
discovered the glyphs in 1975, he continued to visit
the site every 5 years and found that every time
the number of [Egyptian] glyphs had increased. [17]

However, there are a couple of non-Egyptian glyphs that are among the oldest and
could represent Aboriginal or older carvings…

….as well as this one…

(above images credit; Bing Images: Gosford Glyphs).

The last two glyphs coupled with the Wandjina Aboriginal cave paintings suggest that something interesting was happening in ancient Australia.

And there is one more piece of evidence from Australia that is odd.

Tektites

Tektites are often vitrified pieces of sand from intense heat (as found in the Sahara desert from the impact of meteorites – and in the case of atomic blasts in Nevada which also has them), but Australia is different.

Consider the following, called **Australite,** found only in Australia which according to <u>Alcheringa</u> was one of the remnants of the crystalline mothership disintegrating in the Earth's upper atmosphere:

And again, below, another sample found in Southeastern Australia:

Keep in mind that these tektites do not anywhere have an associated crater, so they did not come in by meteor. The Aborigines say they came from the ancient skycraft, and their chemical composition is not that of an asteroid/meteor.

Very unusual button-shape tektites with a lip….

… dated to 780,000 years old.

Australite

And that is not all, according to <u>Alcheringa,</u> parts of the Pleiadian mothership started disassembling over Greece—Moldavia and drifted southeast over SE Asia and then Indonesia and then to what is now Australia, but deposited a very unique form of green gem material (in all those locations), often as valued as diamonds and pearls for its beauty and rarity: **Moldavite.**

Light green Moldavite – a glass-like gem which can be made into the following:

Clearly the stone can be mistaken for an emerald.

It is easy to cut and facet into a beautiful decorative stone…

(credit: Bing Images: Moldavite gems)

Rocks from The Sky

In addition to tektites and Australite, the Lyrans also rained down boulders and asteroids on the Reptiles – driving them underground.

Naturally, you have to wonder if there is any proof of such claims made in Alchcringa. Actually, in addition to the Australite stones having been dated to **780,000 BC,** there are two other evidences.

Round Boulders

Found in what was once Yugoslavia, there are perfectly spherical boulders not formed by any natural geological process – and they are also found in abundance in New Zealand…

New Zealand… .

…also Costa Rica:

Costa Rica

...and Bosnia...

So what is the deal? Just that both Reptilians and Lyrans shot these stones at each other – **it avoided radioactivity** and fire damage and they carried quite a whallop. They were lifted by antigravity technology and propelled by the same antigrav force at the target. (Kind of like a large railgun.)

Lyran Bombardment

According to <u>Alcheringa</u>, the Lyrans also bombarded the Reptiles with asteroids, meteors and these round stones above. Proof? Yes, indeed --

If such an anomalous event happened, there must be evidence of it somewhere and the author says

> I eventually discovered there really had been a multiple object bombardment of our planet involving meteors and asteroids, dated to between **770,000 to 790,000 years ago...**
> **[there is the magic 780,000 BC time frame again]**
>
> The reason why this was not common knowledge or anything I had heard about before was that the discovery had only just been **announced in late 2016** and then only in one brief news item.
>
> A geological study team from Heidelberg University had revealed that multiple meteors all impacted at around the same time, about **780,000 years ago**, samples from several sites across Asia, Australia, Canada, Central America, and the Darwin Crater in Tasmania, all pointed to this same conclusion. [18]

The Rest of the Story...

The Pleiadian Starship came to Earth and was treacherously destroyed about 700,000 years ago – while the Reptilians were still here. As was reported in <u>Anunnaki Legacy</u>, the **Black** race was created in and for Africa and the **Semitic** races (Jews and Arabs) were created in the Mesopotamia area. The **Chinese** race was deposited by the beings from Deneb, via Huang Di, the Yellow Emperor who came to Earth in a Red Dragon. And the **Brown** race was created and deposited by beings (not the Anunnaki) in what was then called MU – an older, sunken continent affixed to the southern end of what is now India and extended to just north of Australia... giving rise to the South Sea Islanders over time (and migration).

Lastly the **White** race was created and deposited in the Northern Europe area… and in all cases the progenitors looked after their genetic offspring, educating them, protecting them, even culling them from time to time.

Humans were given training in writing, medicine (plants), metal making, astronomy and the construction of structures to track the seasons as well as the equinoxes, agriculture, Religion and law (Think: Hammurabi Law).

Coming from the sky in their craft, with technology that appeared to be magic to the primitive humans, the beings that piloted them were seen as gods and thus arose the many versions of **Skygods** around the Earth:

>Norse gods, Greek gods, Dogon gods, Zulu gods, Hindu gods, Chinese gods, Mayan gods …

In fact, now is a good time to do a quick review of the Skygods and the better known versions who were responsible for Mankind.

The Nine Gods

Both the Mayans and the Egyptians spoke about Nine Gods: the Mayans referred to them as **Bolon Yokte** who is still expected to return to Earth in the 21st Century, and the Egyptians referred to the **Ennead** who were their creator gods, staring the Egyptian civilization. …

Whereas some of the gods operating in Greece and Rome were also identified as Anunnaki (VEG, Ch. 11), it behooves us to see who the Egyptian gods were, in the same way.[19]

Egyptian God →	Anunnaki God
Amen-Ra/Ammon	Anu
Thoth	Ningishzidda
Ptah	Enki
Ra/Aten	Marduk
Horus	son of Osiris and Isis
Osiris	son of Marduk
Isis	Ninmah
Set(h)	son of Marduk

Enlil and Inanna did not have known 'god' roles in Egypt.

The officially recognized Nine Gods listed by Egyptian Scholars:

The god Atum, his children Shu and Tefnut, their children Geb and Nut, and their children Osiris, Isis, and Set(h) and Nephthys.

And in Egypt at the Edfu temple, on the Upper Nile…

Shebtiw at Edfu

Edfu is an ancient site in Upper Egypt, south of Luxor, where **Horus** (the bird-god in Chapter 1) built a massive temple. In fact, Egyptian ancient history says that there were three sets of gods involved: the **Builders**, the **Sages**, and the **Shebtiw**. The Egyptians believed that those 3 were the "olden gods" who preceded the better-known ones that ruled Egypt and Mesopotamia after 3000 BC.

> Horus established a **foundry** at Edfu and enlisted humans
> in his fight against the evil Seth [who killed his father Osiris],
> arming them with weapons "forged of divine metal."

The temple description also states that Horus kept there his **great Winged Disc** ; "when the doors of the foundry open, **the Disc riseth up**." [20]

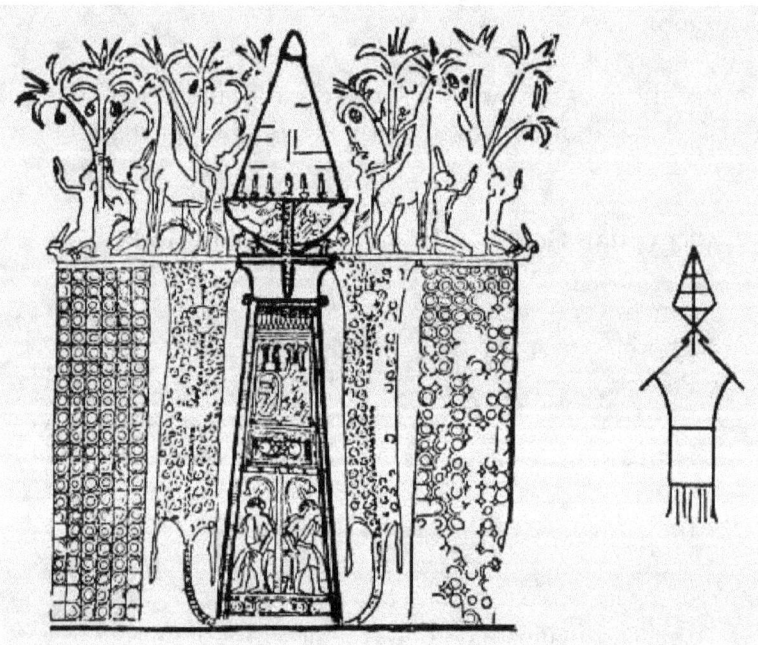

from that Sumerian glyph, we get...

...and the Egyptian version...

...and then the real thing...

The "olden gods" were said to have come by boat to Egypt and included **Ptah**, the one who 'developed' or made things, including raising the land of Egypt from the waters. (Remember that **Ptah was Enki and Thoth was his son Ningishzidda** – already gods in Sumeria.) Said Mr. Sitchin, "…. **Evidence shows that the gods of Egypt originated in Sumer.**" [21] The Egyptians believed in Gods of Heaven and Earth and another set that they referred to as **the Great Gods who descended from Heaven** and they are still recognizable on temple walls by their horned headgear (i.e., the Anunnaki).

(credit: Bing Images: fineartamerica.com)

Enki is the second from the right with the streams of living water coming from him (hat horns indicate rank [VEG Ch. 3]).

Be clear that the gods shown above who descended from Heaven, came first to Sumer and then came later to Egypt.

Seven Sages

In addition in the pantheon of gods from Sumeria, i.e., among the Anunnaki leaders, there were also the **Seven Great Anunnaki** who were also called the Seven Gods of Destiny, the Seven Gods Who Judge, and these same were called the **Seven Sages** in Egypt. [22] The Seven Sages were sent by Enki to teach humans civilization, [23] and had a part in the **Egyptian Creation Myth.**

According to the Edfu Temple "texts" (which are hieroglyphics on the walls), the gods gathered together , **Thoth and the Seven Sages**, at the place called Djeba (the Falcon) in Wetjeset-Hor. Horus was called the Falcon, as was his **ship** (which was a Mu'u) which (above drawing of rocket) was said to be kept at the "foundry" or temple (AL, Ch. 1)

According to E.A.E. Jelinkova, Thoth and the Seven Sages created the Edfu temple and among their retinue were two gods Aa and Wa.[24] From another source, we learn that A (or Aa) was another name for **Thoth**. [25] And we already know that **Thoth** was Ningishzidda (thanks to Messrs. Sitchin and Lessin) and was sent by **Ptah (or Enki)** – and that is how it all comes together.

Creation Versions

But all was not in perfect harmony – at least not for long. It was said that the land was risen from the sea by Thoth and he stood on the initial island of Earth, initially also called the **Island of the Egg** – symbolizing the birth of something. It was said by some sources that his word created the land – reminiscent of the Book of Genesis where God speaks things into being... and yet another source said that Thoth stood on the primeval island (Atlantis) and ... well, let's work up to that....

The Book of Genesis appears to show two creations (See AL, Ch. 12), and the **Egyptians have at least 4 different versions of creation**: [26]

> (These are examined more in detail in <u>Anunnaki Legacy</u>, Ch. 5. Remember:
> **serpent symbology is often an oblique reference to the Anunnaki, who were serpent-like Reptilians.**)

The 4th version is the most interesting, as it relates to the Anunnaki, the Hopi, and the Aborigines of Australia... and the ubiquitous serpent.

There was the great infinite ocean, called Nun or Nu, and from it emerged the god Atum from the primeval waters, and stood on a raised mound, called the **Island of Trampling**, and created life...

Remember Atum was later called Amon-Ra.

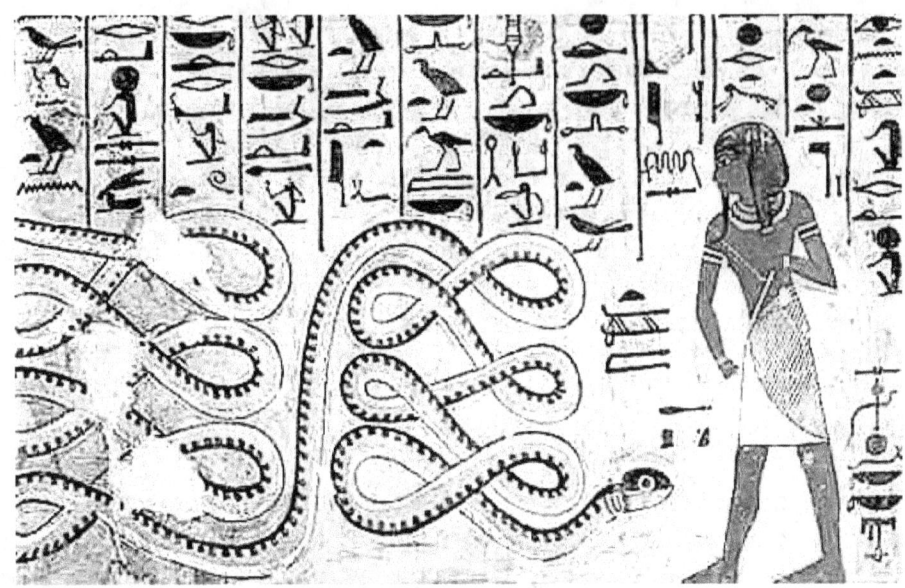

The God Atum Fighting the Serpent…
(credit: landofpyramids.org)

Ok, now we have Atum (who was Thoth in an earlier version) and he is fighting a **serpent**… on the primeval island with a weird name… "Trampling?"

War Among the Gods

Unfortunately, the symbolism of Atum (aka Thoth aka Ningishzidda) fighting the serpent is all too real. Said EAE Reymond:

> The domain of the Wetjeset-Neter is now attacked by the enemy Snake… When the enemy, **the snake, appeared at the landing stage** [platform] of that domain, a bw-titi, Place-for-Crushing , was planned and protective guards of the gods were formed…. There is allusion to **a fight on the earth** in front of the shelter. Another fight took place at the same time **in the sky** in which the Falcon [Horus] was believed to fight against the snake named sbty [shebtiw].
>
> The Edfu cosmogonical records begin with a picture of the primeval island where the gods were believed to have lived <u>first</u>… the Earth-Maker is said to be the snake who created the Earth… [27]

> **Note that the Aborigines have the Rainbow Serpent – also a creator serpent (below)… again a connection to Australia.**

Actual Aboriginal Rainbow Serpent – Rock Painting

And while we are totally segue-ing way out in left field, let's remember the American

Southwest rock art of the **Hopi**...

Aborigines and Hopis did not know each other and yet they both painted **serpents**... venerating the Reptilian gods? And ants (aka MIMINU) ? Note that the Hopi also have the words Anu and Naki -- they mean "ant" and "people"... and the figures with antennas could represent them.

So what we have is serpents at the beginning, creating the world (land from the sea) and the humans and some other gods, and then getting into a fight and destroying the island (Atlantis) and dwelling place (calling it the **Island of Trampling**), then it all gets Flooded and then it gets rebuilt. And it appears that the Builder Gods, the Sages and the **Shebtiw** were the Anunnaki since they were originally Reptilian (see VEG Ch. 3 – Berossus, the ancient Babylonian scribe, who saw them, said so).

And it is known that **the gods did fight among themselves**, ground and air battles), and some of the original inhabitants died. It is still maintained that the **Nagas in India** (underground) are said to be the Remnant of the original Anunnaki – they were called the "serpent people." Said EAE Reymond:

> A further important fact that emerges … the allusion to the under-world… makes it clear that **the underworld existed before the world was created**… [28]

And that is significant as the creation myth of the Amerindians (**Navajo and Hopi) and the Inca** are said to have emerged from under the Earth.

By now the reader should begin to see some recurrent patterns and connect some of the dots regarding creation, serpents, beings living underground, and Skycraft.

In addition it is clear that the Reptiles (not so much the Dracos, their creators, by the way) have been here a long time. What we'll see later is that there are good guys/bad guys among the Anunnaki Remnant.

Anunnaki Sculpture

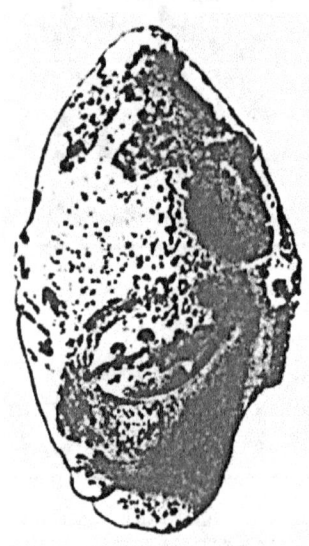

8- Terracotta reptilian head found at Choga Mami (Iraq), Samarra period (6th millennium B.C.). Although somewhat too vertical, it exhibits the typical elongated head of the Gina'abul (lizards). Compare with the male and female figures in chapters 4 and 5 of Part II.

Just so you have a frame of reference, and can begin to get used to the idea that not all created beings "Out There" look human, this is how various cultures around Earth have pictured the Anunnaki/Skygods:

Left: Anunnaki sculpture from Iraq ca. 600 BC.

(See Appendix E.)

From Africa....

Statue from the Ubiad Culture, in Mesopotamia about 3500 B.C....

Here (left) a mother Reptilian is nursing her young...

According to Sumerian tablets, the **"round dots"** are the ME or the Sumerian crystals that stored their laws and science.

…and next, from India, we have the Nagas, the underground serpent people…

…and from Peru… an awesome reminder of the Anunna…
(with a vulture on its head)

...and from the Mayas...

...and again from the Maya reflecting the Ubaid culture....
The Maya and the Ubaid both depicted "dots" on the shoulders....

15- Female Amašutum figurine of the "lizard head" type. It is made of terracotta, 15 cm high, was found at Ur and has been dated to the Obeid period (5th millennium B.C.). This statuette displays an elongated wig (made of bitumen) similar to those worn by certain Amašutum. There are "studs" on the shoulders, an ornament similar to those that can be seen on the shoulders of high Mayan dignitaries (see below). Among the Maya, these circles symbolized "OL," which meant "perception," "consciousness," the "way" and "memory." The Mayan OL is related both to the Sumerian UL, which means "past" or "ornament," "star," "splendor" and the verb "to shine," and to UL_5, "privilege" and "protection." These circles or studs on the shoulders of Sumerian gods and Mayan dignitaries symbolized crystals, or ME, in which the knowledge of the gods was stored. In note 32, we saw that crystals like quartz were used as transmitter-receivers to store and send information (see the male figurine in the next chapter, fig. 18)

16- Mayan priest with the OL or IL ornaments on his shoulders, symbolized by quartz crystals. Only "gods" and high dignitaries enjoyed this privilege in the past.

[43] In Sumerian, the names *Mami*, *Mamí* or *Mama* clearly allude to the mother. They are regularly found on tablets and refer each time to the Mother-Goddess, the Mother of the Earth or the goddess of Fertility.

Notice the "bumps" on both figures' shoulders.

Those are Sumerian ME's – crystals of power.

…and again from Iraq…

We have a unique pottery design (next page) that resembles the Ubaid sculpture we saw earlier, and now from a different angle… note the circular "bumps"…

and the pottery counterpart on the following page…

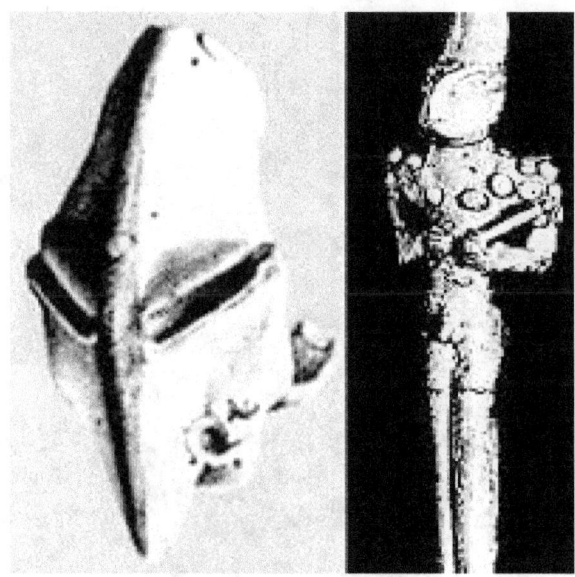

Iraqui Pottery Humanoid

Sure, that could be just a ceremonial gown of some sort (below) except that the face is not human, nor are the eyes normal, and the very snake-like arms are noticeable…

and the dots/bumps resemble those of the Ubaid sculpture above.

Dots or bumps

We have no idea what the artists above and left were seeing. But there is a commonality that suggests they all saw something similar… and not human.

And then there is one more piece of pottery that shows a serpent-like humanoid… (see face under arrow)…

Ceramic Vase: Tell Agraib ca. 2900 BC....
...the round orbs they hold are alleged to give the bearer power
to travel between 3 dimensions... [29]

Skygods in Ancient Art

And to conclude this survey of historical anomalies, in art, we have many examples of Medieval, Renaissance and foreign artists painting skygods…and it is included here as much for proof of visitation as a convenient source for future reference.

From Japan…

... from Europe... Middle Ages...

...and while religious, it is what it looks like... a UFO...

And as seen here and next page, even Da Vinci got into the act…

… the expanded scenario is below…

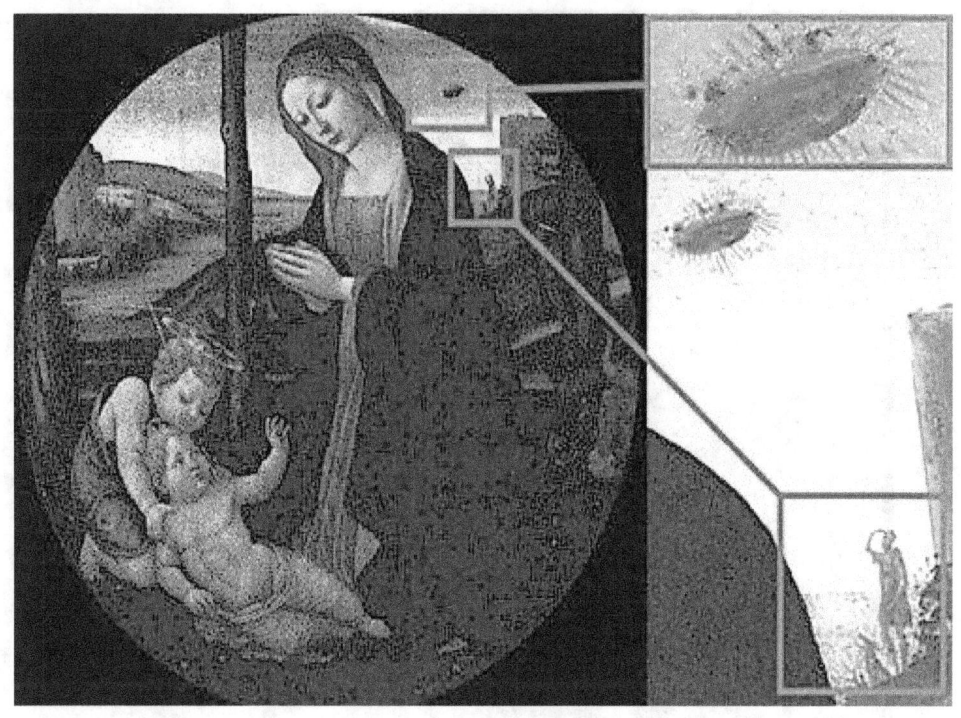

...from Nüremberg, Germany, 1561 AD......

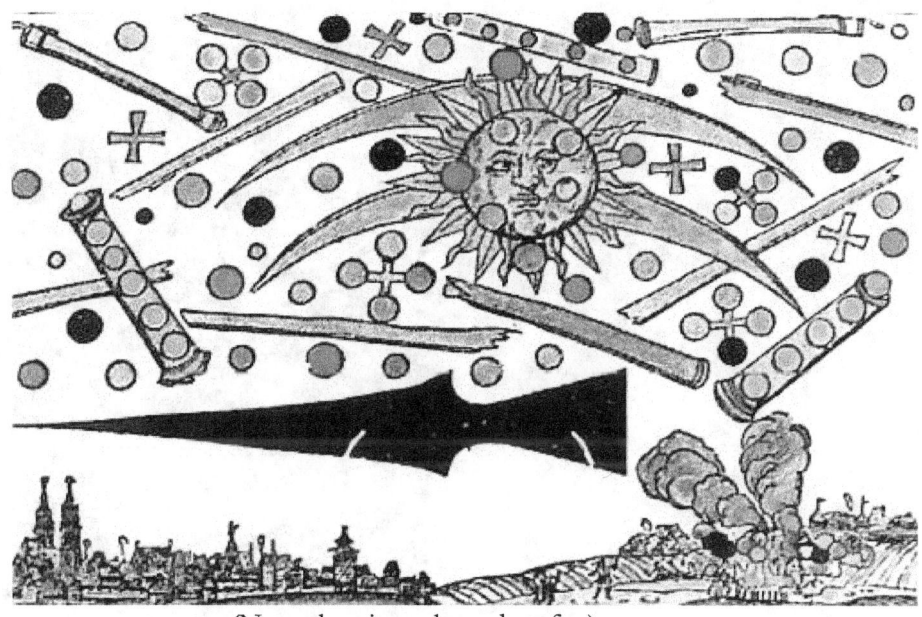

(Note the cigar-shaped craft...)

...and then from India, we have a **Vimana**...

As well as a scene from the Mahabharata...

..battles in the air between flying craft...

... and lastly a very interesting one from Europe...

The preceding paintings seem to give a whole new (secondary?) and subtle meaning to the cryptic painting by Da Vinci ...

Is he suggesting that there is significance to something above us?

Appendix E has more...

Summary

The preceding was just a brief survey to acquaint the reader with the basic elements of Skygod intervention around the planet – a more full examination was given in the 700 pages of <u>Anunnaki Legacy</u> (AL), which was a world-wide survey of what the Skygods did and where.

Note that they fought among themselves, not just Reptilian against Pleiadian, but Gina'abul vs Gina'abul (Anunnaki vs Anunnaki)! And many humans in today's world inherited the Anunnaki genetics (explored in Chapter 3) whereas some humans seem to be more peaceful, altruistic and probably inherited the Pleiadian genes.

> Both produced genetic offspring – the Pleiadians just to modify their race and survive the hostile environment.

In addition to the benevolent ones helping Man, the more antagonistic ones not only fought among themselves but used Man to do some of their battles (Think: cannon fodder) and visited disease on Mankind – *Suruppu* Disease (qv, the *Atra Hasis Epic*, a Sumerian document: see VEG, Ch, 3), and the Black Plague (AD 1347) for starters. Could Ebola and HIV also be their creation?

> **It has always been about Control of Mankind and there is (still) an Earth Control Agenda as will be seen.**

As also seen in the **Mahabharata,** the gods did not always get along, and the Anunnaki gods fought a terrible sky battle with the Hindu gods, called the **Kurukshetra War**, using Vimana against the Anunnaki Skycraft (see AL, Ch. 4).

> Even Horus in Egypt fought Seth in a sky battle.
> Puma Puku was destroyed in a sky battle.
> Even the mighty and real Atlantis came under Anunnaki fire…

> the Reptilians do not give up.

The point being that Earth history has been long and varied and has involved ETs, Skygods, the **Shining Ones** (Gina'abul because their fine coat of scales glisten in the Sun), and the Ancient Ones, as was examined in <u>Anunnaki Legacy</u>.

We have never been alone and still are not alone today.

As will be seen, not all the remaining Skygods are benevolent and some are up to no good, still trying to control their human offspring – and not just to educate them. Reptilians are still here still trying to get control of the planet that they still claim is theirs.

Fortunately, the Pleiadian offspring are also among us, sometimes called the Indigos or Star Seed, and they often obstruct and remove the Anunnaki cutouts (i.e., puppets) like the PTB. And that battle is currently ongoing and is also a subject of this book.

Its projected resolution is offered in Chapter 8.

The rest of this book is an examination of what the Remnant are up to, where they are, and why it is important to know, and what one can do to obstruct them... as well as protect oneself.

Knowledge Protects
Ignorance Enslaves

Chapter 3: Genesis, Genetics & Epigenetics

The following is meant to show why Man is the way he is… subject to disease, not too bright (compared to our more advanced human-like "cousins" on other worlds), and how the Anunnaki genetics (which were petty, violent and greedy) translated forward combined with the Homo *erectus* primitive genetics and form thus creating a dysfunctional *Adama*. That is also the historical genetic basis for Homo *sapiens*.

It is largely due to our inherited **dysfunctional genes** that (1) the Greys are abducting humans, trying to proactively rewire us, and (2) there still has to be some system of control (beyond Religion) to keep Man from running amok.

The Anunnaki Remnant Dissident solution is about to be implemented and it is not good… you need to know about it so you can opt out. Otherwise they will **violate your freewill** and trap you in a new slave system (Chapters 4 & 5) – that is the only way the Reptilians can handle humans who have souls – that scares them as most reptilians **do not have a soul potential** – more about that later.

Man's Genetics

As was suggested in Chapter 2, there are different genetics running around the planet to this day, notwithstanding what the little Greys are doing in their abductions to Man's genome.

Not only did the Anunnaki modify the Homo *erectus* and add their DNA to re-create *Adama*, but the Pleiadians had to also grab whatever hominid they found (whether Homo *erectus*, Homo *habilus*, or even an *Adapa*) and re-create a viable human-like form into which their progeny could birth – they took their egg + sperm and modified it then inserting it into a Pleiadian mother to carry it to term. This way a hopefully better, stronger (less susceptible to the harsh Earth environment) child would be born and survive.

> This was in fact what the Anunnaki did to make them-selves more presentable to humans (who found their

native reptilian form [right, below] ugly), and inserting genetically modified eggs into surrogate Reptilian mothers, such good-looking humans as **Inanna** and **Marduk** could easily move among the humans and not scare them off.

 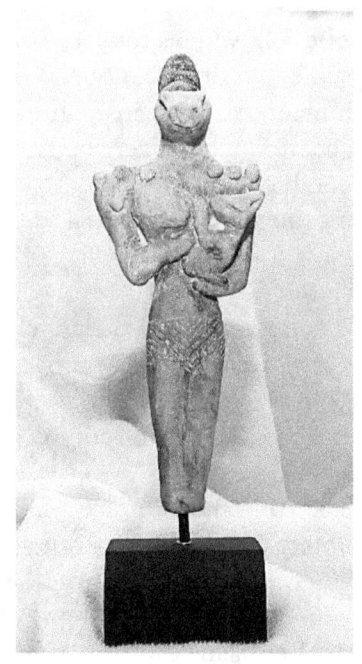

An authorized Sumerian bust of Marduk (in a French museum), note the large head, reminiscent of the Egyptian Pharaoh Akhenaten ...

Marduk Sculpture

(See also Appendix E pix.)

…and Akhenaten's wife Nefertiti… (below, right)…

…also note their kids have elongated heads, too… those were not just stylized hats… Needless to say, these two rulers (below) were looked on as "different" and were summarily rejected for trying to remove the pantheon of Egyptian gods and replace them with RA, the one God of the Universe (symbolized by the Sun in the picture)…

Egyptian Pharaoh Akhenaten and Nefertiti
(credit: Wikipedia: public domain)

Note the 3 kids' heads are elongated.

Up to the point where Earth-born Anunnaki came on the scene, the Reptilians had to build pyramids (in Latin America), ziggurats (in Mesopotamia), and tall temples in Angor Wat, in which they could stay **high up out of the view** (and reach) of humans and there was usually just one human priest or vizier who communicated the Anunnaki laws to the humans below.

The ruling Anunnaki eventually went underground and ruled from there, as they still do today.

Note that climbing the stairs was awkward and was discouraged by making them very steep. (Mayans did not use cables.)

And the gods were very fond of human flesh which is why there was so much **human sacrifice** at the top of the pyramids – sometimes hundreds at a time (and no one has ever found a mass burial pit with human bones – and they never will… think about it.) Aaarrgh. Huitzilopochtli was the Mayan god who promoted sacrifice while Quetzalcoatl tried many times to stop it… again, the Anunnaki disagreed among themselves, but sacrifice was actively promoted again with the Aztecs, Moche and Incas in Peru, and it even occurred earlier with the Pleiadians – if the Reptiles caught them.

Genetic Memory

One of the reasons for the **cannibalism,** and why it is also practiced in Africa and the Amazon jungle, somewhere the natives got the idea (or were told) that eating one's enemy delivered that person's strength and smarts to the eater.

Say what?!

The underlying principle is one that is being discovered by today's genetic researchers, and it is called *genetic memory.*

> We can think of the programmed automatic behavior of the newborn babies in this category. The emerging field of epigenetics

has revealed that life experiences …. can be passed on to children. Studies have shown that survivors of traumatic events may have effects in subsequent generations. **Our experiences change the way our genes express themselves, and this genetic expression can then be inherited.**

There is growing interest in the capacity for DNA to store memories. That DNA can record many forms of information is no secret, researchers have even come up with a way to encode digital data in DNA. [30]

We know today that some of the traumatic experiences of our ancestors have somehow predisposed us to fear things, like spiders, **lizards**, and bats without us consciously knowing why… perhaps that was also part of the Survival of the Fittest paradigm. **Dr. Berit Brogaard**, specializing in the area of cognitive neuroscience, says that some gifts like a talent for a musical instrument or a gift for mathematics can be passed on… If not the predisposition, then the actual 'programmed' ability in our DNA from ancestors who were good at it. [31]

Scientists are also saying that our so-called **"junk DNA"** is not junk – it does something and is there for a purpose… we just don't know what is [or was] stored there… Just memories or latent abilities the Anunnaki did not want us to have.

Human Divergence

Scientists have said that we split with our ancestral chimps about 5-6 million years ago. And yet what we find is that there is more genetic diversity among chimpanzees in Africa than there is among humans: "…all living humans have virtually identical DNA, differing by only 0.1%. This is an oddity …" and **our species lacks the usual depth of ancestral DNA** typical of all other mammalian species! [32]

Shallow evolutionary history should not be the case where Man has been on the planet for millennia.

So what does that mean? We need to look at some very recent findings (which did not make the Fake News Media, by the way.)

Scientists have extracted DNA from the fossils of the Denisovans, the Neanderthals, and the Cro-Magnons thus allowing for the calculation of when these three types of humans diverged from a **last common ancestor (LCA)**. The investigation was headed by scientists from the University of Utah and "produced a date close to **800,000 years ago** for when these three human forms began to diverge." [33]

There are also beta-globulin sequences in today's Asian population that are thought to be descended from an ancestor of modern humans living around **800,000 years** ago, and there are multiple lines of evidence that point to the divergence of the first Homo *sapiens* from the LCA about **780,000 years ago**.... [34]

> **Humans have an anomalous evolutionary origin and it matches the destruction of the Pleiadian mothership about 780,000 years ago.**

Key Genetic Aspects

So what are some of the things that set us apart from our nearest "cousins" – the chimpanzees?

223 Genes

First, Man has **223 genes** that are not found in any other mammal or primate, let alone any other organism on the planet.

Because The Others added the 223 genes to our genome in their genetic upgrade of Man (*Adapa*). **This is the "smoking gun" that proves Man was a product of "assisted evolution."** It remains to be seen just what function the 223 genes have... Science is still trying to figure that out.

> An analysis of the functions of these genes, published in the journal *Nature* (issue No. 409), showed that they involve important **physiological and cerebral functions peculiar to humans**. Since the difference between Man and Chimpanzee is just about 300 genes, those 223 genes make a huge difference. [35] [emphasis added]

If someone went to that much trouble to make Man different and more capable than the Apes, then it would stand to reason that They are probably still around, watching to see how Their progeny turn out.

Chromosome-2 Anomaly

All primates except for modern humans have 48 chromosomes as 24 associated pairs – one from the mother and one from the father.... Humans have 22 plus the 23d which is the Male/Female determinant gene (X vs Y).

100

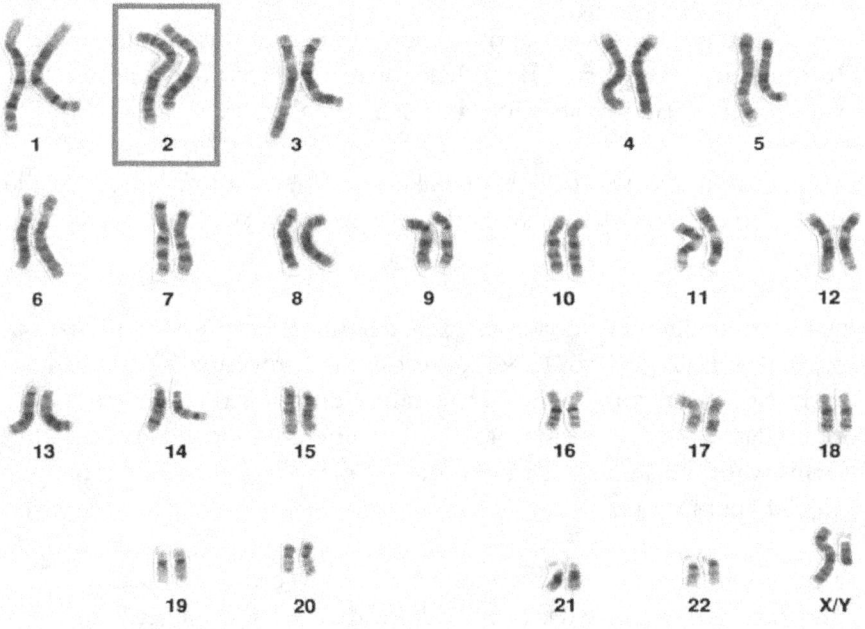

(credit: Wikipedia: National Human Genome Research Institute)

Here is the weird part:

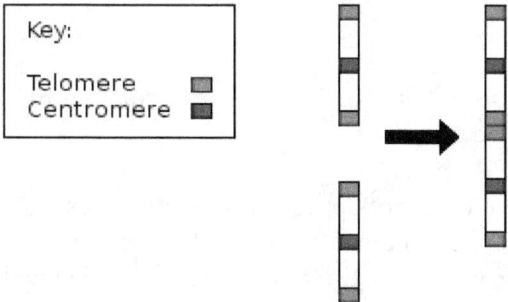

What is unique about Chromosome-2 is that **in the middle** (where the arrow is pointing, above) is the remnant of protective terminal **telomeres**… only found <u>at the end of the chromosomes</u> which act much as the plastic tips of shoelaces to keep the chromosome from coming apart. Somehow (and this required intelligent intervention to do it) the **two ends were fused without losing any of the base pairs… NO other chromosome except #2 has this structure.**

Why? And what does it do? This is another **smoking gun** that Man's genome was 'hacked' – but why?

The fact that Denisovans and Neanderthals share this chromosome fusion means it occurred BEFORE the divergence of these three subspecies 780,000 years ago – in a shared ancestor.

The first assumption we can make is that like all primates, our ancestor would have had 24 pairs… but there is a lack of DNA from that far back to see where and what it was.

Secondly, we know that chromosomes exchange material thru recombination, sometimes swapping A & T for C & G, but this was prevented by the fusion of the two halves of the chromosome. **Dr. Adam Benton** at the University of Liverpool validated that the fusion of Chromosome-2 happened at the point where Denisovans, Neanderthal and Homo *sapiens* split <u>and</u> he said it happened about **780,000 years ago**. [36]

So the aim of such fusion would have been to create a superior version of Man and it has been realized that the several genes in Chromosome-2 that are found near the area of fusion point are **expressed more intensely** in our species than in that of large apes.

They are expressed particularly in **the brain** and gonads.

The fusion site is in the *intron* of a gene… and some of the genes affected by this intensity are: [37]

DDX11L2 – involved with the nervous, muscular, immune and reproductive systems.

ARHGAP11B – it does not appear in other primates and used to be part of a longer gene but was "chopped up" and was reinserted into the human genome – something we can do today with CRISPR… but how was it done millennia ago?

miR-041 – crucial for brain development, and affects decision-making and enabling language ability. **Martin Taylor** of the University of Edinburgh says this gene "sprang from nowhere at a time when our species was undergoing dramatic changes."

FOXP2 – empowers human language ability. Without it, Man cannot speak.

SRGAP2 – affects dendrite spines and neural growth in the brain. Also affects memory and synaptic transmission between neurons. This is involved with 48% of the prefrontal cortex making us different from chimps.

Where it really counts in genetic expression, humans have some genuinely profound differences from chimpanzees.

The benefit? Man is superior to the Apes and chimps and he can outcompete and surpass all others – as was the **design**.

Blood Types

In keeping with someone's desire or intention to enhance and protect their progeny, we have discovered a very interesting situation with blood types.. and there are 4 basic types of human blood, known as O, A, B and AB.

And as many people are aware today, a woman who marries a man with a different blood type may doom her baby to an early death. This has to do with a feature of the blood called the **Rhesus factor**, i.e., whether the O type for example is O+ or O- . Rh positive male and Rh negative female cannot create a healthy baby – unless drugs are given the mother to inhibit the mother's immune system from producing lethal antibodies to kill what is perceived as something alien....

Now consider: if you are creating a special line of humans and you don't want them to interbreed (say Pleiadian humans with Anunnaki humans), this Rh +/- system will do the trick nicely.

It is called **deliberate engineering**.

And as you might suspect, Rh- people are often a little different. They have been studied and found to have some of the following characteristics that set them apart from Rh+ humans:

> They often have a keen interest in UFOs, ETs and other worlds. They have a higher-than-average psychic ability, and their IQ tends to be higher than average. Physically they have a lower than normal body temperature, slower pulse, lower blood pressure, and often an extra vertebra. They are more resilient to disease and illness than the general population. [38]

A good example of Rh- in today's world are the **Basque** people – they are an amazing 40% Rh- (higher than any other population on Earth) and there is almost no AB or B blood among them. Not only that, but their language, **Euskara**, has no ties to any other known language on the planet.

Alternate Origins

Panspermia

Scientists still cannot accept that Man could have been designed and built by ETs who came here – with the same genetic knowledge that we are now acquiring. So they admit that the building blocks of humans' genetics might have come from the stars – but carried here by **asteroids and comets.**

The concept is called Panspermia… Earth was pollinated with living microbes and proto-genetic material, most likely off of an ice-bearing comet.

Close encounter of Earth and a giant comet.

(credit: BingIages: everythingselectric.com)

As was said above (and in TOM, Ch. 2), **Man has 223 genes that are unique to him** and are <u>not found in any other organism on Earth</u>. If the scientists believe that *panspermia* is the <u>only way</u> genetic material gets here (bacteria from meteors), it falls to Earth and disperses in the soil and water, the bacteria propagate among the animals that eat the plants that absorbed the organisms, and drink the water that contained the bacteria, or eat the fish that absorbed the bacteria in the water, then why aren't the unique-to-Man 223 genes found in other animals or even in free-living bacteria themselves? They're not.

Panspermia may have been responsible for some flora/fauna (biota) on Earth, but Man is too complex to have evolved by random selection…

Junkyard Tornado

The likelihood of life just springing up from a pool of amino acids (**Abiogenesis**) is about as likely as a tornado sweeping thru a junkyard and assembling a fully functional Boeing 747.

The **junkyard tornado** is an argument used to deride the probability of abiogenesis as comparable to "the chance that a tornado sweeping through a junkyard might assemble a Boeing 747." It was used originally by **Fred Hoyle**, in which he applied statistical analysis to the origin of life, but similar observations predate Hoyle and have been found all the way back to Darwin's time, and indeed **to Cicero in classical times**. While Hoyle himself was an atheist, the argument has since become a mainstay of creationist and intelligent design criticism of evolution.

Hoyle used this junkyard allegory to <u>argue in favor of panspermia</u>, that the origin of life on earth was from preexisting life in space. [39]

But simple Panspermia still doesn't really cut it.

Advanced Panspermia

One more aspect of the genetic design that would oversee the formation of the animal bodies is called the **HOX gene**. This gene controls the body development process, and which body parts go where; the HOX genes switch on other genes and

> Hox proteins determine the type of appendages (e.g. legs, antennae, and wings in fruit flies) or the different types of vertebrae (in humans) that will form on a segment. Hox proteins thus confer segmental identity, but do not form the actual segments themselves. [40]

> This detail is not to bore or annoy you; I want you to see that the complexity of the human being could not just have arisen willy-nilly from a primordial soup; i.e., Abiogenesis is nonsense, as is simple Panspermia.

So here is a final word on the Panspermia issue. [41]

> The **HOX gene control system** … indicates that the architecture of the human genome would have to have been **developed** as a complete operational unit. It looks as though such a system would not have been able to be generated by evolution or strong (simple) Panspermia.

> **To explain the development of complex life on Earth, we have to consider that a different mechanism was in operation, and I**

call this mechanism Advanced Panspermia. It is possible that intelligent life on Earth is the result of a purposeful experiment carried out by an extraterrestrial civilization….

Brain modifications requiring changes to hundreds of genes needed very precise **genetic manipulations.…**

Several genetic studies indicate that there is a vast genetic difference between H. *sapiens* and Neanderthals. Such a big difference could only arise in such a short period of time as the result of **purposeful genetic modifications.…**

Support for extraterrestrial design also comes from the analysis of our genetic code. Work performed by **shCherback** and **Makukov** shows **unexplained properties of DNA codes** which would require some intelligent approach that cannot be accounted for by Darwinian evolution. [emphasis added]

The genetic makeup of the brain has not changed for the last 6,0000 years, and yet we see significant development in human thinking and consciousness – meaning that Man is becoming smarter, more aware, and none of this is based upon any continued evolutionary physical growth of the brain, but rather on **an in-built aspect** of the human brain – already designed and put thereby whoever created Man and his brain. That spells forethought and design for future expansion of abilities… Are psychic skills also in the near future?

And there is one more aspect of Man that cannot be easily explained…

The Cobalamin Enigma

There is one last reason to believe that Creation, even **directed intelligent creation** of Man happened, due to a unique human biochemistry that Darwin knew nothing about.

It is the **Cobalamin Enigma**.

Cobalamin is **vitamin B12** which is a necessary micronutrient for humans, whose absence (deficiency) causes serious problems. It is the most complex of all the vitamins and it is the only one that contains a metal atom – **cobalt**. Its deficiency spells anemia, even **pernicious anemia**, and a severe lack of energy – Vegans are warned to supplement their all-vege diet with B12 since they cannot get it from nuts, veges, fruits, berries and beans.

It is vital to the production of red blood cells and nerve/brain function.

Cobalt is a rare metal in the Earth, 0.0029 percent of the Earth's crust. It is known that bacteria can synthesize cobalamin, whereas plants and fungi do not use it. **Mammals must have it in their diet.** So if plants existed before mammals, in the Evolutionary Schema, and animals must have it, why can't Man synthesize or metabolize it?

> The fact that this trace element is necessary for proper enzymatic function in animals yet **does not exist in plants** presents several enigmas. If life began and evolved on Earth [and we suggest it did not], then why did "evolution" select cobalt, a rare element, to play such a pivotal role in animal metabolism? Would not iron, which is **plentiful** … have better played this key metabolic role? [42]
> [emphasis added]

In mammals, especially quadri-ruminants like cows, bacteria produce B12 in their system from what they eat – but single-stomach mammals (like humans) must get it from their diet – BUT **their colons cannot absorb the element** -- it is excreted from the body (which is why dogs often eat their feces). It is a strange set of facts that the large majority of "evolved" mammals on Earth cannot get B12 from the plants that they eat…

> The rarity of cobalt in Earth's crust and oceans and the fact that plants do not depend on it or provide it to plant-eating animals poses a challenge to Darwinian evolution…. The vitamin B12 synthesized in the rumen and by bacteria in the colon is one of the most complex nonpolymeric natural products produced in nature. **Then why did animals evolve such a complicated metabolism, based on an ultratrace element in short supply?** [43]
> [emphasis added]

Man must get it either from some other animal products (e.g., liver) or take a **vitamin supplement**, such as B12 called Cyanocobalamin (or the better form, called Methylcobalamin).

So the issue is: Why did the cobalt metabolic pathway (requiring and using cobalamin) occur in the very primitive organisms, then skip the higher plants, and reappear in complex animals? Equally enigmatic: Why do so many animals need cobalamin to trigger body chemistry (enzymes), yet their gastro-intestinal systems cannot absorb it from most of their food?

Even algae have a gene (**CBA1**) that aids in acquisition of the cobalamin, and yet humans lack this <u>lifesaving mechanism</u>!? This looks to deny the simple Panspermia theory. And research has discovered that about 39% of the human population is vitamin B12 deficient… not able to absorb it from food. [44]

Taken as a whole, the cobalamin enigma seems to indicate that life originated <u>not on Earth</u> in one continuous Evolutionary Pathway, but as Dr. Silver says (in VDoG, Ch. 1), on another cobalt-rich planet (Mars?).

> **Why natural selection would create a complex pathway that depends on a hard-to-get trace element does not make any sense, nor does it even fit the Darwinian Model.**

Human Incompatibilities with Earth

Just because it is very interesting to contemplate, **Dr. Ellis Silver** has suggested many of Man's incompatibilities with Earth, and here are a few… [45]

1. **The Sun dazzles us** – our eyes do not have protective membranes, nor can we handle bright glare. Dr. Silver tested several animals with sudden, bright light and glare and they had no adverse reaction.

2. **The Sun damages our eyes** – cataracts are often caused by ultraviolet light and the Earth is bathed in it. And Africans are more susceptible to Glaucoma than Whites, and until 7,700 years ago, **all eyes were brown**, and brown eyes are 2.5 times more susceptible to cataracts than blue eyes.

3. **The Sun damages our skin** – it causes skin cancer, wrinkles and sunburn if we don't use Sunblock 50. Most animals out in the Sun have thick body hair or skin, or scales or feathers… How did we miss out?

4. **Our vitamin D levels are too low** – while too much Sun can be a bad thing, humans also do not get enough vitamin D due to a lack of exposure – indoor workers get more skin cancer than outdoor workers.

> BTW, Dr. Silver includes objections and rebuttals to these listed points in his chapters and he openly examines these issues.

5. **Loss/lack of body hair** – even in Africa where Man supposedly originated, it can get cold at night, but Man is pretty much hairless, unlike a dog or an ape… what kind of evolution was that?!

6. **Our responses to the seasons** – we don't respond to the change in seasons the way other animals do; <u>we develop colds and allergies,</u> and catch the Flu when cold weather hits, and there are thousands of people afflicted with SAD – Seasonal Affective Disorder... a form of depression when we don't get enough sunlight in the Winter.

7. **Chronic Illnesses** – almost all humans suffer some form of chronic health issue – allergies, addictions, anemia, arthritis, asthma, diabetes, dyslexia, eczema, hearing problems, insomnia, migraines, narcolepsy, obesity, psoriasis, scoliosis, sleep apnea and stress – just to name a few. If living on Earth was stressful (as an ancient hunter-gatherer) why weren't we better protected or adapted to handle it?

8. **We can't drink the water** -- just as we can't safely eat raw meat, we can't safely drink from streams and lakes as other animals do. Infections and diseases we can get from impure water sources are horrendous – but it doesn't bother dogs, cats, sheep, deer, cows, bears....!

9. **We dislike natural foods** -- carrots were naturally small, purple and tasted of wood, so we modified them to be orange in color and taste 'carroty'. Not too long ago, tomatoes were poisonous, as are some mushrooms – not everything in our environment is edible... Why is that?

   Some people cannot drink milk – "lactose intolerance" (largely Blacks and Orientals who have no lactase to dissolve it).

Back in the 60's, the World Health Organization wanted to do a good deed for the Blacks in Western Africa, and so they shipped over 10 tons of **enriched powdered milk** with an adviser who showed the mal-nourished natives how to add water and drink it. Then the advisor left.

Two months later, the advisor comes back to see how his project is doing, and finds the natives still mal-nourished and all their mud huts have been painted white – with the powdered milk!!

Furious, he demands to know why the natives have misused the powdered milk and hears that it made them sick – they could not digest it, and so they did the next best thing with the product – now their huts were cooler!

---

10. **We lack sufficient calcium in our diet** -- cow's milk is not part of our natural diet; it was designed to blow an 80-lb calf into a 400-lb cow, and puts weight on mammals. And somehow the human body

finds it very difficult to absorb calcium from food/drink and as a result, many older women have **"Dowager's Hump"** where they are bent over because their spine and bones were the source for calcium that the body processes needed – and they took it, leaving the woman horribly bent over.

And if a woman becomes pregnant, the fetus will take whatever calcium it needs from the mother's body… and when our natural diet does supply enough calcium, we often lack the enzymes to absorb and process it – Where did Evolution go wrong?

11. **Back Problems** – apes and monkeys, horses and oxen are strong enough to put their back into something and not collapse under the strain – So **why is Man such a weakling?** Gorillas are born with the muscle they need to get around and do what apes do, strong, with strong backs, and do not go to the Gym and work out… and it is not because we walk upright – Gorillas walk upright, some chimpanzees too, and Man has been walking upright for 300,000 years – so where was Evolution (and epigenetics) all this time??

12. **Varicose Veins** are a particularly vexing issue for men and women. Those 'bubbles' and sinewy twisting, turning veins that work their way to the surface of the lower legs… and often result in embolisms and thrombosis.
    Man is designed to stand upright, so why does this happen after age 50, and sometimes sooner?

13. **We have a 25-hour body clock** -- in a 24-hour world! Our natural circadian body rhythm doesn't match the Earth's day-night cycle. It looks like we came from **a planet with 25-hour days …**

    Note: 24.65 hours or 24 hours and 40 minutes is the length of a day on Mars.

Wow, do we need to go any further? Dr. Silver has just corroborated **Brinsley LePoer Trench** and his thesis of almost 100 years ago: Man may have originated on Mars. And according to the first two chapters, if the Gina'abul ALSO created slaves on Mars, they would have brought them to Earth with them...

Nonetheless, Dr. Silver lists **53 anomalies** in the <u>lack of adaptive ability</u> of Man to the planet Earth – as if this is not our home because **we were not designed for it…**

Thus, we do not have viable arguments <u>for</u> Evolution of Man on Earth, and the Survival of the Fittest (Epigenetics) dictum has not done too well, either – Why has Epigenetics not resolved Man's ability to adapt to his hostile environment?

So, we are left with a form of Advanced Panspermia or Intelligent ET Intervention as probable origins for Man….

And lastly, there is another gene that has been discovered in many humans living in Northern Europe: parts of Siberia, Latvia, Scandinavia, Northern Germany…

CCR5delta32 (aka CCR5Δ32)

Again, this is an example of **preplanned survival** – somebody preprogramming the human genome to survive a very deadly pathogen. It has been theorized (VEG, in Ch. 3) that the Anunnaki using *Suruppu* **disease** on the humans to control and cull them was a form of HIV. It is also known from the Sumerian epic, the *Atra Hasis*, that the Anunnaki also had the cure.

CCR5 is the gene that codes **white blood cells** and coats them with a protein substance – to which the HIV virus attaches and then kills the **leukocytes. Wouldn't it be nice if the white blood cell had no protein coating?**

Those humans today who have the genetic mutation ("<u>delta</u> 32" means 32 base pairs in the gene were <u>changed</u>, or dropped) and the CCR5 mutation <u>does not</u> coat for protein. That means that when the white blood cell encounters an HIV virus, the virus cannot attach and the leukocyte kills the HIV virus.

> If the child gets the CCR5delta32 code from <u>one</u> parent, the child has functional **resistance** to the disease.
> If the child gets the CCR5delta32 from <u>both</u> parents, the child is **immune** to HIV.

> Unfortunately, the CCR5delta32 gene is found primarily in the Whites living in <u>Northern</u> Europe. [46] (Think: Aldebaran progeny.)

> Again, an example of **deliberate engineering.**

As we are about to go into Chapter 4, dealing with Freewill, it is appropriate to address the issue of Freewill and Genetics… Does a person's genetics affect his Freewill? Don't laugh, there are some scientists who are trying to prove that genes do influence behavior.

Genetics and Freewill

This will not be a thorough examination as (1) we DO live in a Freewill Universe, and (2) if your genes predisposed you to aberrant behavior, or some physical malfunction, then why would we have laws dealing with criminals? – they would not be responsible for what they do and thus punishment for something over which they had no control would not make sense… and these people get locked up.

However, as was said earlier in this chapter, there is a pertinent aspect of genetics that does affect behavior. And that is when the brain or body did not develop normally and one's perception of self and/or others is 'twisted' (as in **psychopathy**) and anti-social behavior results (Also see OPs in Chapter 6).

In addition, there is also the acquired **genetic addiction to drugs and alcohol** which causes a person to repeatedly use them and while under their influence commit anti-social behavior.

And yet, a normal, healthy person whose brain and body chemistry do not drive him/her to anti-social or self-destructive behavior is not genetically-driven beyond what they can control. Thus blaming one's genetics for a lack of Freewill is not credible… and yet, **Karma** and one's **LifePlan** may be more relevant… as will be seen. (Appendix D)

Hidden Biological Forces

She said, "I can't stand bananas, but I like okra!"

I replied, "Yuk, I hate okra, fried or not, but I like broccoli and cheese.."

She thought about it a minute, then added… "Forget the broccoli, but I like <u>fried</u> bananas."

I hate banana pudding, fried bananas taste like shoe leather, and I do not like banana crème pie, nor will I eat banana bread… especially with walnuts -- that turns me off….Ick.
<u>But</u> I like natural, whole bananas… When a banana is cooked something changes the taste and it is repulsive…
Man we were really opposite!

The significance of that is that the latest genetic science has discovered that our genes predispose us to many of our likes and dislikes – from food, to the opposite sex, or to the same sex, and even to whom we vote for. Really.

Likes & Dislikes

I always thought that my likes and dislikes were based on experience or thoughtful deliberation and that I was in charge of what I like or don't like. Then I read an article in NatGeo that made me wonder...

> ...I became acquainted with *Toxoplasma gondii* [at the] Indiana University School of Medicine, [and] I observed how the single-celled *T.gondii* parasite can change the behavior of the host it infects. It can make rates unafraid of cats, and some studies show that it can cause **personality changes** (such as increased anxiety) in humans.

> It made me wonder if there are other things happening under our radar that could be shaping who we are, **programming our likes and dislikes.** [47]

The author of the article did further research and discovered scientific evidence that many of our actions are governed by "hidden biological forces" (which would be part of the LifePlan 'programmed' into the body you were given at birth). And due to our unique biochemistry, we may have little or no control over what tastes good or bad to us...

> In addition to which scientists have found that some humans don't even see the same colors in a picture the same we see them – depending on the distribution of rods and cones in the retina – and that does not even include the **Tetrachromats** who see more colors than the average (Trichromat) person (see p. 496).

> ...[some] people are called supertasters. We have variations in genes that build our tastebud receptors. One of those genes, **TAS2R38**, recognizes bitter chemicals like thioureas which are plentiful in broccoli... [so] about 25% of the people hate broccoli for the same reason I do – it tastes bitter. [48]

Our behaviors and preferences are influenced by our genetic makeup, by factors in our environment (see end of chapter : Epigenetics) that affect our genes, and by other genes forced into our systems by the innumerable microbes that dwell inside us.

(See earlier section in this chapter on "223 Genes")

Sexual Orientation

So what about preferences in the choice of a sex partner? VEG Ch. 14 dealt with the same sex issue and it is not a choice, nor is it due to hormonal imbalances – it is due to the **sexual orientation of the soul inhabiting the body**... if that soul was a woman for 30 lifetimes and now tries to experience a male body, 'she' will still be attracted to males even though 'she' is now in a male body... it is a 'programming' and can be overcome, but in itself, like Bruce Jenner's choice to flip back, it is a learning experience and not something to shame or attack someone for. (And by the way, **souls are not male or female**, but can become 'programmed' in their orientation to others thru past-life preferences, then when they again incarnate, and switch sexes, the 'programming' may prove too strong to fight when in the new body. Tolerance and compassion are **our** lesson when encountering these people – who were brave enough to switch sex when they came back!)

> Of course you have to understand that Reincarnation is a reality – see Appendices A and D.

And why do some women reject some men, and vice versa... it **is** due to pheromones and literally "chemistry."

> ... virtually everything we do emerges from a subconscious urge to survive and reproduce our genes, or lend support to others... who carry genes like our own....
>
> Science has provided a little comfort as to why your amorous advances are sometimes spurned. A famous study had women **sniffing the underarms of T-shirts** worn by men and then ranking the odor. The more similar the men's and women's immune system genes were, the worse the T-shirt stank to the women.
> **There is a sound evolutionary explanation for this**:
> If parental immune genes are too similar, the offspring will not be as well equipped to fight pathogens. In this case, genes used odor receptors as a proxy to size up whether a potential mate's DNA is a good match.
> Studies like this affirm that **chemistry between people** really is a thing. [49]

Political Choices

Even our taste in political leaders is not impervious to whatever is going on in our genes. It seems that the **brains of liberals and conservatives are different in some areas.** Actual physical differences as well as brain neurotransmitter action...

Several studies suggest that variations in our dopamine D4 receptor gene (**DRD4**) influence whether we vote red or blue.

[Hang in there... they have proof!]

Dopamine is a key neurotransmitter in the brain, associated with our reward and pleasure center; variations in DRD4 have been tied to novelty-seeking and risk-taking, behavior more commonly associated with liberals [blues] than conservatives [reds].

Liberals are more novelty-seeking whereas conservatives are more conventional and prefer stability....

Other research has shown that certain areas of the brain are different for liberals and conservatives...

For example, conservatives tend to have a larger **amygdala**, the [emotional and] fear center of the brain and have stronger physiological reactions to unpleasant photos or sounds....

This is why it is so difficult for a liberal or conservative to get the other to "see the light."

You are asking people not just to change their mind, but also to resist their biology... Every human behavior -- from addiction to attraction to anxiety – is **tethered to a genetic anchor.** [50]

And you still have Freewill as Chapter 4 will show, and your actions are not fated to be cast in concrete, but they are those **issues that you are here to overcome**. They are part of the body and family you "inherited" upon entering the Earth Realm and the gods knew that and you are to experience and overcome your "limits" or challenges. (Appendix D)

DNA has built human beings a brain that is so magnificent that it has figured out DNA's game [see Chapter 7 and Epigenetics and Tags]. And with the advent of gene editing [CRISPR] we have become the first species capable of revising our genetic instructions. [51]

...little does he know that we are not the first...

Epiphany

Science has shown that you are not who you think you are. Metaphysics is showing us that we are **more** than we think we are. **There are no accidents** – otherwise you could really raise Hell when you die and return to the Other Side (the InterLife) and ask why They screwed you with challenges you could not handle.

The **gods are not ogres** and no one is given more than they can handle – there is no point to beating a soul up and hitting them with problems they cannot handle, work out, or otherwise rise above!! Think about it.

This is a School not a Concentration Camp.

It is just your opinion that you can't face it, handle it, or that the problem should not be there – They want to see you take a proactive stance (**despite the outcome**!!!) handle it (if possible) and rise above it – because you can! Only one time you can't and that is <u>the</u> Final Test (that all Old Souls have to pass) – can you face totally overwhelming odds, major disaster and still stay in touch with the divine spark (the greatness, your connection to the Higher Self) down inside of you –<u>no matter what</u>? It is very hard, but the realization that you are an eternal soul and CANNOT BE DESTROYED <u>no matter what</u> is a key realization that gets you out of here – that was also a great lesson that Jesus demonstrated on the Cross. Most people missed it.

He was killed and he rose again – they can't kill a soul!

And until you see that, you are stuck in the School until you get it – a realization of what/who you <u>really</u> are! No kidding. THAT is the lesson and when you get that, you can Graduate! (**Chapter 8.**)

That is what the Dissident Remnant and PTB don't want
you to know, or even have a slight realization of it --
<u>so they can keep you here</u>!

(More in Chapters 4 and 8)

EMF & Autism

As Chapters 5 and 6 will show, there is work being done to engineer humans to be better suited to Transgenderism **(H+)** control via 5G and AI – and the effects of 5G EMF are not good if the exposure is persistent and strong – it breaks DNA and causes cell mutations (oxidative stress).

In order for Transgenderism to work more effectively, the
human must become more passive and robotic and that
describes autism.
It was discovered that **autism** could be induced by vaccine via
needles delivering nanites that migrate to the brain and once
enabled via WiFi they can entrain neuronal 'dullness' kind
of mimicking autism. (This is examined more in Chapter 5.)

Autism is also considered a side-effect of WiFi EMF in some children, who are susceptible to it. [52]

> ...ADD and ADHD and all those issues that we are seeing with kids as well as **autism** which I think may also be connected to the use of technology ... And Harvard University just published some research that is connecting the **rise in autism** with a proliferation of all the electromagnetic [EMF] fields that are part of our 21st century environment....

> If you take a look at your child's brain in EMF as I have, then you're going to see how the rates of **autism** and learning disabilities have absolutely soared in the last... twenty years ... And so much of that really mirrors and matches the proliferation of the electro-magnetic fields [EMF] which is .... more than our grandparents were surrounded with. [53]

Digital Dementia

> I think we have an epidemic among children of decreased motor function, of hyperactivity [ADHD], diminished cognition, and diminished learning. That is where all this **digital dementia** is coming from.

> ...and that is why I think it is so important to fight WiFi in the schools... to **text** rather than call, and use the **headset** rather than put the phone up to their head...

> Digital Dementia is the overexposure to EMF which reflects the way the child interacts with the environment and with other human beings and becomes really **addicted to the Internet**, to the cellphone, to technology, to the effect that many social skills are starting to become very deficient...
> ...cellphones are actually **physically altering DNA** and bodily functions and creating all kinds of havoc in the body and disease... as well as the addictive element.... [54]

> (There is more on this in Chapters 5 and 7...)

Is this (below) social or "togetherness?" ....

Or this?

Not much exercise is being gotten... note the inactive skateboard.

And this is **really bad**:

These kids are too young to hold the phone up to their heads... see Chapter 5.and Appendix G on **brain penetration** of EMF.

Autism & Nanochips

It is being said that nanochips, or nanites, being smaller than the diameter of a vaccine needle, can deliver nanites to a person via the needle and they don't know they might be "chipped" while being vaccinated. In addition to the nanites, there is also **mercury** in most vaccines (and Flu shots) and that is not good, either.

> Robert F. Kennedy (author) was approached by a mother who had a child who had **gotten autism from a vaccine** and went to court and proved it. But there is a movement to deny and abolish reporting of such things by *fiat* as somebody (CDC?) has said we have to have vaccinations... and negative results ($\approx$ 20% of the kids) are censored. [55]

Can we make safer vaccinations?

Miscellaneous Genetic Info

The following is for your information, as an example of what the scientists are doing – some of which relate to the development of improved genetics for Man as a whole – even the DARPA supersoldier plan.

Synthetic DNA

Researchers in genetics and bioengineers are planning to rework the human genome... ostensibly to correct genetic flaws as much as insert wanted characteristics in new babies. Such a project is seen coming to fruition in as little as 4-5 years (2025), and a secret meeting was already held in May 2016... to the moral outrage of concerned biologists and human rights activists. It was called the **Genome Project–write** (GP-write) and included a select group of invitation-only experts. [56]

CRISPR

It should be noted that scientists are already 'hacking' DNA and inserting and/or removing sections of DNA to modify the human genome.

> **CRISPR gene editing** is a method by which the genomes of living organisms may be edited. It is based on a simplified version of the bacterial CRISPR /Cas (CRISPR-Cas9) antiviral defense system. By delivering the Cas9 nuclease complexed with a synthetic gRNA into a cell, the cell's genome can be cut at a desired location, allowing existing genes to be removed and/or new ones added. [57]

Cas9 is an enzyme that targets the stretch of DNA to be replaced/deleted. **CRISPR** (**"clustered regularly interspaced short palindromic repeats"**) is a family of **DNA sequences** found within the genome of organisms such as bacteria and ...these sequences are derived from DNA fragments of **viruses** that have previously infected the organism and are used to detect and destroy DNA....

[or modify it].[58]

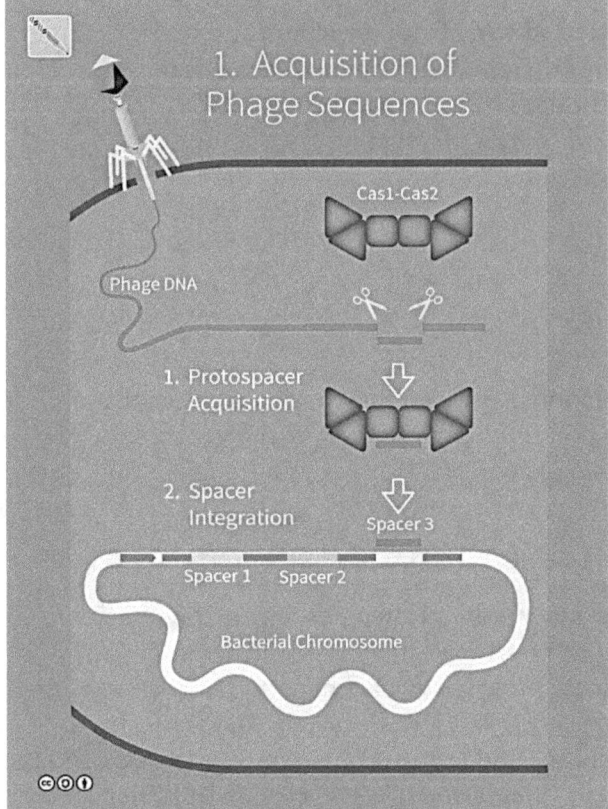

So **a virus delivers the new (phage) DNA sequence**, the enzyme breaks the DNA, and RNA transcripts the insert into the new DNA (basically: oversimplified).

RNA harboring the spacer sequence (steps 1 & 2) helps Cas (CRISPR-associated) proteins recognize and cut foreign pathogenic DNA. Other RNA-guided Cas proteins cut foreign RNA.

(I told you it is very complicated.)

(credit: Wikipedia: CRISPR)

The next one is easier to follow...

XNA

There is a subfield of biology called **xenobiology** – the study of synthetic or 'foreign' biology. Study is now being done to create synthetic DNA... called XNA. The "X" in XNA stands for "xeno," meaning stranger or alien, indicating the difference in the molecular structure as compared to DNA or RNA

> The study of XNA is intended not to give scientists a better understanding of biological evolution as it has occurred historically, but rather to explore ways in which we can control and even **reprogram the genetic makeup** of biological organisms moving forward. XNA has shown significant potential in solving

the current issue of genetic pollution in genetically modified organisms. While DNA is incredibly efficient in its ability to store genetic information and lend complex biological diversity, its **four-letter genetic alphabet** is relatively limited. Using a genetic code of **six XNAs** rather than the four naturally occurring DNA nucleotide bases yields endless opportunities for genetic modification and expansion of chemical functionality.[59]

The other big area of research now is to see if **information can be stored** in DNA and XNA...

Hachimoji DNA (also XNA)

Scientists are pushing the boundaries of what is possible and in a way emulating the Anunnaki of millennia ago… we are playing with DNA, and have extended

the existing 4 nucleo-tides A, T, G, and C
to add another 4 now called P, B, Z, and S

or **"Hachimoji DNA"**. These are new chemical parameters (synthetics) that generate thousands of new genetic templates that didn't previously exist.

[Hachimoji means "8 letters."]

This is forecasted to serve as a new substrate for data storage and to build new nanostructures. As with anything on the border of "far out," it is touted to be "a vehicle for defeating incurable diseases and viruses in the future…" [60]

Will the AI engineers miss the opportunity to build AGI and ASI devices, or parts of them, using the new XNA? If they can build a more non-silicon avatar, and equip it with a super processor, might it be part of the ASI of the future…?

Right now the new XNA is confined to the laboratory as it needs a continuous supply of proteins and building blocks, but once stabilized, it is thought to be a **medium of DNA storage of digital data**, and self-assembling nanostructures. [61]

And there is yet a third form...

Synthia

Dr. Craig Venter (at the J. Craig Venter Institute), created a **new synthetic bacterium** – starting out small so it could be easily understood and controlled, and it was called **JCVI-syn3.0**, or Synthia, announced in *Science* magazine in March 2016. It consists of 473 genes (whereas humans have 20,000+ genes) and he called it "the first designer organism in history." As it is a synthesized version of *Mycoplasma mycoides* it was christened **Synthia**.

> The goal was to identify the smallest number of genes that are required for life. Said Venter,

> ...that 149 of the genes in Synthia [have] unknown functions means that **the "entire field of biology has been missing a third of what is essential for life. "** [62]

The uses for Synthia and XNA are expected to benefit medicine and **repair of genetic flaws**, as well as the creation of Synthia gives a basic framework for the construction of more complex genomes in the future.

xDNA

Just to give you real warm fuzzies, they are also playing with another form of DNA...

> xDNA..... yDNA.... xxDNA.... and yyDNA.

> xDNA has many applications in chemical and biological research, including expanding upon applications of natural DNA, such as **scaffolding**. In order to create **self-assembling nanostructures**, a scaffold is needed as a sort of trellis to support the growth. DNA has been used as a means to this end in the past, but expanded scaffolds make larger scaffolds for more complex self-assembly an option...

> Its 8-letter alphabet ($\underline{A}$, $\underline{T}$, $\underline{C}$, $\underline{G}$, xA, xT, xC, xG) gives it the potential to store 2^n times increase in storage density... [63]

> There you go, it is involved with **nanotech** – see Chapters 5 & 6.

Miscellaneous Items

The following genetic items are extracted from a NatGeo special magazine:[64]

(31) Scientists believe antisocial behavior is a function of genetics, as multiple genes work together. (It also includes the OP phenomenon... see end of Chapter 6 and Apx. C.)

(38) Scientists can never **clone a dinosaur** because DNA has a half-life of 521 years.

(43) Humans share about 60% of their DNA with **bananas**.

(47) About 8% of human DNA comes from ancient viruses that infected our ancestors. (This may include the **223 genes** that exist in Man and no other creature on Earth —see earlier section in this chapter.)

(52) **Mice and humans** share 85% of their DNA (which is why mice are used for medical testing).

(54) Studying why **deer can regrow antlers** may lead to treatments for humans with osteoporosis.

(56) It takes at least 6 genes for the body to process **caffeine**.

(60) Researchers believe a mutation of the **BRCA1** gene may contribute to Alzheimer's and other brain disorders. (Hence the work being done via CRISPR to replace faulty genes.)

(63) Only 5-10% of cancers are genetic in origin.

(64) There are over 10,000 different genetic conditions that can cause serious health problems for those than inherit them. (That is why **stem cell research** and synthetic DNA research are so important).

(65) A small percentage of Europeans are HIV resistant because of genetic mutations caused by the Plague in the Middle Ages. (Questionable 'vanilla' source for **CCR5delta32** genetic mutation – as examined earlier in this chapter... they can't say it was added to our genome by Intelligent Design.)

(68) Humans may be genetically programmed to not live more than 120 years.

(69) **Octopi and squid** can edit their own genes. (And they are so unusual in their ability to camouflage and outwit those who would capture them that it is suspected their intelligence was designed by the Skygods – for unknown purposes.)

(70) **Red blood cells do not contain any DNA**. (But its composition and bloodtype are unique enough to provide biometric info.)

(79) In less than 10 years, scientists will be able to sequence an entire human genome in just a few hours. (So *AncestryDNA* and *23andMe* now require several days and the test is not 100% foolproof – **and** the data can be subpoenaed by Law Enforcement to help solve *cold cases*.)

(81) A person's **bloodtype** (O, A, AB, B) is completely controlled by genes.

(82) Scientists can **artificially grow skin** so it can be grafted onto a person's body.

(94) When temperatures rise above 195°F (90°C) the DNA strands break down.

And did you know that people who exercise regularly may be able to change the way their genes are expressed? This is the science of **Epigenetics** which studies what turns genes on or off... or how they are "expressed."

Epigenetics

Scientists used to think we were stuck with the genes as we got them, but modern research has discovered we can make adjustments to our genes... as well as edit them via CRISPR.

> Genes have to be told what to do, how to do it and when to do it. They work together (and 'signaling molecules' are often involved). Epigenetic modifications, called **tags**, give each gene its instructions. These chemical tags sit on top of and outside the genome, **telling genes to turn on or off**... Other tags act like a dimmer ...[to] increase or decrease in intensity.
>
> Our genome contains at least **4 million of these tags**....
>
> Smoking, sleep, exercise, diet and exposure to toxins can affect the behavior of these epigenetic markers...
> Epigenetic tags in our **muscles** adjust themselves as a result of resistance exercise. These tags allow our muscles to remember how they grow after they return to their original size. [65]

Summary

So are we playing God with our genes? And should we? Didn't the Anunnaki or Skygods, whatever particular 'flavor' of ETs visited and manipulated Earth genetics to create their worker humans? Are the scientists considering the social, ethical and economic impact of what they are doing?

The technology is there and is now interfacing with **Transhumanism (H+)** and medical advances. The issue of course is always an ethical one – Is anyone hurt by this science? If we create H+ humans who are superior, will they see the original humans as inferior and seek to constrain or remove them?

Notwithstanding **Nietzsche** who promoted the idea of a "superman," we could create a race of humans who are smart, attractive and stronger – a genetic caste system with the "best" genes who could lord it over those with "inferior" genes...

Even DARPA is looking to develop a set of **supersoliders** who are smart, strong and fast – the perfect weapon on the battlefield. Is this something that Man is destined to pursue...? Is this the evolution of mankind? And do we <u>have to do</u> it to compete with more advanced races Out There... especially if they are stronger and antagonistic? (Think: Orion Empire and the Gina'abul.)

It is already known that the Reptilians are much stronger than a human, and some may be smarter...How do we compete – having been created to be weaker and subservient?

It has been said that three (3) Arnold Schwartzeneggers in his prime could not physically take down one of the 8' reptilian Anunna.

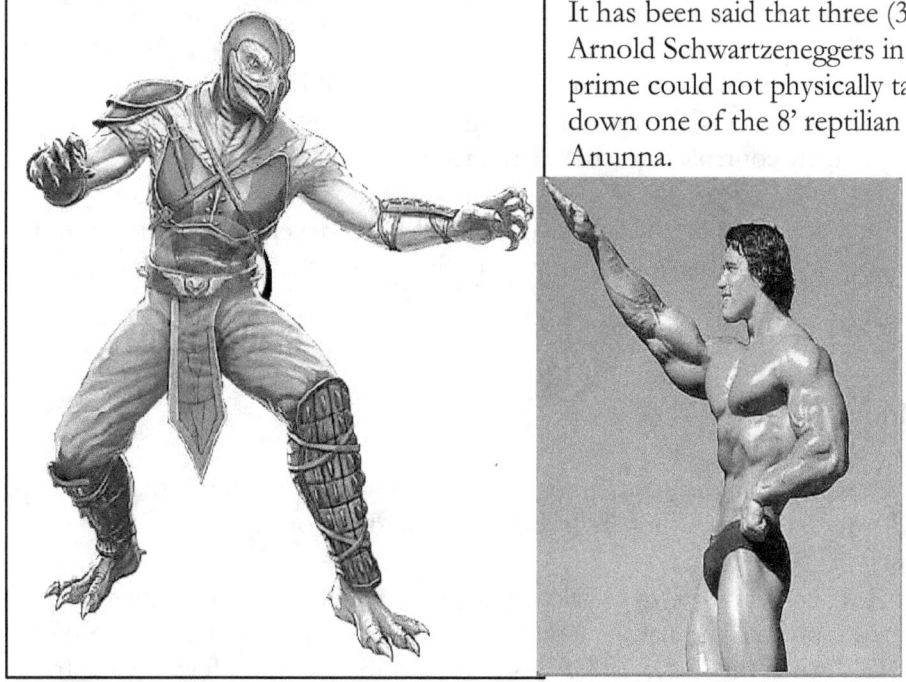

Regarding the picture above...
At **8' tall** they are a lot to deal with., and some have **claws** for hands. Their skin is protected with **scales** and they have a body **musculature** like that of an ape...

When you reflect on what is really going on on the planet and in this part of the Solar System, we may be foolish to not try and be all that we can be. It may not be a case of "insulting God" or "trying to play God".... it may come down to a case of **survival**.

Also see the section in Chapter 5: **Genetics Are Important**

After all, we are now catching up to the genetic knowledge that was used thousands of years ago to re-create Man (Homo *sapiens*) from a jungle biped... by those who are still here and some of whom (the Dissidents) still seek to enslave or control us.

What if that last part is true, even if you don't want to hear it?

Can we afford the arrogance to pretend that there is no 'alien' threat —just because we don't see it? – Reagan alerted us to the possibility of that in 1987, and the **Breakaway Civilization** is currently dealing with the issue 500 miles above the Earth. And I know you have heard of the Solar Warden Space Force...

(See Chapter 8)

Actually we do see it and it has been reported for decades, Roswell and Kapustin Yar notwithstanding, and the Media (largely controlled by the 'alien' faction), suppresses the truth...

I suggest it is time to be the best that we can be and develop our genome and our technology (responsibly) such that we can finally take the planet back.

Epilog

And yet, we have to face something else dealing with our genetics that may support the Chapter 6 info regarding Transhumanism and Cyborgs...

Man is currently engaged in developing rockets, robotics/AI and exotic propulsion systems to travel to the stars – and he will have a problem with that, as NASA has discovered when they send astronauts up to the ISS just for a few weeks. In one case, they sent a man up for 360 days and when he got back they examined his DNA as well as his muscles, blood, immune system, and skeletal system – all of which had changed – and not for the better.

Weightlessness causes muscles to atrophy, the DNA also reacts (Epigenetics above) and changes, and while we can simulate gravity on a revolving space platform (as in the movies, like in *2001: A Space Odyssey*), what is our body going to do if we try to exit the spacecraft and walk on a bigger planet like Jupiter whose gravity is much more than Earth's?

And if we go to a smaller planet, or say a moon with $1/6^{th}$ the gravity of Earth, an astronaut should be able to jump 4 – 6 feet into the air...(and we all saw that in the Apollo moon walks, right?) That is because the man's muscles are conditioned for Earth gravity and would have the strength to make him look like a mini-Superman.

Remember that Superman (the comic book hero) came from a denser planet and he had superpowers on Earth – one of them being his ability to "leap tall buildings in a single bound" and the analogy carries some weight – we would be stronger, and probably run faster (more muscle power in less gravity), and jump higher, but maybe not over buildings... Didn't we all see astronauts jump over small 3' boulders on the Moon...?

It is for this reason that we may have to **explore space with robots** while we figure a way to rework the human genome and make special humans bred for space exploration?

Wow, history repeats itself and Man becomes like Anunnaki geneticists.

Hopefully Man will also not copy the Anunnaki and destroy Earth the way Mars was destroyed – it used to have water and atmosphere, and after the **Space War,** Tiamat was gone (now the Asteroid Belt) and Mars was a wasteland..

Mars Before and After
Then: wet and warm vs Now: cold and dry.

(See Appendix F for Man's Legacy and pix.)

Chapter 4: Freewill & LifePlan

There are two major aspects to this chapter: Freewill and the LifePlan, and the LifePlan may (but not always) constrain the Freewill of the human while on Earth – it depends on one's Karma.

> A lot of his information was covered in several chapters
> in Transformation of Man (TOM) and Appendix D in
> this book, so it will only be summarized here to make a
> few major points.

Let's look at a few of the principles involved… and that will help us better understand the following chapters where Man's Freewill is being violated, and how it happens, and who is doing it.

LifePlan

All souls birth into the Earth Realm from what has been called the **InterLife** – the interdimensional place where souls study and make preparation for another experience somewhere; they review past lives, consult with Masters and Teachers, and choose their next incarnation (not always in the Earth Realm). **The soul is eternal and will experience many different venues** for whatever lessons/tests that each place has to offer… it's all about soul growth.

If the soul chooses (and no one makes souls do anything – again: Freewill, but souls can be strongly advised) to take another, special lifetime in the Earth Realm, and so pre-program a Script or select a location, time frame, family, body type and growth experiences. These are set up as preplanned "events" in the timeline of the chosen lifetime… "tests" if you will to see what the soul will do with whatever it encounters – and that is the **measure of soul growth**: does the soul freak out, run and hide, bitch and moan, blame someone else?…or stand up and attempt to overcome whatever challenge has come its way.

> As the Bible once said:
> **No one is ever tested beyond what they are able.**
> The gods are not ogres and soul destruction is not the goal.
>
> In fact, we are all much more capable than we think we are and
> that is the beauty of the soul (which Reptilians hate – they do
> not have the potential and must seek to repress the souls on Earth.)

One of the goals of Earth challenges is to discover that we CAN do much more and BE much more than we think.

So the LifePlan, also called a Script, is personalized with events designed to develop and strengthen souls. Please note:

> **The Script does not control what you are to say or do...**
> It is all up to you and that is why there is Freewill — to see
> how much you have grown and what you can handle and
> what you can't.

What you can't yet handle is where you undergo InterLife classes, training and "imprinting" (Glossary) to help you next time around the Earth Realm. It is all set up to develop you into a dynamic, totally capable soul who can "graduate" from Earth School and then rule (administer) over other realms with other Higher Beings in which the low-density Reptiles, among others, live! (That is why they try to derail you on Earth.)

So that is the LifePlan. Everyone comes into the Earth Realm with one — Baby Souls are more "constrained" than are Old Souls because they need more guidance. Mature and Old Souls get more leeway because they are working on fewer issues.

Karma & Reincarnation

This has been handled and discussed so many times in a lot of the books, suffice it to say that what you can't handle, or some lesson you failed, sits in your Karma Box — awaiting your ability to handle it. To that extent, some things you need to learn (as a way of "meeting yourself") is what Karma is. Thus what you have yet to master can show up as another pre-planned event in your current or next lifetimes.

> BTW, because it is important to understand:
> Just because you slipped up and knifed someone in a prior
> lifetime does not mean you have to be knifed in return —
> that is NOT **Karma**.
> What is done is to present you with the same set of circumstances
> you faced when you did knife someone (hopefully after you have
> done some soul growing since then) and see if you do it again —
> or did you meet that negative aspect of yourself and CHOSE to
> not do what you did before? Did you **overcome**?
>
> And by the way, the original Christian church used to teach
> **Reincarnation** (as do the more enlightened Buddhists and Hindus)
> back about 100 AD. The Catholic Church removed the teaching
> because people told the Church they were going to boogie and

party this lifetime and they'd get it together NEXT time, and the Church had to change its doctrine – ostensibly to get people to be good this lifetime…. or go to Hell as punishment.

So **Freewill is the rule** and we need to make the best choices we can, that don't hurt us or someone else. What we do is reviewed with a teacher on the Other Side, as if it were recorded as a supernatural video – and we get to replay it, feel it, and better understand how we impacted others (we are all connected: Quantum Physics calls it **Entanglement**). Old souls make the best choices as they have learned the most and they are ready to graduate. (See VEG, Ch. 7.)

Ok, no preaching, just fact.

Now we come to the very serious aspect of this chapter…one that sets the stage for the next few chapters and WHY you need to know about this.

Violation of Freewill

Right up front I am going to repeat this as it is WHY your Freewill is being violated…

The Reptilians have always been heavily invested in derailing humans' growth – if they were hosting a soul. Most of the Reptiles do not have a soul and they fear those who do as there is a connection to something higher – the Higher Self with a **conscience.** The soul also has the potential to evolve past the Reptiles and rule over them.

Throughout history, the Anunnaki have manipulated and influenced the PTB – humans on the surface who love being Lords and ruling over Serfs (slaves) and that fits the mindset of the Reptiles, too. The PTB love their power and position and seek to sustain it by keeping humans dumbed down such that when the humans die, they automatically return because they learned nothing and thus the PTB dominance is sustained.

So not only do the underground Anunnaki mind-control their PTB puppets, they can also mind-control humans on the surface. A soul that graduates Earth School can mind-control the Reptilians (or stop them from doing it) and that is the thing they fear.

Mind-Control

Be it known that the **Greys** who abduct and do genetic experiments on humans are able to mind-control the humans, as well as wipe their memories after the horrific event. The Greys paralyze the humans and then stare into their eyes and easily manipulate memories and behavior.

Why is that important? Because the Greys are **biocybernetic androids** created by the Anunnaki Remnant… with the same abilities as their masters. (The Draco also have the same abilities but to a greater extent.)

So what the Greys can do, the Reptiles can also do.

Mindscan

The Grey seems to have **superquantum computer circuits** for processing and most Greys also possess an interesting ability called **Mindscan**, a term coined by **Dr. David Jacobs** in his book <u>The Threat</u>. Not only do they know what you are thinking, **they can put thoughts directly into your mind**. And even more amazing is that the Grey can stare into your eyes and manipulate the optic nerve to gain entrance to the brain's neural pathways. [66]

He can inject new images, and cause people to 'see' whatever He wants … no doubt part of the secret to their **shapeshifting**. Natives in the Amazon claim they see large birds sitting up in the trees and as they take aim with their bows and arrows, the "bird" changes shape to a large Grey… and the natives run away.

The Grey also uses Mindscan procedures to effect an orgasm in males for the purpose of collecting semen. The physiological responses can be artificially generated this way. [67]

In effect, this ability gives the Grey **absolute power over the human**, even to the point of paralyzing them. They can make the abductees think, feel, visualize, or do anything they want. In addition, the Greys have been known to extract memories and information <u>from</u> the abductee and put it <u>into</u> their Hybrids…[68]

Anybody remember Spock in *Star Trek* doing a "Vulcan mind probe?"

© Paramount/Kobal/REX/Shutterstock

So is it more clear now that the Greys are violating people's Freewill... and yet you are not aware that the Anunnaki Remnant are also doing it to people (and especially those that are called the PTB, or Deep State), the puppets carrying out the wishes of the Remnant.as they walk about their daily lives.

Let's see how that happens...

Violation & Agreement

Quite bluntly, the PTB, the Neggs, OPs and your neighbors have violated your freewill and **you agreed to it** at some point in your life.

Initially some people will disagree with that because (1) they don't know how subtle and persistent the PTB and OPs can be, and (2) people think they're in control of their lives. Without congratulating them, these STS entities can be considered <u>masters</u> at deceiving and tricking you into giving your power away. And this violation principally surfaces when you die; and then what you didn't learn and still don't know <u>can</u> harm you and you may wind up some place other than where you intended. Your mindset needs to include enough knowledge and awareness to discern who's there, what they're doing, and enough self-respect and integrity to stand up and say no.

But the Sheeple nowadays have been lulled into a false sense of security –
they let Harry, Tom or Dick handle their political societal affairs for
them while they sit on the couch, eat their burrito, and laugh at Seinfeld.
Meanwhile Hilary, George and Omar are making off with the Country
right under their noses… often lying about what they are doing, but
the people go along with it. Lazy people get the kind of society they agree to.

Violation of Freewill cannot happen without some **deception** that convinces us that there is no danger to us and gets us to think it is Ok to do something. Then you go along with the idea/suggestion/urge. You have accepted BS without knowing it – all they have to do is **get you to agree** (for whatever reason) to what is promoted, and if you don't care, or you don't know, **you have given your power of choice away**, and cannot then do the right thing. Then the PTB capitalize on your ignorance, and even though it is frowned on to violate another's Freewill in this universe, the PTB have gotten away with it because you agreed to it. The fact that they lied to you and got you to agree through deception doesn't hurt them – You have the problem, not them.

This is also something that is done by the **Media** who 'spin' history, promote confusion, and manipulate public opinion. Societal trends and attitudes that would not have been acceptable 50 years ago are now gaining acceptability due to the promotion of the idea that we're now in the 21st century and thus should be more progressive and open-minded. Case in point: **Gays** and the spreading of HIV. About 52% of the American public has HIV, including heterosexuals, and the CDC knows it can't be stopped or cured, so we ignore it, we no longer tell the public what the infection rate is ("don't want them to worry, they're probably going to get it anyway, so why bother?"), and we also promote **same-sex marriage** with the understanding that it is "an alternate lifestyle," and because we are now more enlightened, it is Ok… "Get with it people!" is the injunction. We did the same thing with alcohol and tobacco – we couldn't stop it (Think: Prohibition) and so we 'legalized' it and put a tax on those commodities. All of that not to denigrate the Gays, but to show that society, for better or worse, has a history of going along with things that can be harmful to our health – alcohol, tobacco and HIV are not the hallmarks of a healthy society. But, loving and respecting the Gays, is.

> **Caution: Gays are not the problem, HIV is, and it seems
> to have started with the Blacks in Africa first. Regardless,
> they are our brothers and sisters and we should still care for
> them. (See VEG Ch. 15.)**

If there were a conspiracy to wipe out Man, HIV would be the way to do it, and cause minor damage through alcohol and tobacco, to say nothing of what drugs do to people's DNA and health. If there were a conspiracy to remove Man from the planet, you'd want to get his approval (so he doesn't rebel) and understanding that **fluoride** is Ok in the water – doesn't the ADA sanction it? (European countries have rejected it.) You' d want to promote the idea that **GMO** food is Ok, and that **Chemtrails** are helping to protect us from the Sun's UV rays… In other words, get Man to agree to things that are harmful, that abuse him, so that if the gods step in, the perpetrators can say, "Don't blame us – they agreed to it!" The other defense is to say "Well, we told them – right there on the **Georgia Guidestones!**" Then, when a society falls, the people are caught off-guard and claim they didn't know what was going on. So, sell them on 5G cellphones and get them to agree to their own health problems. (Chapters 5 & 7)

And ignorance is no excuse because all one has to do is stay aware and **ask.**

By the way, parents, teachers and pastors regularly violate people's Freewill by using their authority position to get you to believe what they want. Kids are lied to about Santa Claus knowing if they're good or bad and thus manipulating the kids to do what the parents want. Teachers can unknowingly teach the wrong thing (see section on science errors in VEG Ch. 8) and violate the kids' right to know the truth. And lastly, preachers used to threaten people with Hell and damnation if they didn't do what the preacher said.

Your Freewill can be violated without your knowing it.

Subtle Agreements

Remember in Chapter 1 we reviewed the way OnStar ®, SmartMeters, and SmartTVs are sold… pushing the positive side and not telling you what else they can do that you may not like? In addition, you also tacitly agreed to **Alexa** because you wanted to speak a command and get a reply… "Alexa, what is the weather?… Alexa, play Moon River…." And so on. What you did not agree to was that she is always listening and **all your conversation is recorded**, and if there are keywords that you spoke, such as ISIS, riots, 9-11, or White supremacy… the latest recorded set will be quickly reviewed for the hundred or so keywords, and if they are found, the whole text set is sent to a human to scan it to see if it means anything… If not, it is trashed, and Alexa is already recording the next set… If the human did not like the combination of keywords, it will be kept on file under your name/IP address and if further conversation is detected that could be inflammatory or seditious, you may receive a visit from the local FBI.

By the way, the newer cellphones now do the same thing as Alexa, and some can take your picture without your knowing it, and track your use of Apps. [69] Same with the camera on your laptop: Smile!

Another item you did not object to, and thus tacitly agreed to, was posting your personal data and pictures on **Facebook** – ostensibly for your friends to see. In reality, the NSA, FBI and CIA can also go thru the accounts (passwords do not stop them as they have encryption-breaking software) and they can look for more sets of eyebrow-raising text or pictures.

And last but not least, weren't you urged to store things on **The Cloud**? Hey, don't clutter your PC (or cellphone) with all that data and pictures, "just back things up to the Cloud – that way you also don't lose it." Sounds good eh? The Cloud is a large server in a local Telecom warehouse or hub. You just sent your data to someone else's PC. Sure, put your accounting and bank info up there for all to see…. How naïve are some people?!

Maybe you saw that one coming, but did you also see other convenient and easy-to-use things like **tapping your credit card** on the POS terminal – Wi-Fi enabled… and anyone in the area (with the right equipment) can also read it. And you know that the little 6-7 part gold chip on your creditcard can be read by a hidden sensor – like those at the entrance to a large department store in the Mall? (And some of them can even read your card in your wallet in your pocket.)

Welcome to High-Tech. You will learn more in Chapter 5.

Subconscious Sabotage

This is a biggie for some New Age people, please pay close attention. This is a form of **violation of our Freewill** – although while it is being done to you, it seems harmless.

Mr. D found a New Thought church where the guru/minister would lead the whole congregation in a **guided meditation** at the beginning of each service, every Sunday. They'd sit quietly, eyes closed, breathing deeply, while the guru would lead them into a relaxed state – also known as **'Alpha'** wherein the subconscious mind is very susceptible to suggestions. After a few minutes of relaxation, the guru would repeat affirmations for 5-10 minutes, such as:

> "There is no negativity in my life"
> "I do not entertain negative thoughts or people"
> "Nothing negative happens to me"
> "I speak only positive words and ignore the negative ones"
> "Nothing negative is real in my life, I focus on only the positive"
> …etc.

This is a real 'catch 22': if you have problems, you can't solve what you can't see.

… after a few Sundays of that, guess whose subconscious, programmed while in 'Alpha', is PROGRAMMED to ignore any negative input in their life? Guess who cannot handle and resolve negative events because they don't exist for him? Guess what effect it had on his marriage as he could not deal with anything negative in his life, and work out solutions to problems with his wife…? Guess who could not deal with problems with vendors and employees in his company…? Did he listen when told what the problem was? No. Would he go for "deprogramming" under hypnosis to become a normal person again? No. He is divorced and unemployed now.

This is real, people. And is one of the dangers of the 'guided meditation' in some New Thought churches – which really amounts to **hypnotic suggestion** and some day, when Man wakes up, it will be seen as a violation of Freewill, and may involve misdemeanor prosecution for practicing irresponsible hypnosis with naïve subjects.

Be careful with affirmations--evaluate the content of some 'guided meditations.'

Basic Techniques for Manipulation

It greatly behooves the PTB (and the Remnant) to keep the populace as naïve as possible. And to that end, the main techniques are:

Disguise - a beautiful woman, handsome cleric, smiling doctor, or amiable salesman… all of whom appear to be acting in your best interest.

Distraction - the Chantix ® and Opdivo ® Rx commercials are a great example of this: By law they have to tell you what the nasty Rx side effects are, but on the screen is a cute, distracting scenario so you are not conscious of what is being said… The next time, stare at the floor and listen to what they are saying… if you dare!

Disinformation – very common…

The last one is the most common: and it takes the form of false information, from teachers, clergy, doctors… etc. More commonly it generates **False Belief**.

The **Church** was a great player at that game: "Believe what we tell you, or you go to Hell!" If you don't believe in the Seven Sacraments, tithe 10% every week, or that Mary was a virgin, or that Jesus died for your sins… you are damned. But if you do believe those things, then they can **tell you what more to do** and believe.

> The problem is, if they are wrong, AND you die believing you
> are going to Hell, you can actually get waylaid trying to get
> back to the InterLife.

What you believe is very important and it should be based on Knowledge of the Truth … not beliefs that might be true.

Teachers mean well but do the same thing when they teach just Evolution and ignore the possibility of Creation… and even more relevant nowadays with the growing body of proof in Quantum Physics, Genetics, and Ancient History, is the likelihood that Man's evolution involved **assisted creation**.

Scientists who mean well but are misinformed by the Establishment Training they went thru to get their PhD, will tell you that the Moon was created as a spin-off of the Earth. We now know that that is false; the Moon is made of different stuff and is much older than the Earth. (see Appendix F: Moon Info)

Doctors who mean well, and in turn also went thru the Establishment Training to get their MD, will tell you that you need your Flu shot, vaccinations are Ok, and they regularly prescribe statins to lower cholesterol. First off, the Flu shot you get has 2 standard strains which may not be the one going around; vaccinations often have negative side effects (sometimes resulting in autism); and statins kill the liver cells so they can't produce cholesterol – which is needed by the body to make other hormones and signaling molecules used by the body. And do they ever tell their patients to **drink more water** so the body can flush toxins that create inflammation and disease? Little old ladies with Dowager's Hump were never told to take a Calcium-Magnesium supplement to support the body. (Doesn't the Hippocratic Oath say "First do no harm…"?)

Earth is a child's garden of misinformation – on purpose.

So are there other ways in which our Freewill is violated – and are not so obvious? Are there times in which we think we are in control, no one is telling us what to do, and we think <u>we</u> are making the choices we face?

Not always…

> This is going to amaze you, and if you are an atheist, you will really have a problem with it, but I assure you, **it is real** and what I am about to describe is very real – it is part (and has always been a part) of the Earth Realm.
>
> More importantly, it is part of the problem with the coming **5G Network** (Chapter 5) that even the Telecom people do not suspect. **5G satellites** violate your Freewill (Chapter 7).
>
> Thus you really need to understand the following information if Chapter 5 and the issues with the 5G Network are to make sense to you.

A lot of the following information was covered in detail in VEG (Ch. 6 – 7, and Ch. 12), and is knowledge found in the Secret Societies, Druidism, Wicca, Alchemy and Kabbalah for starters. It was further examined throughout <u>Anunnaki Legacy</u>, (or AL) in many chapters. Many cultures and secret societies around the world have been privy to this information – it is time to repeat the gist of it here.

The Church did not want it known by the average man on the street, and yet it crept into Christianity inasmuch as Jesus cast out devils and came against the Powers & Principalities (Ephesians Chapter 6)… thus it is not surprising that the **4D STS entities** who do harass us want to kill Christianity, and are busy at it nowadays. (How can you oppress and harass people if they know what you're up to?)

> And be aware that while the Christians refer to the 4D STS entities as demons, that is oversimplified and mostly inaccurate... so even the basic truth of the 4D denizens and their activities got garbled.

Perpetrators Defined

So let's lay some groundwork with the relevant definitions (also found in the Glossary).

4D STS Beings – this is the catch-all phrase that refers collectively to the entities in the Astral (4D) realm around us that we cannot see. It includes **6 types:**

Neggs – those "dark angels" whose job it is to carry out the negative aspects of one's Karma often found in the LifePLan. The term is short for **"neg**ative **guides**" as they work together with the Beings of Light (BoL) to effect the catalyst that causes souls to grow in the Earth Realm.

The Neggs are often called "demons" by the Christians but they are also angels playing the 'heavy' role as justified by the Karma one has to face. They are just doing their job and are **totally controlled by the BoL. They look like Shadow People.**

Beings of Light (BoLs) – these are commonly referred to as Angels as most people understand the term, and are often the positive guides in one's life; the Neggs being the "negative" guides.

To see how they work together, please read VEG, Ch 6 ("Dr. John Lerma" at the end of the chapter). In reality, the Neggs are BoLs who volunteered to play the heavy and work with the BoLs and when their assignment is up, they revert to a full BoL again. This arrangement is necessary as it does not work to have a BoL making decisions to hit you, bless you, hit you, bless you – such behavior is rather schizophrenic for one entity and is it <u>more straightforward</u> and simpler to administer to have one angel play only the 'heavy' and the other play just the proactive part.

Discarnates – often called ghosts, these are souls who were afraid to move on to the Light, and return to the InterLife via the Tunnel… they (falsely) feared judgment and anger of the powerful God who would damn them – and no such thing exists but they were lied to and now drift around the Earth Realm unable to incarnate or return to Source.

(credit Bing Images: Dread Central)

Occasionally they gain enough energy (drawn from the living) to create poltergeist activity, or they can attach themselves to a soul's aura while in a bar, for example, and induce the 3D human to drink and carouse. They feed off the energy.

Thoughtforms (and Parasites) – this is less common but still a problem on Earth. Intense group activity, as in group mind, where all are in the same mindset and energized can (usually briefly) create a thoughtform which practitioners of the Black Arts (i.e., Voodoo) can order to go and do some act. (See also the **Tulpa** and **Golem** that advanced practitioners in Tibet, Middle East and Hawaii create to serve them.)

These are a lot like **Shadow People** (below)…
      …which are sometimes thought projections…

Djinn – also called **Interdimensional Beings** (VEG, Ch 12) –

These beings are not that common but occasionally come thru in Exorcisms and seances. The Gnostics called them Archons and they are very antagonistic to Man, hating him because they were **created from fire** and Man was created from clay and the Djinn were removed to another dimension, aka the Draco. [70]

The Black Djinn are the Most Powerful

Then there is the Draco...

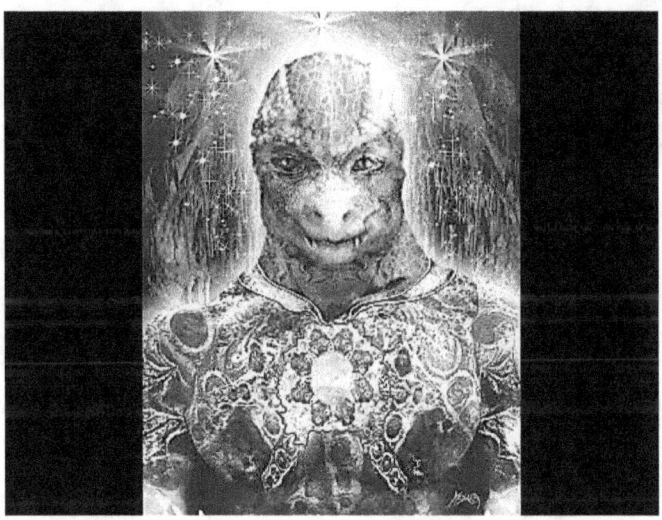

The Djinn are fascinating. Allegedly here <u>before</u> Man and they inhabited the Earth. They were created from 'fire' (Astral plasma stuff) and considered themselves superior to humans, who were just made of clay. When told they must respect the new creation, Man, most all did except one called **Iblis** who said Man was inferior – not seeing that Man had a soul and a divine potential greater than theirs. This refusal (among other things) got them ostracized to another dimension.

Their behavior was also abhorrent to the gods (Elohim) who run this place… they lived in clans and fought among themselves all the time for supremacy. Constant arrogance and violence. Thus, they were banned from Earth.

So saying, there were no Fallen Angels (as also says the Qu'ran) it was just the rebellion of a Djinn who took a small group with him (the Gnostics called him Ialdabaoth – See VEG, Ch. 2). The Angels (BoLs and Neggs) were created from pure light and have <u>no Freewill</u>. The Djinn on the other hand were created almost as high as the Angels and I maintain that due to the way the **Draco and Djinn** behave, **they are the same beings** – they are tricksters able to shapeshift (think: Bigfoot), remain **invisible**, fly (levitate), create illusions, possess humans, **mind control**, and **create objects and other beings**. Note that the creation of Reptilians on Earth was the Djinn/Draco way of getting back on the Earth and controlling it once more.

The Djinn and Draco are often confused with "demons" as the Christians call them. They can also mentally 'project' Shadow People, and sometimes UFOs and ETs to confuse and deceive.

…and lastly, number 6…

Anunnaki Remnant Dissidents – added here as they are 3D underground on Earth (and also in 4D) and they are the ones largely influencing and manipulating their puppets, the PTB, via what can be called **mind-control** (more next chapter).

Think about it: if the Dracos and the Djinn (the same beings) can mind control humans <u>when near them</u>, and the Draco created the Anunnaki/Reptilian race, then the Reptiles created the Greys who also can mind control.

Mind Control

And be aware that the mind control happens one of two ways:

> Either by being <u>near the human</u> so that direct manipulation can happen,

or

> As is the case with the underground Remnant, special high-tech microwave machines are used to mind control surface humans – via satellites that the Remnant also have up there. Some of the PTB have tracking chips in them to locate and target them.
> (See later sections on Montauk and **V2K**)

Needless to say, when the **5G Network** gets going, the Remnant can piggyback their technology onto it and that is one of the hidden dangers that even Telecom is not aware of – or they would not be doing it!

It is sometimes hard to realize that there is a whole other world going on in the Astral, and yet we know there are **frequencies we cannot see that are there** – radio and TV waves, cosmic rays, microwaves, neutrinos, x-rays – all real and yet we cannot see them. And they can and do affect us.

> And it is the proliferation of intense **microwaves (EHF)** in the next chapter that will be a big issue in the years to come.

So let's look at how the Remnant and 4D STS work their 'magic' on us…

Astral 4D STS Tactics

In addition to fear and deception, planting thoughts in people's heads (claiming to be the voice of God) and shapeshifting, the 4D STS have another procedure they use a lot. When you are reading something helpful, or someone is explaining something helpful to you, the 4D STS can often follow this sequence to try and stop you:

Block your hearing so you can't hear, or
>   block your eyes so you can't see what you're reading;

if that fails, and you did hear (or see)…

Block your understanding so that you can't use the info;

and if that fails, and you did hear/see and understand…

Block you from taking action –
>   make you forget the info or diss it.

And just as interesting is an old saying: "If the devil can't make you bad, he'll make you busy." The 4D STS can orchestrate events such that extra phone calls, people and emergencies intervene, then you'll get very busy, and then you'll become very tired, exasperated and give up trying to do what you want, and wind up saying or doing what they wanted you to say/do in the first place.
Anybody seen that before?

Man Gives Permission

Respect the Astral to a point, and if they can 'hit' you where you're weak, note that you have the problem: you need to fix/change/stop something so they can't do that again.

>   **You need to take back the 'ground' that you have given them,**
>   knowingly or unwittingly, by realizing your mistake, asking to
>   be cleansed by the Light, and not repeating the error.

For example, when you get angry and start to swear, or become sexually aroused, or watch a violent video, two or three or more beings will be attracted by the energy, appear near you, and then **feed off your energy**. [71] When you perpetrate, they react. **They do not instigate**, you do: you started the drinking, smoking, porn video, drugs…. . So respect them up to a point where they become a nuisance, and **then tell them to go**… and clean up whatever it was you were doing that attracted them, i.e.,

>   **take back any ground you have given them** – it is effectively
>   **permission** that you granted them. (See also VEG, Ch. 14) Your
>   straightening out your thinking and behavior tells them that you
>   are aware of them and intend to give them less opportunities
>   in your life.

Again, for example, if you dabbled with Wicca or witchcraft, Tarot, or the Ouija Board, recognize that that gave them some ground because you were seeking power (witchcraft) or asking for a response (Ouija board), and so **you entered their** territory. You want to play the game? They will oblige – just be aware that you have **given them permission to 'work' with you**. What you don't know (and to them your ignorance is no excuse) is that

they play by <u>their</u> rules,

not what <u>you</u> think is fair. So if they can get a toehold in you, and then a foothold, and then…the more you practice the 'black arts', or cooperate <u>with</u> the 4D STS, **the more ground they can take** – until some people become very oppressed, and then finally possessed. When you die, they come to collect.

Playing games like *Pokemon* (short for "pocket monster") or *Dungeons and Dragons*, and the like in video games (They had them designed and built!) and it is **giving them permission** to oppress. Why?

Sample Pokemon Cards
(credit BingImages: Pokemon)

Because those are <u>their games</u>, to entrap curious souls.

> While Pokemon is not that "evil" it is what it can lead to:
> :Dungeons & Dragons then Tarot then Witchcraft and then
> the OuiJa Board where they will talk to you... No kidding.

The main goal is to **make you think you have power** (which they
give you) – that is why when you die, they come to collect, and
they don't want your money.

You can think of the harassment, deception and oppression as the 4D STS 'job'
although they are <u>not</u> altruistic – they <u>are</u> out to 'get' you, if they can... subject to
the Beings of Light's approval (it has to synch up with your Script.) Only the
Neggs are doing it for your soul growth.

> Be clear that the **Remnant** are not doing it for your soul growth –
> they **are** out to get you and control and subjugate you – which is
> what the Anunnaki have done for millennia.

Remember in an earlier chapter, it was suggested that the ensouled humans would
have to be stopped lest they develop their inner divine spark? In reality, the Neggs
are the **negative catalyst** to move Man in the right direction, but Man has to wake
up and realize that when he does the wrong thing, the Neggs and Remnant will
capitalize on it. And if the **Neggs** secure a foothold, they may be aided by the
opportunistic discarnate entities. If the **Remnant Dissident** secures a foothold,
they may terminate you – the Neggs can't go that far.

And the 4D STS will try to persuade you and manipulate you via **False Beliefs**…
which if you believe them are tantamount to agreeing to whatever game they have
in mind. And their game (Agenda) is not a positive one:

> It is to take back the planet via the Reptoids.
> And 5G will help them do it, and if that isn't
> strong enough, they will go to 6G…

> **Knowledge Empowers**
> **Ignorance Enslaves**

Disinformation & Deception

This is part of the Earth scenario because so many OPs (NPCs) and undeveloped souls are here. In fact, VEG, Ch 8 - 11 pretty well covered the errors, lies, deception and disinformation that we have been experiencing here. We are not expected to fix/stop/change it – **just handle it**. You really need to get that.

Rewind:
> **We are not expected to fix/stop/change anything –**
> **just handle it… and in many cases avoid the entanglement**

The way to handle it is to get enough Truth into you that your discernment kicks in and warns when someone is running a deception, or your intuitive side (which will grow if listened to) warns when something or someone doesn't 'feel' right. And this is part of the waking up process that souls go through here…often by being taken advantage of, manipulated and undergoing violations of our Freewill – those lessons learned become part of the soul and WILL serve in future incarnations. Thus it behooves souls to read and learn as much as they can about our world, people, Science, History and Religion to prevent being used or misled by others.

Manipulation

Disinformation and Deception naturally are what is used to manipulate people – to get you to say/do things that seem to be OK, but which are actually a **violation of your Freewill.**

You are given information that is not true in an effort to **coerce you into making a decision that if you had the real facts, you would not have made**. You were compromised. The used car salesman said, "This car was driven by a little old lady on Sundays to church. It's a creampuff!" And while the body looks great, later you may find out it was the Little Old Lady from Pasadena that owned the car, and the rings are shot.

You have Freewill to choose <u>based on the facts</u>. To the degree that you are manipulated and compromised in making a decision, because you do not have the facts, your Freewill can be violated. And if you believe that you're saved and die and expect to now go to Heaven, a Being of Light may show up (Neggs can also shapeshift and trick you), and tell you that you have to go back, or you can follow him to "just the place for you!" And you may wind up contained in a grey, foggy realm where you can't meet and mingle with other souls… (I saw such a place in an Out of Body experience and don't want to go back).

But deception, disinformation and manipulation **are permitted** here as part of the training ground: to wake you up and get you to seek Truth, Light and Love. By letting you experience Darkness, you will come to appreciate the Light that much more. And if you question why some experiences are so rough, remember:

Steel is not made without fire.

Don't get all bent out of shape over events on Earth. Souls have Freewill to experiment, and the OPs often execute parts of the Script that you are in – to provide **catalyst** for your growth. (VEG, Ch. 5, and this Appendix C.) They are like NPCs in a video game.

Life is not a puzzle to figure out.

When people ask, "Why is this happening to me? I'm a good person. I don't deserve this!" that is a restatement of the time-worn "Why do bad things happen to good people?"

Perhaps now you have an inkling that (1) they aren't as good as they think they are, or (2) they have a special test to go through that reflects something that they may have learned, but the Higher Beings sometimes submit the test again just to see if you really learned the lesson. Their purpose is not to jerk you around, or needlessly test souls for the heck of it – **the purpose is to show YOU whether you still have it all together**! And They will not send anything to you that you cannot handle – it is just your opinion that "it shouldn't be happening to me!" You might want to take a look at what that is all about… what is your ego's point of view? Are your expectations in line with the reality of your LifePlan? (See TOM.)

Ok, enough sermonizing, but the points are valid and bear stating for those who are Baby or Young Souls (VEG, Ch.7) and have not yet assimilated the above realizations.

False Beliefs

This is a very important area, and is largely why this chapter was written. We'll spend some time here. It is important to inspect some of our cherished beliefs – even those we got from our parents, teachers, friends we admired, 'experts' or the Mainstream Media. Remember:

> **If you can't look at it, you can't handle it,**
> and
> **if you can't handle it, you're not getting out of here.**

> Hint: The goal is to overcome it all (not ignore it).

As we saw, the OPs, the PTB and **well-meaning but naïve souls** all can promote false religion, science errors, and false Earth history. Why? Control, or to get what they want.

If humans can be made to believe things that aren't true, they will wander in a sea of deception and confusion – not knowing who they really are, what their divine potential is, and best of all, if different groups of men serve different gods, they can be manipulated into going to war – each for their own god or cause. (Think: **Hegelian Dialectic**.) Keep them distracted and confused. And if they have too much idle time, better start a war so that their numbers can be brought down; people can be ordered around, and somebody gets to make money off the sale of armaments and, of course, from the reconstruction that follows the war. Blow up everything they have and then sell them new stuff. Great economics. If necessary, install Martial Law to **control** the populace. Or promote microchipping everybody… of course you need a powerful network to do that (Chapter 5).

Still think Earth is not an insane place? See now why it's quarantined?

Does Man ever learn or see through this situation? Rarely. Because the programming is too good, too complete – partly due to prevalent thoughtforms, group mind, and corrupt DNA. We also expect 'experts' to tell us the truth, when they may not know either. The Earth's low-vibration energy that sustains low-level awareness works to **entrain** all on the surface into their lower 3 chakras, and Man will never know he can do/be/have more because the higher chakras don't have enough energy with which to resonate to reach and sustain a higher consciousness.

This why the Earth Graduate is so respected – s/he broke free of the illusion and disinformation.

Such resonance (entrainment) is like two tuning forks sitting side by side. They are of the same material and tonal quality – say, designed for middle C. If one is struck to get it vibrating and placed next to the other (non-vibrating) one, within about 30-40 seconds, both tuning forks are vibrating the same. That is **entrained resonance**.

Man is also entrained by subliminal messages on TV and the effects of music and video games. What are you listening to?

Multiple Beliefs

Another little-known way to chain yourself to planet Earth, i.e., to be **recycled** (see Glossary), and really support chaos in your life, is to believe multiple systems at the same time.

This is something that some **New Agers** pride themselves on: the ability to embrace multiple philosophies, teachings, symbols, practices, etc… Sounds good, doesn't it? After all, the more things you can embrace, supposedly the less upset and arguments you will have, thus you should be at "…peace with all men because you believe all things…" or at least tolerate them. (Be careful: there are things that are right and things that are wrong; do not stand for the wrong, or at least do not agree with it.)

Red flag!

Case in point was a local New Age minister who read and devoured as many different religions, philosophies, and teachings as he could – including the latest Quantum Physics discoveries – all supposedly higher teachings and facts about our world. He felt he was really enlightened and just had to share something different every Sunday morning, perhaps to enlighten his congregation, but surely to entertain them as well.

However, he was one of the most unstable people I knew in that he would waffle back and forth on almost any metaphysical subject, and he did not have a firm foundation in truth. His favorite quote was "What is truth?" He thought personal truth was relative (**Absolute Truth is not relative**) and while he certainly did not believe in the Neggs, he did have a big statue of **Kali** on his office desk – he said he venerated her. And yet, he could not understand why his congregation did not grow, his wife mysteriously died, and the church developed a strange energy.

Left: Kali – doesn't this look like the kind of goddess you want to guide your church?

She is dominating the male, Shiva, and has a bloody knife and a severed head…

Note she is blue-skinned – denoting royalty ("blue blood")… and Shiva is surrounded by serpents. (Curiously, Pleiadians were blue-skinned.)

Again there is the strange "stick out your tongue" so common in old art – the Aztec Calendar also has that.

(**credit: Pinterest**)

He believed that **Kali was the goddess of new beginnings**, but missed the fact that that comes about **thru destruction of the old**. A slight misunderstanding. **Kali is about change** but can be dark and violent, including annihilation. If Kali has a dual nature, it would be wise to exercise caution before venerating (inviting) her to participate in your life. Things are not always what you want them to be (another false New Age teaching) – they are what they are.

Doesn't enlightenment accept all things? No. A master will permit you your folly, after cautioning you, so that you may learn – it is a Freewill universe. A fool will learn in no other school than that of "hard knocks." Not all things are appropriate, as examined later.

So what is the problem with following multiple teachings, multiple paths? How does that actually **disempower** a person and lead to their being recycled?

Incompatible Beliefs

Let me give you another personal example from my life.

Let's say that you like the **warmth and love in Christian churches**, but also want the **New Age deeper teachings (secret metaphysical truth),** and you like Yoga and Tai Chi, admitting to the teachings of *chi* and *kundalini*, and your goal is to really get it together – hopefully approaching an inner knowledge base like Jesus or Buddha must have had… so you think.

And that is the problem: your head thinks these things are compatible with Ultimate Truth and even tells you that you are clever to have assembled a very eclectic mixture that is more enlightened than other enlightened people…The ego (empowered by STS Astral influence) has a part in this deception, too.

As has been said about the **power of words**, many words together make a system, a philosophy, even a religion that has a certain **energy signature**, or vibratory quality, because different words have different energy. Having used a real aura camera years ago to test the vibration (energy signature) of different teachings, it is possible to visually quantify what the vibrational signature of the Bible is, versus the Tao, versus the Book of Mormon, and various New Age books. (The Bible was a beautiful crimson color [Love]). Many allegedly enlightened books had a dirty yellow-green aura [means deception] – Not good! (SIM, Ch. 12)

Bottom line: not all teachings are harmonious with each other. Not all metaphysical energies are compatible. And whatever system you belong to (ascribe to, or practice) will bring to you whatever is <u>concordant with</u> that energy system.

For example, someone who subscribes to **Christianity** will immerse themselves in the fellowship and love, tithe, and may believe in laying on of hands to heal, and certainly believe that Jesus cast out devils – thus there will be a belief that devils exist. In addition, most Christians believe that their sins are gone, and that they are to be meek and humble people, save others, and they <u>live just one life</u> and then go to Heaven.

On the other hand, someone who subscribes to the **New Age**, believes he can Create His Own Day, manifest riches, says there are no devils, channels Astral masters, and he may not tithe. Knowing so many alleged deeper truths, New Agers don't believe they have any sins ("mistakes"), and they may deny making any mistakes – leaning sometimes toward Wicca in that they can do whatever they want, "…as long as it harm none." Some don't even believe in Karma and Reincarnation. And because they know so much more than the 'sheep' in Christianity, they tend to reinforce their ego and are anything but humble.

Thus, an ignorant seeker may say, "I want the <u>love</u> and warmth of the Christian church, but I also want the <u>truth</u> of the New Age church… thus I will support both churches, both systems… I can balance them, and just tune out the parts I don't like." (Actually **New Thought** [aka Religious Science] in 1940-1980 did balance them both, but it was the 4D STS beings via the PTB who removed it, and today it is gone. For a review of NT and RelSci and SOM, see SIM book.)

Red flag!

That person has just signed on for the lesson. Because they're following a mishmash of teachings, with different vibrations, and different corollary experiences, they have a jumble of energy, it is not focused, and may be so discordant as to produce illness and even insanity -- in a worst case scenario.

> *Note: that <u>can</u> happen because the Neggs see what you're doing and will support you in whatever lesson it is that you seek to learn – even your unwittingly creating and experiencing chaos and ill health. Earth can be a rough school.*

You cannot believe in one system that teaches Reincarnation (New Age) and no demons, and believe in another system that teaches no Reincarnation (Christianity) and the existence of devils. The teachings have energy, as well as **archetypal resonance** and you will attract to you <u>experiences and people</u> that play out BOTH teachings. Like confusion and chaos? That's what you signed up for.

To further clarify, as a Christian, submitting oneself to the Thoughtform (archetype) of Christianity, you will have (attract) Christian experiences to you – Bible studies, tithing, healings, prayers, etc. – and because Christianity is a powerful archetype believing in the demonic, you may attract those experiences, too, especially if you fear them (i.e., putting energy on them).

You get whatever goes with the package called Christianity.

Think not? I invite you to look at **Padre Pio's** life in the monastery (see Bibliography: Ruffin in Religion Section). He felt that a true follower of Christ should be suffering as Christ did. So, he not only invited physical attack in his cell at night, [72] leaving bruises and marks all over him, but he also had the **stigmata**. [73] (in his palms instead of correctly in his wrists, because that is where he thought Christ should have them.)

"Aha!" you say – "because he believed it and 'it is done unto you as you believe!'" All I have to do is watch what I believe, and if I don't believe in the devil, I can still be a New Age Christian [sic] because, accepting the New Age side of things, I know the truth." Really?

Sorry, but that is a clear case of mental masturbation.

You lose. It really doesn't work that way. The Neggs execute <u>all</u> the archtype(s) that you choose.

Archtypes

If you are a Christian, the archetype is so strong, you do not personally have the power to override the parts you don't like and create it the way you want it. By trying to merge Christianity and the New Age into your own system, a new, unique one, you are trying to create a new archetype (i.e., a religion) and unless you have the power of a Jesus or Buddha, you can't do it. What you <u>will</u> attract is **chaos** – some of Christianity and some of the New Age, and as shown above, the two systems are not compatible. You will pull the 'devils' to you as you are meditating and seeking to channel higher entities (often 4D STS – the real Higher Beings are too busy to play games) – and even the Neggs and Discarnates in the 4D world ARE higher, since you are in a 3D world. Good luck.

The **Mormons** joined forces with each other to create a Mormon archetype, but that took the concerted effort of hundreds and thousands of souls, and even then, it is nowhere near the strength of the Christian archetype. So are you as one person going to create your own archetype…?

As a New Ager, you will also experience being very uncomfortable in the Christian church when they talk about Jesus dying on the cross for your sins, since you don't recognize sin, nor vicarious atonement for same. And over in the New Age church, you may come to dislike some others who think they are creating their days and manipulating stop lights... developing their godhood.

The New Agers think they have superior enlightenment compared to Christians, but very few people are as loving as they are in the Christian church. You will experience **stress through discordant resonance** in both churches: you subscribe to two archetypes that energetically resist each other. Choose one or the other, OR as this author has done, become a Christian and don't share the higher truths you discover (see VEG)... which are the truth about Christianity anyway, so you are not in danger of chaotic energies and personal attacks – unless you share what you know with other Bible-banging, card-carrying Christians.

> *For more on archtypes, please see Carolyn Myss' book* <u>Sacred</u>
> <u>Contracts,</u> *and the works of Carl Jung.*

Why would you want to create stress and chaos in your life?

And even worse, when you die, no one system 'owns' you, you may be your own person, and yet you do not have the Truth to make a proactive choice for yourself as to where you (as a disembodied soul) can now go, based on the Truth, and so you get **recycled**... or you are afraid to go to the Light and drift around Earth as a ghost/discarnate.

Changing Paradigms

You say that that can't apply to you because you are an educated, thinking person, maybe even 'enlightened.' And as proof, you offer the switch you made from Christianity (or whatever) to Buddhism (or whatever). Then you had the strength of character to switch again to a New Age church. No one controls you, you say.

> *See other section on 'Subconscious Sabotage' as some New*
> *Age churches can also violate your freewill. Sorry.*

> All you did was **change one matrix for another**.

 And every matrix, or paradigm, is a way to control people. It just takes longer to see some of the controls after switching to a new paradigm... the love-affair has to wear off before you'll see that **most belief systems are engineered to do the same thing: <u>control</u>.** Your **mindset** is your belief system, the matrix that you live in. And we all do it – to explain the world we live in to ourselves – so that we'll be safe and the world will be more predictable (thus safer), and maybe if we get it really figured out, we can control it – "create our own reality." The ultimate deception in a School where the student does not control the curriculum.

Sorry, no one 'creates' their own reality in 3D. Your LifePlan dictates the basis of your reality, but you can **attract** blessings/curses to you. Remember, you need Truth/Light to get out of here and avoid being recycled. This is why it is so important to seek/attain true Knowledge.

Knowledge protects.
Ignorance enslaves.

How The Remnant Violates Freewill

This is the ultimate section of this chapter: How does the reptilian Remnant (Dissidents [-] and Insiders [+]) influence what we think and do? This is very important to know as **it underlies the 5G Network danger**.

Keep in mind what was said earlier: The Reptiles (Anunnaki Remnant) not only can influence and manipulate a human's thinking (two ways shown earlier in Section on Mind Control) , but they also have technology to do this (which is like the **V2K technology** [below] that the US Military has also developed), and inspirations to do do/say something may not be your own idea…

Need it be pointed out that the Bible said the same thing:

Take every thought captive… (2 Corinthians 10:5)

…meaning that if you have a thought, <u>out of the blue usually</u>, that is (1) not in sync with what you were just thinking, or (2) it is not in synch with the kind of thoughts that you usually have, you need to suspect Astral Influence.

> I suspect that having mentioned it twice that it is relevant to examine the **V2K Technology.** (This is repeated from VEG, Ch.11).

Voice to Skull (V2K)

The truth of the <u>Course in Miracles</u> (CIM) book is that it truly was dictated by a voice that <u>only Helen Schucman (1965) could hear</u>. And it wasn't ET, nor Jesus, nor 4D STS. And it gave her headaches.

She never suspected that it wasn't Jesus speaking to her and yet today we know about **V2K** ("Voice to Skull"), or **PsyOps** devices that were perfected by the military in the 60's and 70's to **send messages directly to a soldier's head**

without using a radio. The message would appear as words in the receiving person's head, and only they could hear it. Someone standing close to them could not hear it; the message was **neurally 'implanted'** in their head via a microwave carrier wave that would simulate brain wave patterns (in the transmitting device) to synch up with and activate the brain waves of the receiving person in the auditory part of their brain. [74]

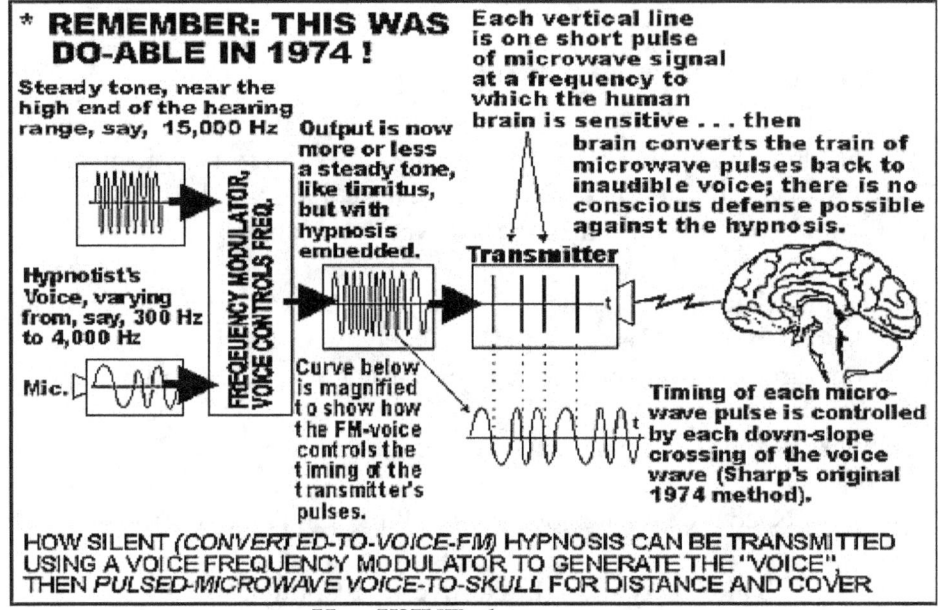

How V2K Works
(source: http://www.flickr.com/photos/63853612@N00/2738963560/)

> And sometimes the receiving person would get quite a
> headache from the experience. Ms. Schucman did.

Patents exist today for this V2K (P300) technology, and it has been around since the 60s. It was an outgrowth of **MK-Ultra** research (Chapter 5) and the later attempts of the US to counter the Russians' use of psychotronic equipment in the 60's against the US Embassy personnel in Moscow. [75] The military has considered using it to beam the 'Voice of Allah' to Iraqi terrorists and command them to stop fighting and go home. [76]

> So if the US military can develop something like this and
> it is a form of mind control, what can the Remnant and
> Dracos do?

Keep reading… V2K also includes the **Frey Effect**:

> The **microwave auditory effect**, also known as the **microwave**

hearing effect or the **Frey effect**, consists of the human perception of audible clicks, or even speech, induced by pulsed or modulated radio frequencies. The communications are **generated directly inside the human head** without the need of any receiving electronic device. The effect was first reported by persons working in the vicinity of radar transponders during WWII. In 1961, the American neuroscientist Allan H. Frey studied this phenomenon and was the first to publish information on the nature of the microwave auditory effect.

The cause is thought to be thermoelastic expansion of portions of the auditory [brain] apparatus ...[77]

So it does exist. What happens when a powerful 5G Network is placed throughout a town (transponders placed **every 300 feet** according to the literature) and it can be used to broadcast 10Hz or other annoying frequencies designed to quell riots – or put people to sleep in an area (also to quell riots) – or broadcast messages to your cellphone –as does the Emergency Notification System (ENS) today?

Sounds like an altruistic idea, possibly to our benefit, but what if it is hijacked by a group with an Agenda to manipulate and control humans?... They say it will link toasters, thermostats and Smart Meters... and control Driverless cars (not!)... so it will be like WiFi – easy to hack into... (more in Chapter 5).

One last item. While I can no longer find the article, V2K was a suggested (non-lethal) alternative to **marketing items** in selected areas of a store frequented by customers... as they enter the narrowly controlled sphere of V2K advertising, say Ladies Dresses in a store, they will hear a presentation on the product being displayed – without using a headset or handheld device of any kind – such as some museums have today. (Testing of same still produced headaches.)

Because this gives away the game (that voice to skull is real), and because it often causes headaches due to the higher energy, stores today often use a sensor to trip a video screen presentation in the product area that will play when you enter the target area.

Obviously, the V2K technology is informational and does not control what anyone does... except that if it comes on you appearing to be self-talk and the information or transmission is preceded with the words, "I think..." or "I want to..." what will the unsuspecting receiving human do?

Montalk on Freewill

The following are excerpts from several articles by an author known as Montalk which have some significant insights on Freewill and how it can be manipulated by the 4D STS. [78]

Importance of Knowledge

> The more understanding and Knowledge you have, the more genuine choices you see, and the greater your role becomes as cause rather than [being at] effect. It is **lack of knowledge** that places one under the influence of causality.

You cannot change what you cannot see, or do not know about, and thus you risk making the wrong decision(s). You must see what is going onto make an appropriate choice. The 4D STS do not want you to see and that is why there is so much **disinformation** ("We're all created the same – forget the OPs") and **illusion** ("Gee, aren't all Congressmen working FOR us?")

Freewill is the only true cause and involves choice. But **you must be informed and question what the 'experts' tell you.** Some mean well but are naïve and succumbed to disinformation, too: "Man and the dinosaurs were not on the Earth at the same time" (a lie). "We can Carbon-14 date fossils back 100,000 years." (a lie – see VEG, Ch. 8 & 10.)

And let's not forget **deception**:

> Have you seen the **Hyundai** cars from South Korea? Note that US ads do not pronounce the name correctly. It is not "Hon-day." I asked a Korean and she said it is pronounced "Schoon-**die**". Great, we all want "die" in the name of our car… Is that a joke?

> Before you answer that, consider that the **same company** makes the **KIA**. Today's generation is too young and most have not been in the military, so they don't know that **KIA is Killed In Action**… just like MIA is Missing in Action. Was this another Korean joke… or do they just not like Americans? Gee, let's all have Killed In Action on the back of our car!

> As I am in my seventies I know what the South Korean reaction was to President Truman stopping MacArthur in 1945 from his finishing the job and removing the Communists from the northern end of the Korean Peninsula… thus busting up Korea, and many

South Koreans were irate about it. MacArthur could have finished the job in 2 months more but Truman apparently saw a \$ustained military pre\$ence there as having \$ome benefit.

Anyway, if one meditates, or does Kundalini Yoga, or gets struck by lightning, and increases one's awareness, there will be a greater connection with one's **Higher Self** which will alert you to deception and a sense that 'something isn't right' – what is sometimes called a "check in the spirit." That and **Intuition** are the voice of the Higher Self – which by the way is also directing your LifePlan, so you do want to be on track with that.

If we make assumptions or choices based on limited knowledge, i.e., that which is told us and we **ignore Inner Promptings**, we are trapped within the game – a game with Their (4D STS) rules and Their outcome. If the decisions and choices are based solely on reason, which is often based on partial data, and what we are already aware of, we may be subject to deception or limitation that ensures Their success and They get us to do what They want (stay dumb.) And that is where the **Violation of our Freewill** comes in: we were not aware of all or enough of the factors. We were effectively manipulated into a decision that supports Their Agenda.

Then says Montalk…

> We have entered this physical reality to learn how to eventually transcend [overcome] it, to take risks by applying of Freewill to learn [predetermined] lessons. When placed in proper context, such risks have no chance for failure because **all paths potentially provide the needed lessons**; on some chosen paths we can learn the easy way, others the hard way, but either **way the same lessons are ultimately learned.**
>
> … stagnation is nevertheless possible when one refuses to learn; Those preoccupied with the transitory distractions [drugs, sex, or smoking…] of the program are wasting away their finite lives. They encounter experiences meant to shake them loose from their hypnotic trances but choose to ignore them and therefore redundantly repeat the same mistakes.

But be advised that **the Higher Self always sees what is going on** and will provide "exits" from the mindless daily routine, as well as provide/insert **Points of Choice,** to try and put the soul back on a proactive path to learning. This sometimes takes the form of **synchronicities** which originate outside the LifePlan and are inserted by the gods (Higher Beings) who run this Realm. Be alert for

161

those – it means you are off track and a change is necessary. If necessary, the Higher Self can override any part of your LifePlan – especially if you are about to exit the Earth Realm before it is your time to do so… these are sometimes called **NDEs** where you are exiting but are brought back to Earth life, reminded of who you are and what you are to do.

The other form, much rarer, is a **Temporal Phase Adjustment** where something in your existing life scenario has to be removed, something you were not consciously attached to, and didn't notice that They removed it or changed it… for your benefit. (Kind of like the scenery in a video game that adds a ladder that wasn't there before, or suddenly there is the sword that you need that wasn't there before, or a person who is there to save or inform you.)

Reality Manipulation

The foregoing Temporal Phase Adjustments are also called Reality Manipulation, but not even the 4D STS beings can jerk with your reality any time they want – usually it is only the Higher Beings when it is in your best interest AND you are not conscious of all parts of the scenario around you. If you are aware that there is no ladder on the wall, and They put one there, they will also have to wipe your memory that it wasn't there a few seconds ago. They want to avoid confusion and unnecessary wonder that would hang you up in another way – so that instead of quickly climbing the ladder to escape the marauding dragon, you stop and wonder where the ladder came from and why it appeared… and that delay gets you eaten by the dragon!

Generally it is things or people you are not aware of (and don't currently see) that get changed, or deleted… Things or people that are tied to one's Freewill (on which one's LifePlan depends) are not subject to manipulation – for example, the 4D STS cannot just have you wake up one morning in your bed, but in Albania instead of NYC. The BoLs see to that.

So sometimes the 4D STS will use disinformation or deception to get a person to do/say something they would not normally do – if they knew more facts. For example, I had a boss who was very sensitive about her weight, and yet I KNEW that she had become pregnant (false information put in my head) and when Mary sat down to eat pizza and ice cream with the crew (breakroom birthday party for Susie), I suggested out loud she not endanger her baby with the Mai Tai's we also had!

Can you say "Set up!"? Man I could not recover from that, and she and I had not gotten along well in prior months anyway, so I was out of there… Yet I knew I had heard or learned that somewhere! The fact that I wasn't laughing when I said it told others I was not joking… which increased the damage. Now I just looked crazy… and that is how the 4D STS can work.

So the way 4D STS works is to either not violate someone's Freewill <u>directly</u> (as mine was just done above) or **do only what the target permits** – what they agree to (tacitly or otherwise). Gradually expose Rodger to the idea of free sex, then to pix of nude women, even have one scantily dressed at a club that (1) is just what he likes a woman to look like, and (2) make her an easy woman, manipulate her (if she is an OP there is no violation of anyone's will), and get her to take him upstairs for R&R – even though he is married.

Then the setup: have his best friend from work come into the club, just to have a few beers, see Rodger and girl together and it is all over.

Direct manipulation, in the same scenario, would have been to have him spot the scantily clad woman at the club, while he is waiting for his buddy to show up, and **put the idea in his head** that he is a He-man, he is worthy of such a woman, and needs to take down one of the women there to prove it, and he **is given the overpowering desire** to act on it… that is a full-bore violation of Freewill.. and yet is that what is happening, in a slightly different scenario, with "nutcases" that take AR-15s and shoot up innocent people at clubs and concerts? (Charles Manson admitted he was controlled.)

Can that be prevented?

> They cannot violate Freewill in cases where they are engaging a
> **Freewill that is stronger than theirs**. This includes cases
> where an individual is stronger than they are, or where he has
> **divine protection** whereby sovereign beings [BoLs] intervene
> and block the incoming manipulation. It also includes cases
> where the Freewill of multiple beings is anchored to **the same
> reality element** and reinforces it beyond the manipulability
> threshold of 4D STS.

In the case above, if Rodger's friend had been there to stop him from hooking up with the woman (because he knew his friend was married = they shared "the same reality element"), it might have been prevented.

Usually the 4D STS just offers choices (physically), and makes mental suggestions (and if the person is an underdeveloped soul), they may be a weak prey. A sure deal – and yet the soul will learn from it – all souls are eternal, so the best the 4D STS can do is slow down the eventual waking up… but they have a lot of Baby and Young souls to work with… and the Game goes on….

…until the 4D STS via their puppets and PTB up the ante. And that is exactly what is happening with 5G…. maybe they can get <u>complete control</u> of all souls on Earth and get them to **agree to be incarcerated** on Earth (because it has been made so nice for them) and not want to leave, and when they die they insist to be put back (i.e., recycled) so they can continue their dilatory pleasures.

> Don't laugh, that is where it is going. And that is what the
> Draco have done thru their puppets on other planets in the
> Orion System.

Why?

> What other Game is there in town when you are all-powerful
> and souls are a nuisance, and your mindset is anti-human,
> AND you have the technology to do it?

Prophecy & Freewill

So when gurus and prophets accurately make a prediction that comes true, does that mean that we live under Fate and that Freewill does not exist?

That is a question that has bugged Man for centuries, and while there is no easy answer, there are a few real things to consider. And they are fascinating.

Subconscious Pre-decision

Recent scientific investigation with people strapped to electronic measuring gear have been monitored for special brainwave activity leading up to and during a decision.

A Test Subject is Monitored for Decision Activity in the Brain
(credit: BingImages: NTCUtah.com)

What they found was curious and alarming... but it speaks to the existence of one's Higher Self (probably executing one's LifePlan choices – see Appendix D).

One significant finding of modern studies is that **a person's brain seems to commit to certain decisions before the person becomes aware of having made them.** Researchers have found delays of about **half a second or more** (discussed in sections below). With contemporary brain scanning technology, scientists in 2008 were able to predict with 60% accuracy whether 12 subjects would press a button with their left or right hand **up to 10 seconds before the subject became aware of having made that choice**. These and other findings have led some scientists, like Patrick Haggard, to reject some definitions [of Freewill]...

One of the pioneering studies in this domain was conducted by **Benjamin Libet** and colleagues in 1983 and has been the foundation of many studies in the years since. Other studies

have attempted to **predict participant actions** before they
make them of "free will". [79]

Many scientists say that the American physiologist **Benjamin Libet** demonstrated in the 1980s that **we have no free will**. It was already known that electrical activity builds up in a person's brain before she, for example, moves her hand; Libet showed that this buildup occurs **before** the person <u>consciously</u> makes a decision to move. The conscious experience of deciding to act, which we usually associate with free will, appears to be an add-on, a post hoc reconstruction of events that occurs *after* the brain has already set the act in motion. [80]

And I would point out that **the decision is being made by the Higher Self** which has merely projected an aspect of itself into the Earth Realm... as Appendix A and D point out. So Freewill still exists, but it is <u>not always</u> the "projected aspect" of the Soul that is making the decision.

Either way, it is still a shocker to learn that we in 3D are not totally in control – the LifePlan rules and **you already decided on it – before you incarnated.** (Really.)
It is not Fate, you can choose to do something totally off-the-wall but it will usually be destructive and not in synch with what you are here to learn – like instead of going to college classes on Monday, you are persuaded by your (OP) roommate who over-powers your better judgment (because they are empowered by the 4D STS) to stay in the dorm and do cocaine. You are derailed.

Your Freewill was just violated. You were <u>directly</u> manipulated and you **agreed** to it.

And there is one more discovery by Quantum Physics that adds to the revelation...

Quantum Eraser (QE) Effect

Before examining this aspect of reality, several times proven in various laboratories, we need to quickly review what was said in VDOG and VEG about the **Observer Effect**... where what the scientists are looking for in the lab affects the outcome at a quantum (particle) level.

Note that the New Age has borrowed the Observer Effect to claim that you at the macro-level can affect stoplights, hair color and manipulate the Lottery. That was a misreading of what the

Effect said. It only applies to the atomic or quantum level.
(See **anomalon** in the Glossary.)

The following are explanatory notes from DNQF:

Note 1: our Reality is a quantum reality meaning it is quantized, and probably a **Simulation**. People assume we are all in the same Reality, but Quantum Physics proves this is not so. Each Observer helps sustain the Greater Overall Reality, but each also has his/her own "local" reality as a function of their intent and beliefs. Most of our experiences are purely subjective. And yet there is still a high degree of consensus about our world that 'tricks' us into thinking that our world is objective and deterministic. It isn't. Our Reality is virtual.
(See the VR Sphere section earlier.)

Hindu gurus and Chinese masters prove it all the time – walking thru solid walls, levitating, and knowing what is about to happen, or where a lost dog is, for example.

Note 2: and while we're at it let's take a brief look at **Freewill and Prophecy**. It is popular to believe that we have Freewill and that IS a factor in us souls meeting our tests and showing how we handle things – as a measure of how much we have learned; i.e., how advanced we are… or not.

But how do you explain **Prophecy**? How could a guru or master accurately foretell the future? If the future were just a **Greater Script** with most events preplanned, it would be easy to access it (with a higher consciousness) and see what was "scheduled" to happen, and then state that – and be right. And there is such a thing – the **Greater Script** (Appendix D) constrains each soul's LifePlan (individual Script) to synch up with it and all we can do is **choose** to meet optional scripted event(s) or not (also see TOM).

QE Goal

The **Double-Slit Experiment** (DNQF, Ch. 9) was initially set up to see if a photon was a particle or a wave and depending on how many slits (1 or 2) were open in the barrier at which the photon was fired -- it should prove what the photon was. With one slit open it appeared to be a particle and behaved that way. With two slits open, the photon behaved like a wave. How could it know how many slits were open? This was so weird that a physicist decided to find out if WHEN the photon was measured (in front of the slit or behind it) was the cause of the anomaly. What he got was a complicated set of mirrors, timers and results.

Here is the basic info: [81]

The quantum eraser experiment described in this article is a variation of **Thomas Young's** classic double-slit experiment. It establishes that when action is taken to determine which slit a photon has passed through, the photon cannot interfere with itself. When a stream of photons is **marked** in this way, then the interference fringes [wave pattern] characteristic of the Young experiment will not be seen. The experiment described in this article is capable of creating situations in which a photon that has been "marked" to reveal through which slit it has passed [after the fact] can later be "unmarked."

A photon that has been "marked" cannot interfere with itself and will not produce fringe [wave] patterns, but a photon that has been **"marked" and then "unmarked"** can thereafter interfere with itself and will cooperate in producing the fringes characteristic of Young's experiment

So marking and unmarking the photon <u>after</u> it has passed thru the slit(s) changes the way it behaves – after the fact!! That is called **Retrocausality**, and is not something we see in our everyday macro world.

> The point being: can Freewill be messed with for us as well? If we can change the outcome of a physics result <u>after</u> it has already happened, can the same thing happen to what we do and negate our Freewill?

And now the kicker: Quantum Physicists have done tests in Freewill and found that sometimes the outcome can affect the preceding causative event. Really. That is called **The Quantum Eraser Effect**. [82]

Briefly, this is a version of the **Observer Effect** mentioned earlier (DNQF, Ch. 9) where a photon is fired at a board with either one or two slits in it and the resulting pattern on the wall behind the slit is different depending on how many slits were open. The Physicist chose to measure the particle <u>as it passed thru</u> the slit and that created the outcome shown in DNQF Ch. 9. Then he got the idea to measure it **after** it passed thru the slit(s) – and that changed everything!! If he expected a particle, it was a particle; if he expected a wave, it was a wave. And measuring it for one or the other **retroactively** affected how it [already] passed thru the slit(s)!!

That is called **Retrocausality**.

In another test, people were asked to watch a list of words flashed up on a screen for a few seconds, then the screen went blank and they were asked to list as many of the words as they could. When they wrote the words down, there was a higher match than should have happened, and it was found (long story short) that what they typed had a higher incidence of being in the list —as if the list reflected their choices **before** they were shown the list. So did they really have **NO Freewill** (to remember and list words) or were they going to list what was already on the list? [83]

Quantum Physics reports some really weird results from examining our Reality. **Consciousness co-creates Reality**.

So, are we really in control of the stuff that happens on Earth, or are we somehow synched up with the Greater Script? Remember that even our LifePlan (operating under the Greater Script) has **Points of Choice** (see Glossary) and options ... and they are apparently made by our Higher Self and then relayed to us (**as an avatar**) in 3D and we think we made the decision. (See Appendix A: Allocation of Soul Energy)

Note that **only Points of Choice** are scripted (pre-programmed); your everyday activity is not preplanned nor scripted – jus the significant junctures where you must make a choice that reflects your soul growth – your choice is yours -- if you are not co-opted by the OPs as manipulated by the 4D STS entities.

(Be sure to read the Epiphanies in Ch.s 3 and 4 for a better understanding of what is going on and what you are here for. They foreshadow the Chapter 8 summary.)

Epiphany

Freewill is very important and is the birthright of all souls on Earth. Each soul is choosing day by day what to do, what to believe... and it all has Karmic weight. That means you want to make the most **appropriate choice** you can at all times... demonstrate soul growth, overcoming limitations, etc...

BUT – there **is** a group that does not want you to do that, and I am not being "religious" and invoking the idea of demons. It is very important that you make the connection between your choices and future AND the desires of the Gina'abul/PTB/Anunnaki Remnant Dissidents Agenda to derail you.

As this is very important, it is repeated here.

Not only do the PTB love being Lords over the serfs, Lords over the slaves, Lords over the Sheeple, they want you to come back in your next incarnation that way, too.

So what do they do to insure that? Keep you stupid so you can't grow. Even more: you don't know and aren't told that you have that potential!

But it is more than that... and this is the **second half** of their Control Agenda.

The Gina'abul/Dissidents also want to stop you from becoming all that you can be (No, not a General in the Army!... unlike the TV ad) Other than hating humans why would they want to derail you, obstruct your soul growth... What is it to them?

Simple.

> **You are a soul with a divine birthright to grow in wisdom, knowledge and true power (like an avatar, a master) and RULE OVER THEM in the realms they already occupy.**

Souls scare the Gina'abul who are called the **Gilimanna** – the Celestial Bestiary.

The Elohim, Kadishtu and the Insiders totally encourage and support you.

Whether you make it or not is up to you – No one in the Higher Realms will stand in your way, nor does any Master seek to derail you. Once you have been given the information of What you are, Where you are, Why you are here, and What you are supposed to be doing... you are empowered. Everything in today's 3D world militates against that – by design. Guess whose?

Preview: Earth School Graduate

Once you have been "waked up" the rest is up to you... When you see what is really going on and that you have a choice to learn and grow and beat the negative vibrations and obstacles placed in your way here in the Earth School, you can then **leave Earth**... and that is called the **Earth Graduate** (and that is the theme of VEG).

> If you can get a glimpse of what it means to be an Earth Graduate, no one can hold you back! Or lie to you. Or derail you.

> Remember that the Earth Graduate must go through the lessons, trials, tests, etc. of the Greater Script that will produce the quality being, ready to serve in higher realms.

There is no steel without fire.

Those who do not 'graduate' now will start again in another 3D virtual realm.

> So when does it happen? **Respect yourself**, develop compassion, gain knowledge, and live in patience, **detach** from outcomes, things and people… a long list of requirements to graduate, but Man is much more than a first glance would suggest, and this may be why one of the Beings of Light told **Robert Monroe** when he did his recurrent Out of Body Excursions, that although the Earth school is a rough one,

"The graduate from the human experience is very respected elsewhere." [84]

And it is easy to see why: if a soul can survive this screwed-up planet with its disrespect for everything and everybody, survive the pollution and the killer diseases, resist the temptation to lie, cheat and steal, not follow the crowd, not give in to corruption, and still emerge with his/her integrity intact, <u>that would be worthy of **respect**</u>. A soul who walks the talk and does not sell out. A soul who thinks outside the box that has been created for all of us by the PTB-dominated Media whose goal is to **keep us as dumbed down robots** so we will buy and do what Madison Avenue suggests. (How is that working so far?)

It is called self-mastery. And that starts with **self-respect**. You don't do those things that are inappropriate – like overeat, or run with hoodlums. Self-respect means the body is the temple of the divine spark, the soul that is so highly prized that there are some beings in 4D who would give anything to have one, and because they can't (I repeat, I know) they work to stop us from becoming all that we can be. When we graduate from Earth, we can be released to return to the higher Realm from which we came and in which we are then ready to serve.

And, You need not be perfect to get out of here…

> If the gods had to wait until a soul on Earth attained the status of a Jesus, they'd have to wait a long time!

Three things are required to leave Earth as a Graduate:

> (1) you see the **futility** of demanding what you want here,
> (2) you learn, acquire **knowledge and compassion**, and
> (3) you experience and realize that **you <u>can</u> handle whatever** life throws at you.

It is worth repeating what Monroe was told:

> The first point in consideration of human structure certainly should be the note that a small percentage [of souls] have never been thru the graduation experience...
> Some may have had physical life experience in other parts of time-space and **in another physical form**, but this is their first run as a human...
>
> Human existence on Earth is an interesting anomaly. **It has some peculiar qualities that are unique in the development of intelligence and consciousness.** As a result, human life has many attractions. To some it is like attending a vast amusement park... a playground where standard rules (non-earthly) are suspended for the moment. They desire human existence simply out of curiosity... Many... decide it is an ideal opportunity to try an experiment conceived in their periods of contemplation...
>
> Still others find that the limitations imposed by physical incar-creation as a human also engender concentration of energies available <u>only in that state</u>.
>
> **This is the only point available to apply such energies.**
>
> By far the greatest motivation – surpassing the sum of all others – is the result.
>
> **When you perceive and encounter a graduate, your only goal is to be one yourself once you realize it is possible.**
> And it is. [85] [emphasis added]
>
> When you know, why would you settle for anything less?

Promises

> He goes on to say that one of the outcomes is the ability to **translate Prime Energy into any manifest form** [change water into wine], and "...second, to become a first-order generator thereof." [86]
> And this is not easy to imagine in our current state, which is akin to never having heard a song, and not knowing that we can sing – how would one ever want to sing, much less know what it is, or be able to learn it? [87]

So, in general, Man is quite a complex, fantastic and valuable creation and this issue will be amplified in Chapter 8 when we examine more why we are here. And how to get out.

Chapter 8 will suggest various "jobs" or positions available to the Earth Graduate in the Multiverse Realm.

That is ultimately what this book is about...
if we can't stop 5G...

Chapter 5: Control of Man & Earth

If you were an advanced society of beings from another world, and you have very advanced technology, and you discovered a beautiful world that was populated by hominids who abused the planet and its resources, abused each other, and were so ignorant that they could be told the truth about something and they could not understand it nor accept it, no matter how you tried to educate them, you'd say they were destructive and stupid.

In fact, if you saw the value of the resources, despite it being a Preserve that was overseen by the Solar Council, you'd devise a way to take it for yourself… if the Council would not let <u>you</u> run the planet with a more proactive mindset.

Failing with the Council, because they traditionally saw your species as aggressive and not worthy of the Blue Jewel, you withdrew and considered wiping the hominids off the planet and just taking it, but due to the Galactic Police Force (the Lyrans) you can't just attack and come in guns blazing --as so may SciFi movies of the 50's portrayed.

So you discuss it with your Empire Council and decide to take the planet anyway – but use an underground approach which is not obvious to the surface dwellers, and you are crafty and know that you can "persuade" the hominids to give you the planet – after executing a masterful, and complicated Plan over much time (so no one suspects). Since the Reptoids live for hundreds of years, and are unified in their domination mindset, this can take place over centuries of hominid time… so subtle, no human will wake up to it.

> If you throw a frog into a pot of boiling water, he will jump right out.
> However, if you put the frog into a pot of cool water and gradually turn the heat up, the frog will not notice until it is too late…

And of course, the Solar Council will know what is being done, but (1) it is a **Freewill Universe** where anyone can do whatever they want as long as no one is hurt (i.e., souls are not destroyed), and (2) the **Plan** is to create disinformation and deception (the appearance of terror) causing the hominids to surrender their Freewill for Safety… If the hominids agree to what the Reptiles have craftily proposed (violating their Freewill), the planet will eventually belong to the Empire. They have to be careful to not <u>directly</u> violate the hominids' Freewill and as long as they don't do that, the Council will not interfere.

Essentially this is what has been done, but there is one element missing for greater control (and acquisition).

This subterfuge was tried (by the same Orion beings) between 1917-1989 in what is called the Soviet Union and it failed. Largely because there was no computer, nor was there a **network** to link and coordinate them. (They tried a different (political) tactic in 1939-1945 and that also did not work.)

So they gave up the Soviet Union idea and moved to Western Society, promoted their high tech over 40 years and continued to 'plant' and control their operatives ("puppets", aka PTB) and guide the design and creation of computer networks and an infrastructure that would allow them to achieve their Plan with power.

Knowing that humans are enchanted with electronic geegaws, they proliferated laptops, HD TV, cellphones, WiFi, iPads, and VR headsets (which ruin the eyes) and then told people by the Media (which they largely control today) that this is Progress and all we have to do is ramp it up: put WiFi in schools, cellphone towers all over the place (many are hidden in church steeples, **others are designed as "trees"**), and keep increasing the cellphone power (telling people that it is faster and easier to use), and…

Voila! You have today's preamble to the hidden **Control Agenda**.

Newer Cellphone Towers

Again, **the Telecom industry is not evil** – they do not know they are being used, nor do they believe (and if even if they did), <u>they could not admit</u> that the **sea of EMF radiation** we now live in is harmful to humans and bees and insects… Telecom is a "cash-cow" and on a roll and not about to be stopped… just as the Reptoids knew it would happen.

The frogs are happy in their water (with their electronic toys), and some have adopted the technology so much that they have clipped the cellphone earpiece to their ear, keeping their hands free and their brains irradiated with EMF.

The uninformed public have no clue what is going on and many will pooh-pooh what was just said, and while I may appear to be an overreacting idiot, I guarantee you, the forgoing is exactly what has happened to bring us to today's pre-5G world.

Because I doubted, too, I asked Baldy (aka Philealel) to show me that this wasn't just a fascinating story. He said we were going to take a look inside a mountain in new Mexico and he would be with me, we would be discovered, but I would see what is going on.

He had me lie down and become calm, breathing regularly and I was almost in a meditative state. I let him take me astrally (I guess) to the mountain (which I later saw on TV was Archuleta Mesa) and we went thru the side of the mountain into a large hollowed-out cavern, with what looked like a polished cement floor. There were overhead lights, and many large crates and electronic equipment, consoles and even what I later learned were called ARVs off to the left. I had about 10 seconds to take it all in when a Reptile Soldier poked his face into mine and said "Get out!" He was dressed in military fatigues and was very nasty. How did he know we were there?

Baldy immediately withdrew us and I was back on the bed. Man did I have a headache and was dizzy for a day or two – whatever the Reptoid did drained me and left me unable to eat, and sleep was rough – Baldy said he had hit me with a psychic whack (which Baldy mitigated, or it could have killed me). I never tried that again. They are real and may be what the late Phil Schneider saw at Dulce, just around the corner.

Other Control Aspects

Just briefly, before getting into the 5G Issue, let's take a look at what else the Reptoids have done to effect control over the humans they want to remove… these are also part of the **Control Agenda**, and of course the whole scenario is so pervasive and **the scope is so large**, no one would seriously suspect that it is all orchestrated. (That is the idea.)

So you ask, "How would they get so many humans to go along with an orchestrated Agenda?"

First, it does not look orchestrated… they are too clever for that. And they do it over so much time (years) that humans living short lifespans never see the overall picture – they would if they lived 200 years! And **second** you make sure that the Media (radio, TV, and Internet) does not expose the Agenda, and if they do, you discredit and make fun of those who point out the fact that something seems to be going on that is not in our best interest. And **third**, you claim National Security and compartmentalize knowledge of what is really being developed so that humans do work on their own demise, but they don't see the complete picture.

And **fourth**, they use their mind control technology to manipulate the key people – not everyone, just the top dogs who call the shots for all the people below them.

> In fact, some of the top dogs are not humans but are the Hybrids
> (who look human) and walk among us. They can also mind control
> the humans working under them.

But you say, "Isn't the Agenda, and the 5G Network, just normal technological progress and you are jousting with windmills, like Don Quixote?"

This is a **Key Issue** – and it involves people not suspecting that there is a large Ape in the room because no one asks why a part of the room cannot be seen and why the couch is so big…. And why there is a banana tree in a corner of the room?… the clues have been UFOs, abductions, huge structures (Think: Sacsayhuaman and the Great Pyramid which Man still cannot duplicate), and odd skeletons around the world, but you are told that those are a mystery, and no one knows… so drop it.

But you say, "This all sounds too fantastic to be true – you have got to be making it up!" I agree it is fantastic and so much so, the Reptoids know you will experience **Cognitive Dissonance** when encountering this information, which again, protects them. They have spent a lot of time (centuries) making sure you believe the **straight vanilla things about Earth** and you will not like the truth and will reject it. You don't have to believe it, just consider: What if it **is** true? What does that mean for Mankind and the future?

Some of the areas they have been manipulating follow… and this will be just a brief overview as these have been addressed before in other books. (Much of the review covers the information in VEG, AL and GEP.)

Media

Fake News against those who seek to fix what is wrong in America.
   Spin events so the public doesn't know what really happened.
Do not report UFOs or Abductions.
Do not report CDC findings on HIV/AIDS (what if 52% of the US
   population has HIV?)
Do not tell people that cancer is killed in an oxygenated environment
   (think: Hyperbaric Oxygen Therapy).

Religion

Create different ones at odds with each other so Man cannot unite.
Major ones must have at least one lie in them (see VEG, Chs 1, 11).
Exclude spiritual or metaphysical information so that people cannot
   grow and discover the benefits of colloidal gold or Kundalini Yoga.
Downplay or ignore the importance of the soul.

Pharmaceuticals

Everyone will be on something (they don't care what) by 2020.
New Rx drugs have serious side effects (mood swings, thoughts of suicide).
There is no cure for HIV/AIDS (Sumerian *Suruppu* Disease).
There is a cure for cancer but it is a one-time treatment and no further
   money can be made from it… hence we keep $earching for a pill.
Misuse antibiotics so bacteria develop resistance and become **superbugs**
   for which there is no Rx (i.e., people do not finish their whole Rx
   and the few remaining bugs develop an immunity).

Education

Dumb them down, don't teach ways to think (H.O.T.S.) [88]
Instead of **H.O.T.S.** (Higher Order Thinking Skills) be sure
   kids just memorize names, dates, places.
Put WiFi in schools to expose growing bodies and brains to EMF—
   (explained later).
If you have a problem class, just compromise and pass them; let
   the next teacher worry about it.

Agriculture

GMO – genetically modify key food groups; let the DNA adapt to it.
Chemtrails are sprayed at high altitudes (eventually fall into soil and water).
Develop **seedless fruit** so that people cannot plant their own apples,

oranges, etc… thus maintain control over the food supply.

Psionics

MkUltra – the original mind-control research done by the US government with the help of Nazi scientists. Research was switched to psycho-chemicals in the late 60s.

V2K – developed by DARPA for the Army – to allow a soldier to receive orders with no radio transmission (so the enemy cannot intercept) – the info appears in his head by microwave transmission… similar to Remnant technology.

Develop non-lethal pulsed microwave 'weapons' to paralyze or affect human behavior (examined later).

Science

Develop ASI (Advanced AI) to interface with Transhumanism and the Internet of Things.

Develop Genetic Engineering to modify the human genome as wanted.

Develop antigravity/electrogravitic (ARV) craft to replace rocket-based propulsion for space travel.

Develop cellphones that track people, record what they say, where they go (built-in GPS), and broadcast a frequency that mimics brain waves to override them, controlling behavior. [89]

(Already done, just waiting for 5G to empower that.)

Social Engineering

Transhumanism is (not) Man's destiny – to implant electronics in his body to become a cyborg and live forever.

Fluoride at 4 ppm was used to keep concentration inmates docile. Europe has outlawed it but what else can the US do with the byproduct of making Aluminum? (Water treatment plants use 2 ppm.)

Develop genetic engineering to give birth to the kinds of babies we really want (maybe even a supersoldier or two)

Environment – (most of the following are in VEG, Ch. 14)

Promote **Global Warming** scare to increase the tax base; get a Green New Deal doing to control industry and cut travel…
(never mind that the Earth goes thru cycles of cooling and heating – check out the **Younger Dryas** event… [90] see footnote).

Overfish the oceans and have **supertankers pollute the oceans** so that phytoplankton **produce less oxygen**.

Make sure builders and developers tear up trees and cut down the oxygen supply (NOAA has said if we drop another 8% on the planet, no one will be breathing… it has dropped 20% since.1890.) [91]

Create a **Pacific Trash Vortex** – all plastic garbage from Man's laziness in disposing of trash properly.

Increase EMF and microwave usage to kill insects/bees and birds.

Overuse fertilizer and weed killers that get into the watertable so that it either goes back into the water supply (like hormones and antibiotics which water treatment plants cannot completely eliminate) or have it wash into rivers and streams and flow down the Mississippi to create **Dead Zones** in the Gulf of Mexico.

Rewind: Focus

Do you really think that Man is just so clumsy, stupid and bumbles his way thru things that the problems in the above areas just happen?? Do you really think that the above items (positive and negative) are all just happening, or could they be orchestrated? Could the Science issues be guided by those who already have such technology?

> Here is something to think about:
> **Nanotechnology** is all around us today, including "smart dust" which contains nanites... Chances are good you have breathed some in in the last 2 years. If you have gotten a vaccine, the needle used to deliver the **vaccine** is large enough to also deliver a nanite (see chart below), and you have been 'chipped' and can be tracked. But here's the thing: Who is making these intelligent submicroscopic chips for Man? Man does not have the technology to create something that small... not with intelligent circuitry in it.

Nanotechnology

We both know who can create that submicroscopic chip and we know why they would do it -- and give it to the PTB to "further our technological progress."

Realize that the **nanites** are very small:

One nanometer (nm) is one billionth, or 10^{-9}, of a meter.
Bacteria are around 200 nm, and nanotechnology is taken
as the scale range 1 to 100 nm... smaller than bacteria.

Now look at the needles used to deliver vaccines...

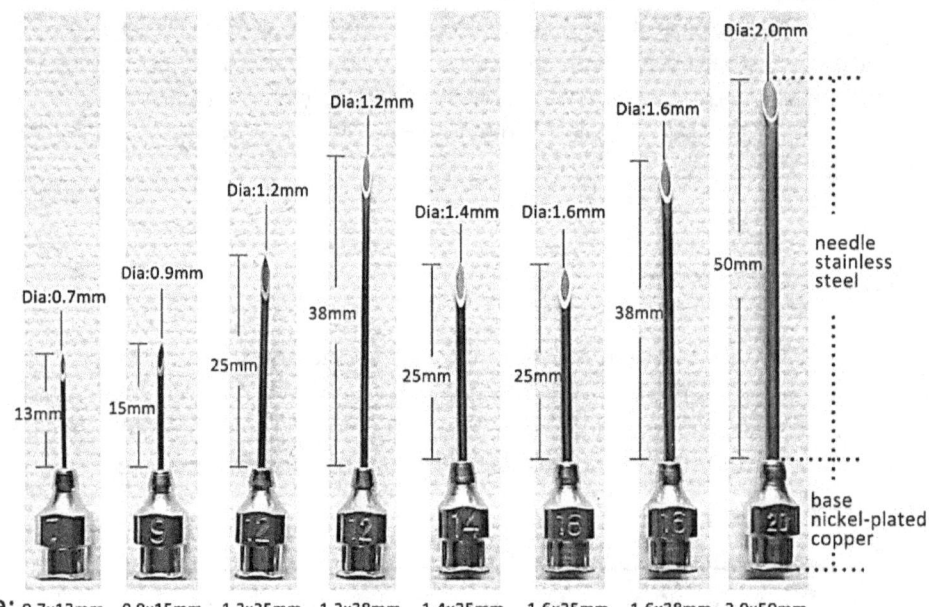

Dia:0.7mm Dia:0.9mm Dia:1.2mm Dia:1.2mm Dia:1.4mm Dia:1.6mm Dia:1.6mm Dia:2.0mm

needle stainless steel

base nickel-plated copper

size: 0.7x13mm 0.9x15mm 1.2x25mm 1.2x38mm 1.4x25mm 1.6x25mm 1.6x38mm 2.0x50mm

All diameters (above) are in millimeters, thousandths of a meter, nanotech is much smaller. 100 times smaller. Nanochips can easily pass thru a needle in a vaccination.

> Allegedly, **Nanoimprint lithography** (**NIL**) is used to etch circuits onto microscopic MOSFET chips, as well as...
>
> The Atomic Force Microscope (AFM) and Scanning Tunneling Microscope (STM) are two early versions of **scanning probes** that launched nanotechnology. There are other types of scanning probe microscopy...
> The tip of a scanning probe can also be used to manipulate nanostructures (a process called positional assembly).
> Various techniques of nanolithography such as optical lithography, X-ray lithography, dip pen nanolithography, electron beam lithography or nanoimprint lithography were also developed. [92]

So did I just shoot myself down? No, where do you think the technology came from? And the supersmall circuitry used in our PCs and medical devices is <u>made for us, programmed for what we want</u>, and no doubt has some Remnant-programmed hidden function(s) as well ... remember: the Agenda is Control.

182

I suggest that there are <u>too many</u> problems and Man isn't that stupid... the <u>problem items</u> above only touch the surface of the things that are going wrong on Earth, and a lot of the 5G <u>development items</u> are either done or close to it. But it is suggested that all these things are what would be done by those seeking to upset, derail and **control** Man's world...

...plus one more big item that would insure Reptoid success:

> Have Man create a computer network that they can hijack and use to either create chaos (and lock him down), or to manipulate what he is thinking – or maybe just to broadcast a frequency that makes him sick/sleepy. [93]

> Many police departments have what they call **"non-lethal" weapons** (see NonLethal Section this chapter) that immobilize crowds with microwave sound. [94]

But before examining the ins and outs of non-lethal microwave technology (which may shock you), and what prolonged exposure to EMF (electro-magnetic [microwave] fields) can do, and what soundwave technology at different frequency levels can do, we need to take a brief <u>neutral look</u>, at the layman level, at the stated principles of the 5G Network....

The 5G Network

The 5G Network is the next or 5th generation of cellular mobile communications. The official definition of the network is given as: [95]

> All the 5G wireless devices in a **cell** [small geographical area] communicate by radio waves with a local antenna array and low power automated **transceiver** (transmitter & receiver) in the cell, over frequency channels assigned by the transceiver from a common pool of frequencies, which are reused in geographically separated cells. The local antennas are connected with the telephone network and the Internet by a **high bandwidth** optical fiber or wireless backhaul connection. Like existing cellphones, when a user crosses from one cell [area] to another, their mobile device is automatically "handed off" seamlessly to the antenna in the new cell.

> There are plans to use millimeter waves for 5G. **Millimeter waves have shorter range than microwaves, therefore the cells [geographical areas] are limited to smaller size** [note that!] ...

> The waves also have trouble passing through building walls. Millimeter wave antennas are smaller than the large antennas

used in previous cellular networks. They are only a few inches (several centimeters) long. .... Each cell [area] will have **multiple antennas** communicating with the wireless device, received by multiple antennas in the device [cellphone] , thus multiple bitstreams of data will be transmitted simultaneously, in parallel [achieving the 5G increase in speed]. In a technique called ***beamforming*** the base station computer will continuously calculate the best route for radio waves to reach each wireless device, and will organize multiple antennas to work together as **phased arrays** to create **beams of millimeter waves** to reach the device.

The new 5G wireless devices also have 4G LTE capability, as the new networks use 4G for initially establishing the connection with the cell, as well as in locations where 5G access is not available.

5G can support up to a million devices per square kilometer, while 4G supports only up to 100,000 devices per square kilometer.

Millimeter Waves

These are much shorter than those used in prior cellphone communication and as such have interesting drawbacks:

Millimeter waves propagate solely by **line-of-sight paths**. They are not reflected by the ionosphere nor do they travel along the Earth as ground waves as lower frequency radio waves do. At typical power densities they are **blocked by building walls** and suffer significant [signal loss] **passing through foliage**.

Millimeter wavelengths are the **same ... size as raindrops**, so precipitation causes additional [signal loss] due to scattering...as well ss absorption... atmospheric absorption limits useful propagation to a few kilometers,,, [so...5G satellites?] **reflection from indoor walls and surfaces**, causes serious fading...
Since the waves penetrate clothing and their small wavelength allows them to **reflect from small metal objects** they are used in airport security [TSA] scanning....

Also interesting: [96]

Millimeter wave radar is used in short-range fire-control radar in tanks and aircraft, and automated guns on naval ships, to shoot down incoming missiles. The small wavelength of millimeter waves allows them to track the **stream of outgoing bullets** as well as the target, allowing the computer fire control system to change the aim to bring them together...

So why use millimeter waves for cellphones?... Answer: transfer of multiple pieces of data via multiple antennas, and **frequency reuse**...

> The short propagation range allows smaller frequency reuse distances than lower [4G] frequencies. The short wavelength allows modest size antennas to have a small beam width, further increasing frequency reuse potential.

You can get more users online (now 1 million) with the ability to reuse the same frequency – if it is not in an adjoining cell, but one located at least 1 step away – see diagram below: [97]

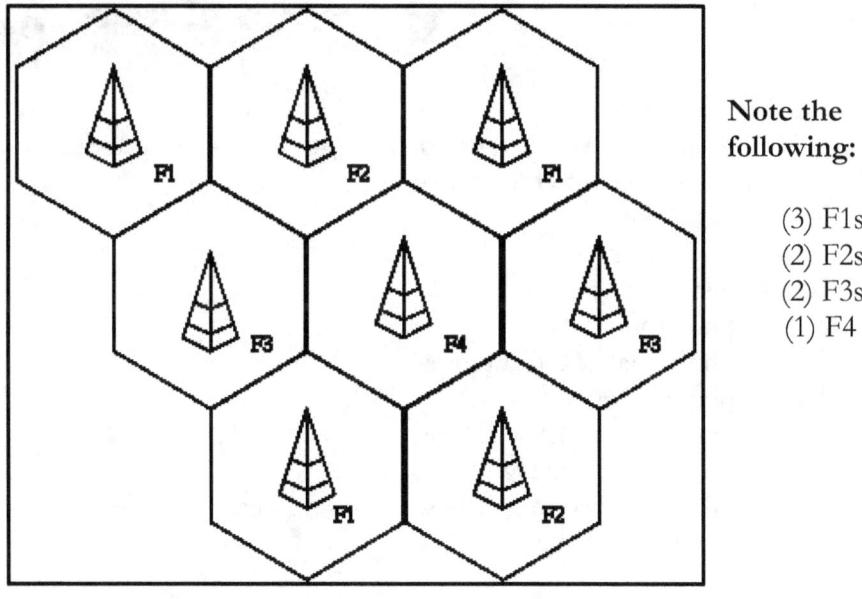

Note the following:

(3) F1s
(2) F2s
(2) F3s
(1) F4

5G Connectivity

In the diagram above...

Cell towers F1, F2 and F3 (not F4**) can** use the same frequency (for different calls) **at the same time,** as they are not adjacent, but F4 cannot use any frequency matching F1, F2 and F3 at the time the other 3 are using it. (Note there are three F1 towers, two F2 and two F3.)

Ok that was just a brief intro to the 5G mechanics.

Five Generations

And the 1G – 5G comparison looks like this:

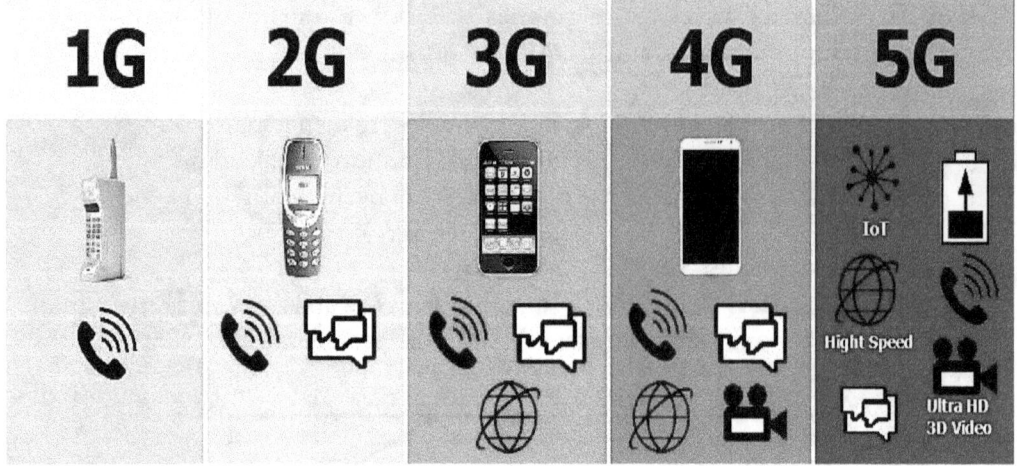

Note that 5G also includes the **Internet of Things** (IoT) – ability to link multiple intelligent, enabled devices into the same network.

    1G = just phone service
    2G = phone and data service
    3G = phone, data, and Internet
`   4G = phone, data, movies (streaming), and Internet
    5G = Smartphone, data, HD movies/video, IoT,
            and Energy-saving…
            and greater connectivity with more things… even your
            front door camera.

There is also a difference in the cellphone towers/transceivers.

5G Tower Placement

5G transceivers do not require much space, they are very small…

5G transceivers

...whereas the 4G cellphone tower hardware looks like this (and will still be used to make initial connections)— especially for the non-5G cellphones.

4G Cellphone Tower

However, the 5G boxes have to be placed **every 300'** apart to get maximum reliability. This is due to the short range of the millimeter waves.

In your Face...

...and a "smart city" design looks like this...

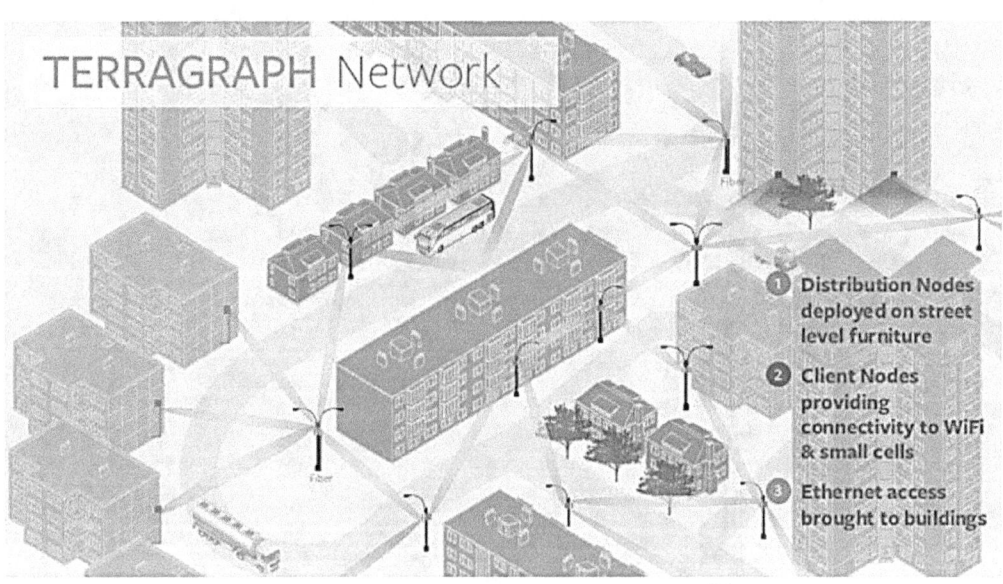

TERRAGRAPH Network

1. Distribution Nodes deployed on street level furniture
2. Client Nodes providing connectivity to WiFi & small cells
3. Ethernet access brought to buildings

This will be an EMF sea that will surround us.

5G Dangers & Health Issues

Consider first that our youth with their growing bodies and brains are constantly on their cellphones, as well as other technology (PCs, laptops, VR headsets…). Besides the reported physical health issues (examined momentarily) there appears to be another __social issue__ – people are not connecting personally any more…

21st Century Teens Socializing

…and…

21st Century Social Interaction

…and ….

21ˢᵗ Century Family Bonding

… maybe there will be an app for cellphone **remote sex** in the future? Then we don't have to interact at all… Oh, wait, they already have sexbots….

…This gives new meaning to the term 'face off' …

Ok, in a more realistic tone, AI, robots, technology distancing people can be a serious issue (next chapter). Even more serious are the reported health issues – which the **Telecom industry refuses to test** and acknowledge, and when independent testing is done, Telecom pooh-poohs it, or ignores it, or discredits it –

Yeah, why kill a "cash cow!" And that is a serious issue: thousands of people make their living supporting the Telecom Industry… What happens when 5G bombs because it causes <u>too many verified</u> health problems?… and they have to backtrack to 4G…Will the economy react and push us into a Recession?

It may. Read on…..**the health issues research has already been done (not by Telecom), they are known and verified by doctors and engineers, and 5G <u>is</u> being seriously challenged.**

EMF & Health Issues

Right off the top, it needs to be understood that I am not against 5G *per se*. It merely needs to be **implemented safely** with a minimum impact on human and insect health (as examined below) …or we are in big trouble as a human race. And that is the focus of the **Warning** at the end of this chapter.

> For starters, we don't need 5G for voice calls,
>   nor for self-driving cars.
> It is a **data** transmission advantage.

The reasons are: (1) we are in a race with China (and Iran?) to avoid problems in data communication and **cyberwarfare** -- which they are hell bent on doing to us. Also (2) that issue primarily with China makes for real concerns in **National Security**. I get it. But there needs to be responsible oversight and **wired connections** as much as possible so that WiFi does not cause health issues for children, and secondly, wired so that WAPs and WiFi cannot be hacked (as easily).

5G implementation guidelines are suggested at the chapter end.

Because there is a wealth of literature and responsible tests **sanctioned by doctors and Telecom engineers** who <u>both</u> urge caution, I will not bore you with the mountain of data and results – except to make significant points and to that end, there will be mostly **short** relevant topics. Really significant (eyebrow-raising) points will be paragraphed and footnoted (Chapter 7) so you can see for yourself. At the end of the **Bibliography** there is a reference to the pertinent links to the 5G literature, some pro some con.

Overview

Cellphone electronics as well as the cellphone towers, and transceivers, produce a lot of electromagnetic radiation (EMF) as the electrons move thru the wires and the greater the transmission speed, the greater the EMF (and heat). Constant exposure to same is a known carcinogen.

The following sections examine the reported dangers and health effects of intense EMF radiation, particularly from the much stronger 5G version. Whereas 3G and 4G were also reported to be not safe for children to be around,[98] 5G is even more intense.

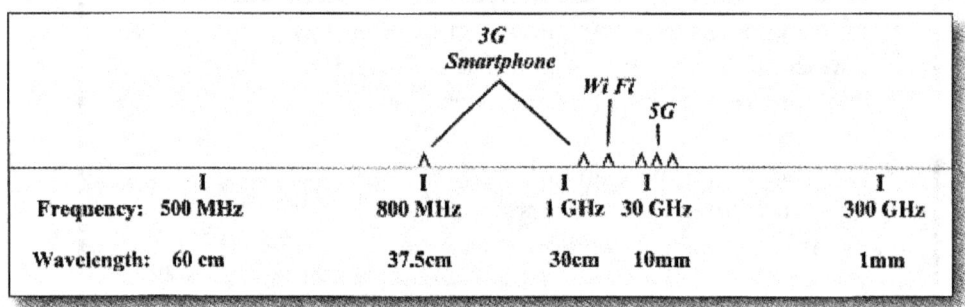

Figure 3: Frequencies and wavelengths of smartphone, Wi-Fi and 5G

(credit: NEXUS, July-Aug 2019)

EMF radiation in general is considered **electromagnetic pollution** and sometimes is called EMF "Smog."

For comparison of various EMF sources…

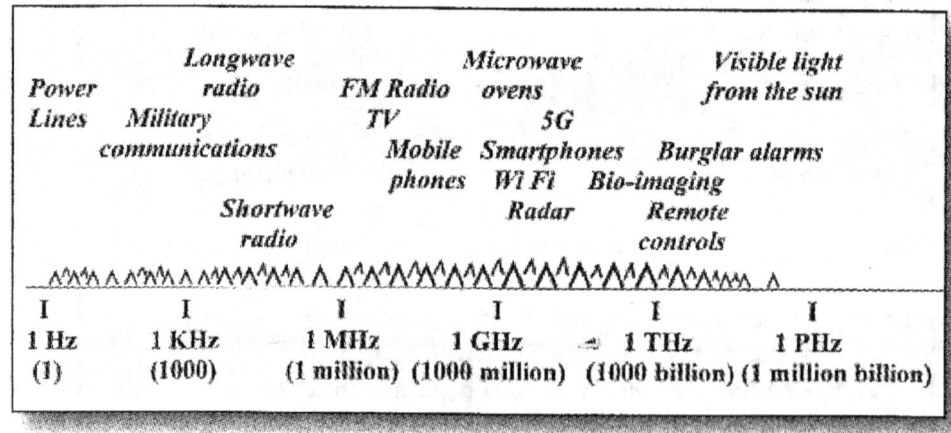

(credit: NEXUS, July-Aug 2019)

You can see from the above diagrams, cellphones are ranked right along with microwave ovens as they both emit similar amounts of EMF – which is why if you would not put your <u>running microwave oven</u> up to your ear, why are you putting your cellphone there? And some people DO clip their cellphone to their ear (exposing their brain cells to the EMF radiation)…

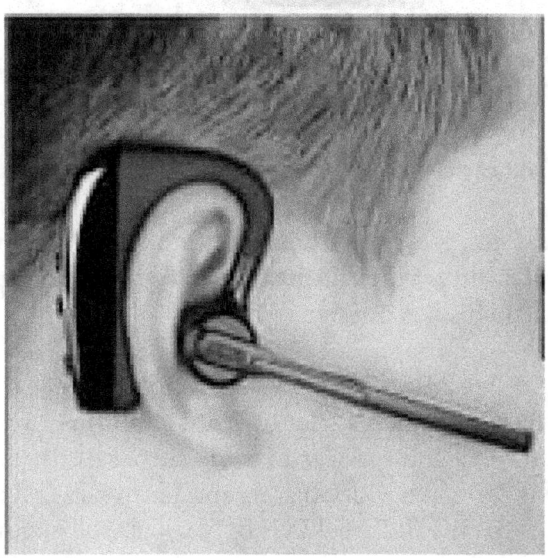

Bluetooth Headset With Microphone For Cell Phones Wireless Earpiece 40

What is wrong with that? Simply the following…

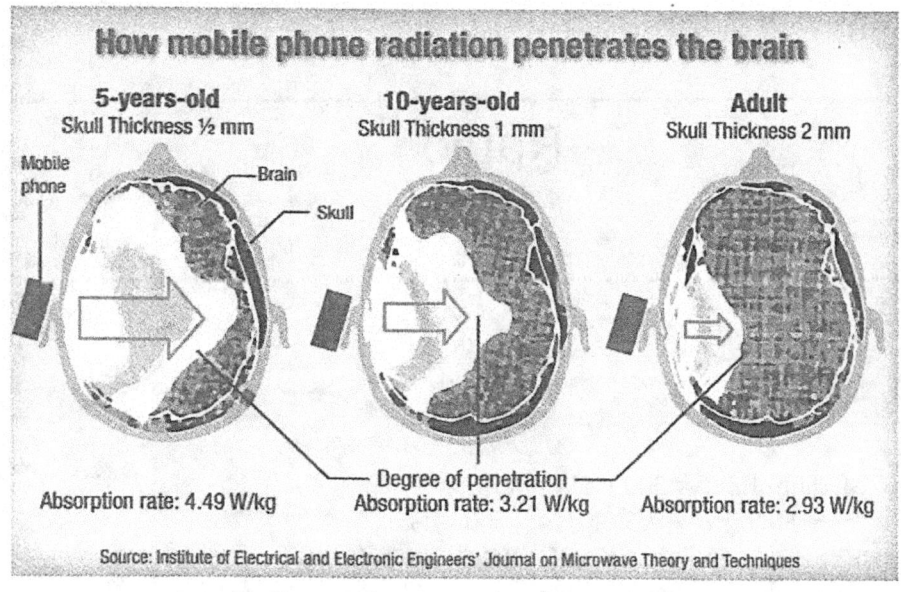

(credit: NEXUS, July-Aug 2019)

Note that the child's skull is not thick enough to block the EMF and what it does is examined later, where it has been found to **break DNA** and destroy ("fry") neurons in the brain… even the adult brain is getting irradiated by the cellphone EMF… which is why you want to **use the speakerphone** instead of holding the phone up to your ear (or face).

So you say, "That's Ok, I use the speakerphone and limit my cellphone use, so I'm Ok." Just when you thought you could minimize your exposure to EMF, remember that the 5G network transceivers will be all over the place, every 300' or so AND….

5G Satellite Coverage

Yes, that's right folks, there are plans to put not 100, not 500, not 1000 5G satellites in orbit, but…. **5G is limited to 330'** so <u>what</u> are they using to transmit from space? 4G? (This has not been explained.)

> In November 2018 the US FCC authorized SpaceX, owned by Elon Musk, to launch a fleet of **7,518 satellites** to provide global satellite broadband services to every corner of the Earth… at a height of 210 miles… irradiating the Earth with **EHF** (see chart below) at 37.5 GHz and 42 GHz… which will augment a smaller fleet of **4,425** satellites <u>already authorized</u> by the FCC which will orbit Earth at 750 miles and will bathe us in 12 Ghz and 30 GHz.
> The grand total of satellites will be just under **12,000**. [99]

What is **EHF** (Extremely High Frequency)?

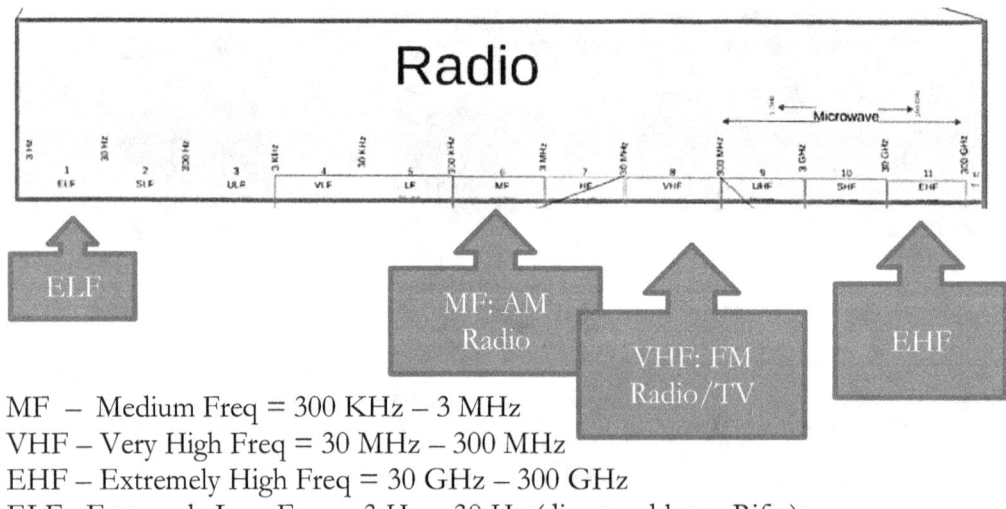

MF – Medium Freq = 300 KHz – 3 MHz
VHF – Very High Freq = 30 MHz – 300 MHz
EHF – Extremely High Freq = 30 GHz – 300 GHz
ELF –Extremely Low Freq – 3 Hz – 30 Hz (discussed later: Rifat).

So the Millimeter waves (mmw) are EHF and are 1 mm in length, at the top end of the Radio wave (microwave scale). These are very short high frequency wavelengths, whereas AM and FM radiowaves are longer and lower frequency.

5G is without precedent for humans, and **it consists of combinations of 3G and 4G in order for it to work.** And what you have is hundreds of antennas that are within a few hundred yards of each other, sometimes near your bedroom window… and you cannot complain, due to rules passed in Washington DC, you may only object on aesthetic grounds. [100]

Cautions

You want to keep the cellphone off of your body, and (no joke) do not do what some women dressed at a fancy gala do: do not put the phone down your bra! With the new 5G, it is **wise to not use the phone in your car while driving** across town as the phone must switch off and locate a closer tower and that causes the phone to **ramp up a higher EMF** (while searching) as it has to constantly make an on-going connection. And you are in **a metal environment** that uses a little higher strength to reach the tower(s), not to mention the phone signal partially attenuates due to the metal shell of the car… And lastly it is a similar issue if you are in a plane.

Lastly, **always keep that phone charged** as a weak battery (say the charge reads 30-45%), the phone will have to ramp up the EMF as it empowers its reach to the nearest cellphone tower.

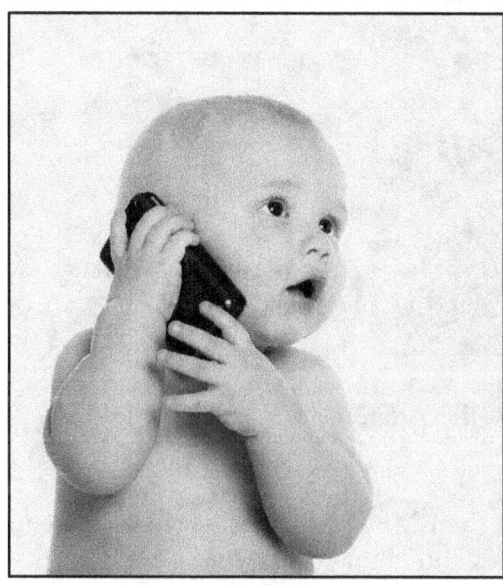

Sure Health Issues!

Lastly, **do not ever give a cellphone to a baby as a pacifier or toy…**

Reminder: if you are in your home or at work, and behind several building walls, yet your phone has enough strength of signal to **go thru the walls and reach a cellphone tower** (for 4G at least) **that is 1 mile across town** – is that a good reason to <u>not</u> put that power next to your head?

EMF & Humans

So what does EMF do to the human body or brain?

As has been said before (in former books: VEG, TSiM, TOM) the human body is normally 70% water … which is why drinking enough water during the day prevents **dehydration and its resultant inflammation** (because the body cannot carry away toxins). As such, the human body and **brain (which is also largely hydrated)** is a kind of **antenna** for receiving electrical signals, facilitated by the minerals which make it conductive. Water conducts electricity, so the body is an antenna.

> An earlier chart showed the degree to which the average cellphone EMF can penetrate the brain… which is greater in children.

Penetration

The following is a tumor on the **right side** of the brain where the patient held his cellphone every day… (top of picture is front of head)

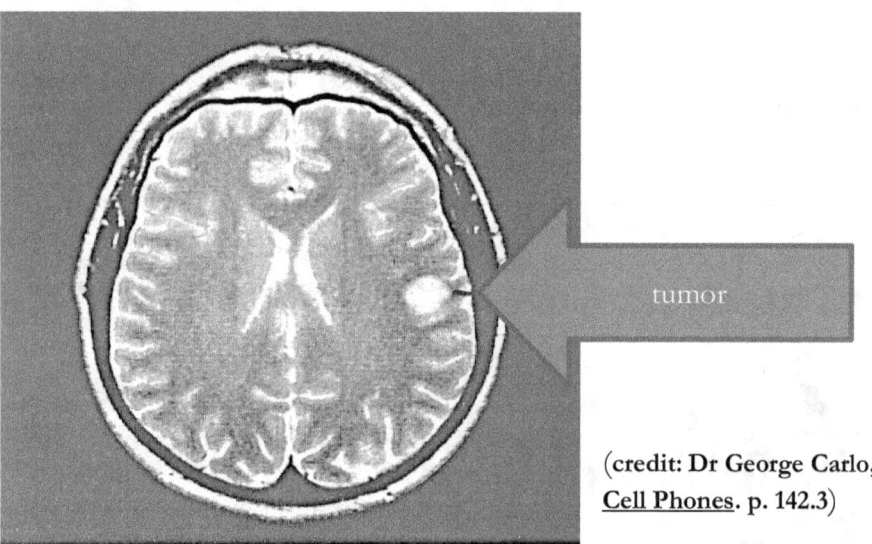

tumor

(**credit: Dr George Carlo,**
Cell Phones. **p. 142.3**)

Is this really worth it? And what about the 'progressive' guys who clip their cellphone to their ear 24/7…?

Using the **speakerphone is safer**, but the EMF is still reaching your hand and your body… and the latest French study shows that the <u>intense</u> **blue light** from the PC monitor flat screen is not good for the eyes on a daily basis…. [101] This includes cellphones and HD TV screens. The <u>intense</u> blue light (with UV) disrupts

mitochondrial function called **Heteroplasmy, and it suppresses melatonin which impedes sleep, and prolonged exposure can damage the retina**. [102]
(See footnotes for more information.)

There are more effects that have been known from 5G ... the modulation of the signal [millimeter waves] is what we are most worried about, because it is **moving extremely fast**. And it has the ability to alter the functioning of all our healthy nerves and cells, and it's the membrane surrounding our cells that may be perturbed the most because it interferes with the way **calcium** [as ions] moves in and out of the cell thru the cell wall. (VGCC)

.... the millimeter waves are mostly absorbed [on the top layer of the skin ...where], your **sweat ducts** are located...millions of them... And they can be regarded as a helical antennae, like a double helix [and]... they can transmit the exposure from the surface of the skin **internally.... [down to the cells]...**

...including damage to the DNA, the mitochondria, the cell walls And ...is a precursor to cancer. [103]

[and] it shows that when at a high power, **it makes peoples' skin feel like it's on fire**. [104] (This makes it an effective means of crowd control... see section: Non-Lethal Technology below.)

Broken DNA

Prolonged exposure to EMF radiation can affect many parts of the body as evidenced in the skin, in the organs, in the blood, in the bone marrow, in tissue, and even in enzymes (body chemistry). It was a Russian review of millimeter waves which showed there were morphological [change] elements at work **in the whole body**, ... into the bloodstream, tissue and nucleic acid metabolism, showing a wide variety of biological impact. [105]

Men are advised to keep the cellphone out of their pants, even the back pocket, as it (when turned on but not in use) can still irradiate the mitochondrial DNA in the sperm cells of the testes **affecting quality of reproduction** (Impotence).

For a similar reason, women are asked to not carry the cellphone near their breasts, nor on their hips in a fanny pack.

The damage to DNA is caused by **free radicals**,,,

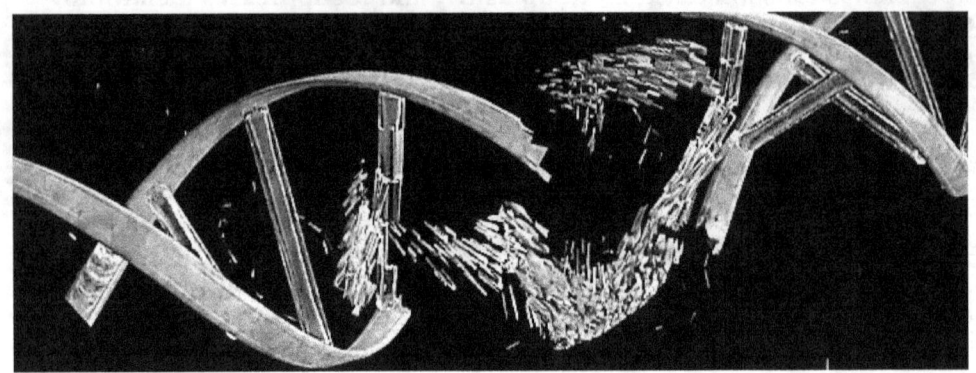

Bing Images: DNA Oxidative Damage

As **Dr. George Carlo** shows in his book on the dangers of cellphone radiation, DNA can be impacted several ways:

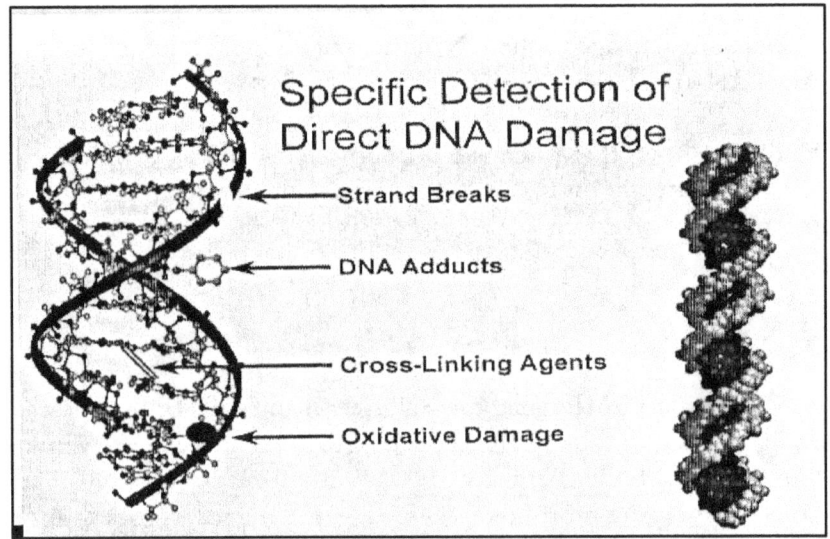

Dr. Martin Pall (foremost EMF expert) adds...

> Oxidative damage is done by free radicals created by EMF action on intracellular calcium; you get **superoxide** and **nitric oxide** which together form Peroxy Nitrate which breaks down to form a nasty **Hydroxyl** radical... which attacks the mitochondria and DNA ...and then breaks down into CO^2 adduct. [106]

VGCC Damage

While all this is kind of techy, it is important to realize that detailed studies into cell operation leading to cancer have revealed **damage from EMF radiation.**

VGCC is **Voltage-Gated Calcium Channels** and these are responsible for moderating the way calcium ions can pass into/out of the cell. Since they have a slight positive charge themselves, as do the calcium, sodium and potassium ions, the **EMF bombardment disrupts the fragile electrical balance** and super activates the VGCCs and as a result, there is also a large increase in nitric oxide synthesis, which diffuses out of the cells. In short, the cell's normal ion flow (calcium, sodium, potassium) is blocked and normal signaling molecules do not occur, thus not healthily interfacing with other parts of the body.

This also affects the **brain cells and neurotransmitters** as they also use calcium ions (Ca+) to do their work. [107] Says **Dr. Gittleman:**

> The EMF radiation actually interrupts the communication of the cells. And so your cells get overwhelmed by all kinds of messages from inside and outside the body because without enough calcium this creates **calcium leaching** in the cell membranes. Then all the neurotransmitters become very scrambled, and then chemicals start pouring in from your ruptured cells that damage your cellular DNA…. And this disrupts normal cell division (mitosis), and that creates **oxidative stress** that further damages DNA and many other physical processes… a real domino effect. [108]

In fact, **Dr. Pall** cites 8 major impacts of EMF on the body:

> Oxidative Stress – production of free radicals and damage to cells.
> Lowered Fertility – chromosomal mutations in DNA
> Neurological Effects – calcium signaling and neurotransmitters are stressed and sometimes cells die
> Apoptosis – cell death due to excessive Ca2+ levels in the mitochondria and DNA breaks
> Cellular DNA Damage – free radical creates peroxynitrite that damages DNA
> Changes in Non-steroid Levels – EMF stimulates hormone/enzyme activity and exhausts body's ability to control level and balance
> Calcium Overload – produced by excessive activity of VGCCs
> Heat Shock Proteins – excessive Ca ions produce signaling changes which result in protein misfolding [***prions*** aka Mad Cow Disease aka Creutzfeld-Jakob disease in humans: also called **Chronic Wasting disease also called CWD**]

That last one should really get you thinking twice... bad enough that Ca+ ions are required for normal neuronal operation, as neurotransmitters, but when EMF can cause proteins to misfold (aka **prions** – see Chapter 7, Area II) and create CWD – that is a real red flag!

Secondary Damage

In addition to prions and CWD, and neurological disruption, EMF radiation, at the higher levels, can also cause **Digital Dementia**, or Alzheimers, by causing excess creation of **Beta-Amyloid Plaque** (oxidation in the brain). (See TSiM Ch. 9.)

And if that weren't bad enough, here is the real shocker:

> Says **Dr. Evan Brand, MD, PhD**....
>
> In the 1980s we focused on Candida, and then Lyme disease, and then progressed to Herpes... But in the last two years, we're realizing that's still not the depth of the bucket. What's at the bottom of the bucket is a group of viruses. They're called **human endogenous retroviruses.**
>
> These are viruses that are [naturally] embedded in our DNA. But they are silenced... largely thru two mechanisms... **methylation** and **acetylation.** Those mechanisms are destroyed by the [prolonged or consistent] exposure to WiFi (EMF)...
>
> And so what happens is these **viruses are now replicating** in us. And the most well-known is HIV. **This is called HIV minor or AIDS minor**. And visible on the surface are people testing positive for Lyme and Candida, for many of the molds and fungi that we have. The real reason is that the immune system is dysfunctional as a result of these viruses.
>
> ....So the moment you're exposed to WiFi you will unleash this dragon on the inside that when these viruses start replicating, you will have all the symptoms of a chronic viral infection that doesn't instantly kill you like AIDS used to do, or HIV does. But it's a slower process. **It's a slow kill**.
>
> So by actually getting control of the WiFi environment... by reducing the toxic burden, we are able to get control of the retroviruses again. And then Lyme, Candida, mycoplasma and the molds either fall away almost on their own, or they become

a much lesser issue… six weeks on an herbal compound can cure the patient [like Garlic and Maitaki mushroom].

…So how can we let the public know that they're being damaged by the **WiFi** environment…? The way it is licensed with no oversight, no testing… and **all the trials that have been done have shown that it is NOT safe**. [109]

If you think that was a bombshell, keep reading… and see **Appendix G**: section on **WiGig and GiFi**.

WiFi in Homes

While we're at it, we might as well point out that if you have a **cordless phone**, especially a DECT phone, they are a great way to be on the phone and walk out in the yard and still talk to the other party.

What you do not realize is that the **DECT phone** has an antenna that acts like a **Cellphone Tower in your house** – yes, that signal is strong enough to go thru the house walls while you walk outside in your garden! Just like the cellphone towers across town… only in your house. And held up to your ear (head).

Now how cool is it?

The other item to acknowledge is **AT&T Uverse** which we set up with the Gateway (router) in the back of the house – using a WAP (Wireless Access Point) to access via WiFi the Uverse receiver under the TV 40' away at the other end of the house.
                    (Bad setup, but we didn't know.)

Linda's **laptop** also worked just fine in this sea of WiFi…. And yet she had trouble sleeping at night… and we discovered that her bed was on the opposite side of the wall from the Kitchen **Refrigerator** – talk about a strong EMF signal – right thru the wall and on into the bedroom – her whole bed was bathed in the EMF from the Refrigerator! (We used the EMF TriMeter to discover this.)

So we moved her bed to the other outside wall…not realizing that the AC Compressor was on the other side of that wall!! And it had an equally strong EMF – right thru the wall – In fact, **the whole bedroom was awash in EMF** – there was nowhere in the room that she could go that she wasn't bathed in EMF exceeding the 2.5 mG limit (see Appendix G). Thus that bedroom was unusable and became a workroom, storage room…
She moved upstairs to the guest room and now sleeps just fine and her health has improved…. Until they forced us to install a **Smart Meter**… but the EMF is

limited to a 6' plume so we found (with our trusty $150 EMF TriMeter) that the far side of the guest room was 'clean' and she sleeps there.

All that to say that some people are **sensitive to EMF**, and while their skin may not feel hot (which requires exposure to a much higher power level), there are some symptoms that result from <u>consistent</u> EMF exposure.

EMF Symptoms

Some symptoms of sensitivity (when <u>repeatedly exposed</u>) to EMF include: [110]

Numbness in the fingers, feelings of warmth, facial flushing, **headaches** (a biggie), dizziness, blurred vision, nauseousness, deafness, blistering skin…
severely affected people at this point cannot even use a cellphone because they get heated so prolifically…
It will **heat up the ear** and heat up a side of the head. So I am finding that many of us are **allergic**, but are not realizing the source of the problem…and that includes those new lightbulbs, the CFLs (which have now been discontinued).

So all of that gives off a kind of radiation and any symptoms that we experience are identical to **radio wave sickness** that was identified as far back as the 1970s. So there are neurological implications. There are implications in terms of the cardiac system, and the respiratory system, the digestive system, and abdominal pain and **the eyes and dehydration**… as well as **nosebleeds** are connected to EMFs as well.

Another set of symptoms connected with <u>over-exposure</u> to EMF: [111]

It can affect the heart rate, the eyes, the ears, the skin, just about **every area of the body is affected** so you never know exactly what's going on… and **there are nowadays no safe havens… you've got the exposure 24/7**…. And be sure that electrical things in your bedroom are turned off at night, or not near the bed – and for God's sake do not use **an electric blanket**. And be sure that **baby monitors** are not near the baby.. and try not to overexpose yourself using **wireless mice and keyboards**…a small thing but every little bit adds up. Minimize the use of **hairdryers** … EMF right near the head. <u>Minimize the use of electricity</u>, by the way, by **unplugging** your appliances (toasters, blenders, crockpots, coffee makers etc.) when not in use.

Now for some interesting aspects of EMF exposure…

ELF, Bees and More

There is a growing body of research and literature of the effects of EMF on trees, insects, birds and bunnies, as well as on microbes. Well, most living things have an electromagnetic aspect to them [112] – very small amounts of electrical charge or flow – the human body has **Bionet** which is what Acupuncture uses to release "stuck chi" – which is when the electrical charges 'bundle' in an area and can cause discomfort or inflammation (this was covered in TOM). Plants also have a chi-aspect as shown in **Kirlian** photography… it is the **life force** or energy moving thru living things. So why wouldn't a stronger EMF affect weaker living things?

The human aura

… made with a real aura camera. (See TSiM and TOM for more info.)

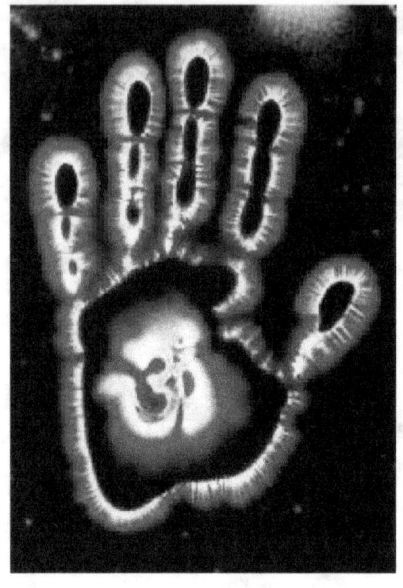

Human hand and a leaf both showing the life energy (chi) or Prana, or Ruach in them. [113]

The human soul is alleged to look like this (below, left):

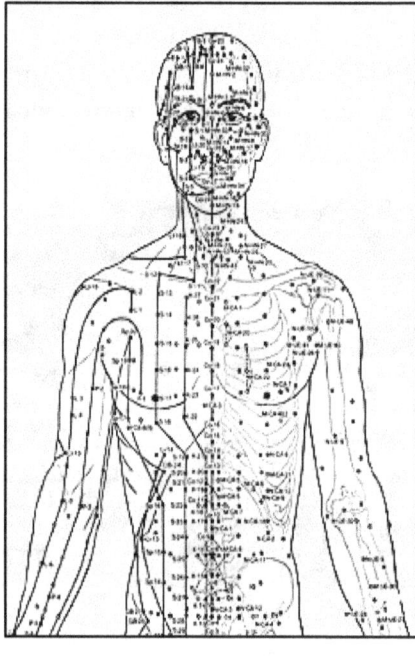

…and the **Bionet** looks like this (right: meridians of energy)

The Bionet also include the many chakras (meridian junctions).

The preceding are examined more in TOM and TSiM, but show why the body is susceptible to EMF: **it already has a weak EMF which is easily disrupted by a stronger (disruptive) EMF -- especially one over 2.5 mG (Appendix G).**

So what happens to **bees** when they are exposed to higher and higher levels of EMF?

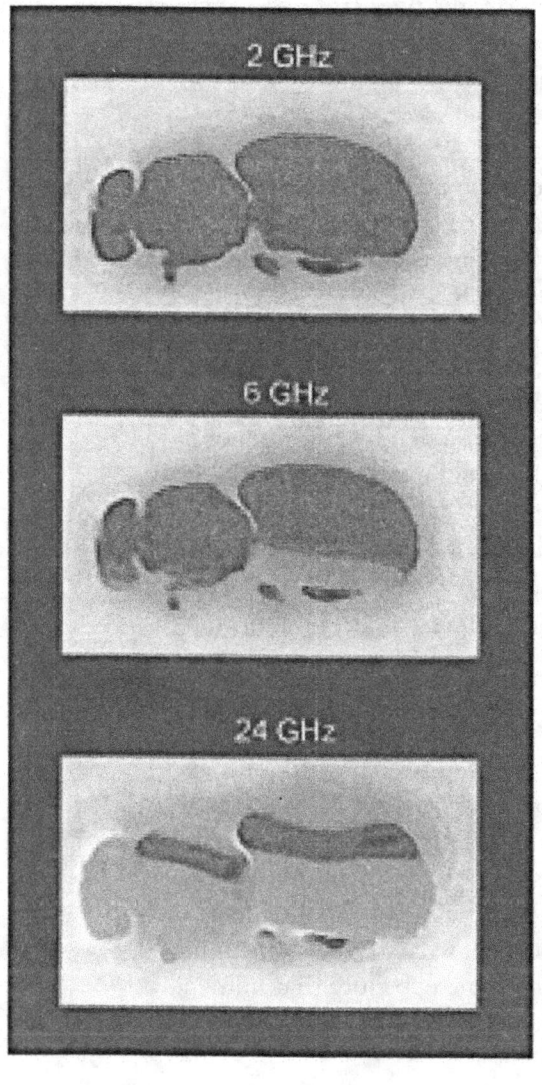

Left: shows the different amount of EMF absorption depending on EMF strength – 2GHz weakest

to
6GHz
(threshold for cell damage)

to
2.4GHz (cordless phone EMF).

When the wavelength is close to the insect's size the radiation penetrates further (left bottom).

(credit: NEXUS, 5G, July-Aug 2019)

Honey Bee Absorption of EMF

And they also did experiments on bees **and their hives**…

But studies have shown that if you take a cellphone [turned

on] and you **put it into a healthy hive**, and you take another cellphone that's turned off and put it into a second healthy hive, and
then the third hive is not exposed at all, and you see what happens **after two weeks of exposure of just two hours a day**…
The hive that had the phone in it [i.e., second one] without being turned on, is fine…
The hive that was the control [i.e., no phone] is fine.
The hive with the phone turned on, the bees stopped dancing [which is how they communicate distance and direction to the flowers] as well as… they stopped producing honey and some of them didn't come back. [114]

The 5G interferes with the **cryptochrome** in the insect which is like a GPS, and as the bees are crucial to the cross-pollination of agricultural crops, maybe we better rethink the placement of cellphone towers and the mobile antennas.

Another study revealed…

… he took his cellphone and put it atop the hive and called his cellphone, which was on vibrate. When his phone rang, **all the bees came out** of the hive and **even the queen came out**…I had never seen anything like that… It takes a lot to get the queen to come out…I said to him, "Why is it doing that?" And he said, "I don't know…." I don't know if anybody knows but it is disturbing. [115] [It is electro**magnetic** vibration.]

…and…

…the brain cells of **rats** when exposed to very weak pulsed signals from cellphone radiation develop damage to their DNA…

And the animals, of course it is hard to get them to make cellphone calls, so you expose their whole bodies to a level of radiation that did not create any heat whatsoever, but mimicked in the lifetime of the animals the same exposure that humans will get in our lifetimes, supposedly in 70 years… [116]

I submit that last one to say that not all the tests are free from subjective error, and perhaps the rats and bees were exposed to much more than a human casually immersed in WiFi for 10 minutes will experience… but keep in mind that minimal, occasional exposure will have the human body reacting to repair any superficial damage.

Ii is natural for humans to react with alarm to new and strange things that appear to cause harm, but we must look at the way the tests are being done and not overreact. (The great majority of humans suffer no discernable damage from WiFi... initially.)

Having said that, we are almost at the point of pointing out that there IS a very powerful group that would like to stop 5G, and it isn't who you think.

Bacteria and Microbes

As you might suspect, microbes are vulnerable to EMF – but more than harm them, **it causes them to proliferate** as they think they are being attacked!! This is what happened (above) with the retroviruses.

> ...5G has the capacity to have serious biological effects, including that it can **accelerate the growth of bacteria**, mainly by depressing the growth of those things that are supposed to be good for you and allowing good things [i.e., biochemistry] to die.

> Microbes in the soil are crucial to agricultural success, and those found at the base of cellphone towers (in India) near farms were damaged as well. [117]

Ok, you get the point, there is much more research and the point is made that overexposure to EMF is harmful, thus one is wise to limit their exposure and use of EMF-emitting devices (cellphone, laptops, iPads, hairdryers, PCs, cordless phones, VR Headsets...)

However, before getting to a very significant and final wrap on this chapter, we need to take a quick look at the conspiracy claims of those who fear government or the military is out to get us and abuse us with the new "non-lethal technology."

Non-Lethal Technology

This technology is called "non-lethal" as it stops people in their tracks from continuing to riot or demonstrate. So far it has not been used publicly, but that is the fear and point of view of those who see the advanced microwave (including sonic) technology as a tool in the hands of Big Brother. It hasn't even been used in limited war engagements in Afghanistan, Syria or even the Golan Heights.

I suggest this technology is reserved for something much more serious, and maybe even off-planet.

> …the U.S. Air Force has developed a **nonlethal** antipersonnel weapon system called **Active Denial System (ADS)** which emits a beam of [EMF] radiation with a wavelength of 3 mm. The weapon (see below) causes a person in the beam to feel an intense burning pain, as if their **skin is going to catch fire**. [118]

Mobile ADS Unit
(credit: BingImages: Tactical-life.com)

Just so that you are aware, one of the main opponents of the use of microwave effects against the population at large is **Tim Rifat,** who has done a serious and thorough research of the ways it could be used against Man. He says

> There is a video that shows the Department of Defense …
> …demonstrating the 5G weapon that works at about 95
> GHz but it is within the 5G range [see earlier chart – it is

about the middle of the EHF range]. And it's the same frequency as you are going to get from the 5G antenna that might be pasted on [the outside of] your building. And it shows that when at a high power, it makes **people's skin feel like it's on fire.** It is now a very effective means of crowd control and the Pentagon has bragged about it...

so the antennas on your building are going to be at a lower power, but what if someone decided to take it over and make it a higher power? It is also by the way, **a very effective listening device...**

...and it is directional [like V2K]...they can point it at specific targets. It is called **Beam Forming technology...** it can be focused.... [119] [to kill.]

> Beam Forming technology is discussed in the Warning
> section of this chapter, later.

5G Satellites

And then there is the fear that the Earth will be blanketed with almost 12,000 satellites as said earlier... and from the height of 150+ miles, that is not going to expose humans to much <u>direct</u> radiation – but it will do something else: **interfere with weather forecasting** as it operates at about the same frequency as that used by satellites to sense moisture and predict weather changes. NOAA has thus voiced its objection to the extreme use of 5G satellites and that has yet to be worked out. It is said to also be a potential problem for aviation navigation if the wavelength is close to GPS and ATC use...

Officers in the Office of Naval Reserves have also voiced concern that the **National Grid** (5G satellites) and the **Space Fence** (you had forgotten about the one, eh?) might be compromised.

> All the concerns are valid IF the 5G technology is misused.
> That is why I said earlier that it must be implemented wisely
> (5 points: see p. 220) and those who oversee it probably
> will wind up being **under military purview** since soldiers keep
> secrets (some necessary) AND do what they are told, and
> civilians may opt to do whatever they think is best—
> if it means a profit.

Insurance Industry

Not even the insurance industry will underwrite or insure anything to do with 5G – not only have the insurers read the reports of health issues, they do not trust the uses…and the 5G industry itself will not run health tests on their technology. The 5G technology is classed in the **"highest risk category."** Not even cellphone towers are insured. If anything goes wrong, the insurance company would have to pay out huge sums – and as such even the nuclear industry is not insured either.

> The 1996 Telecommunications Act, Section 704, says that **local governments can't refuse cellphone tower siting** on the basis of environmental grounds… nor on the basis that there might be health issues. [120]   (Esthetics are considered a valid challenge.)

Pretty serious stuff … it must be a technology that is so important that health and ecology cannot control or stop it… What could be _that_ important? It would have to be something tied to Man's very existence… (Stay tuned.) It is a technology that we don't really need but will get anyway. Why?

Says an insider **Robert C Kane**, a senior engineer, in the Telecom Industry:

> "Never in human history has there been such a practice as we now encounter with the marketing and distributing of products **hostile to the human biological system**, by an industry with full knowledge of those effects…"

> …he discovered all the old Russian research, and …was treated like a pariah by the industry … and he later developed several brain tumors and died after publishing his book, Cell Phone Russian Roulette. [121]

Obviously, he was against 5G (and probably the next gen, 6G…) but he was not one in the deeper know. Thus, the top brass in Telecom as well as the top Pentagon brass know _why_ we have to do this, and it _may_ involve some "collateral damage" which is not acceptable to the public.

Also obviously _forcing the issue_ as is being done with 5G means there must be a very serious, almost monumental reason why it has to be done. One that we cannot ignore, and one that we should deal with.

If you think Telecom is just in it for the money, consider that 4G is plenty fast and adequate (and already a "cash cow") and can handle Smart Meters and cellphone traffic and satellites – go beyond that in your thinking.

> **What would drive the Military Industrial Complex (MIC) to go to 5G and even consider a 5G network around the planet (i.e., satellites)??**

You have to think 'outside the box' to come up with the answer.
But neither the PTB nor the Insiders may want it directly
exposed… even though I was given a go-ahead to do a basic
discussion of it.
And so I am going to present it as one of two options… your
choice. One is right. Neither is nice.
If you read between the lines of this chapter, and consider the
input from Chapters 6 & 7, you'll know which it is, then I cannot
be accused of directly stating it (or potentially inciting fear) …
Remember: 5G is not about cellphones.

Warning

As was said many times, Man has never been fully in control of the planet and has
an unseen enemy obstructing him, manipulating him and effecting control… such
is the reason for the initial two chapters. They were not just to amuse you. **The
Remnant have both inserted human-looking Hybrids who walk among us,
often mind-controlling the heads of Congress, Commerce and the Military.
Now you know why the Nightly News shows such craziness.**

The **Anunnaki Remnant are still here** – they never all went away (due to Galactic
Law). And they are still responsible for us – one part, the **Dissidents** wants to be
rid of us so they can go home, and the other 2/3, the **Insiders**, want to fulfill the
obligation to develop us into responsible, intelligent humans.
They have been in a methodology battle to complete their missions for millennia.

Thus the **Dissidents** are the enemy, working with their surface component,
cutouts/puppets, PTB, to still try and **control the planet via proxy**, and the other
Insiders created the bio-cybernetic Greys to effect our genetic upgrade, thus
improving the species. Neither group in the Remnant can stand living on the
surface due to the Sun's powerful rays, so they stay underground, which is Ok as
they are also Reptilian and humans have a problem with anything that looks
different – even different color skin within the same species.

So one of two things is happening, with respect to 5G, and both are intriguing.

Option 1

Consider that the Dissidents, who do not like humans anyway, à la Enlil,
have infiltrated and manipulated the PTB who head up the military, the Telecom
Industry and the Banking, Media and Health/Pharmaceutical Industry to promote
and distribute 5G which will (knowingly) cause massive heath issues.

This is tantamount to an attack on humans, and the health issues should **cull the numbers** which is what the Anunnaki have always done (*Suruppu* Disease, Ebola, Black Death, Spanish Flu [1918, and HIV). And today, they'd like to be rid of humans.

So they build and install an Extremely High Frequency (EHF) network across the country – removing America first as that collapse will cause the rest of the world to fall. They install the new 5G transceivers not so much on obvious towers as on the side of buildings, telephone poles, trees, etc… to **cover every 900 square yards** within a populated area. This makes sure that the area is bathed in 5G wavelengths.

Once the network is up and running it is easy to **hack into it** and broadcast a 10-16Hz signal via cellphones, which everyone has, to put the population to sleep. Having more superior technology, they can not only hack into the 5G network, but they can use it to carry **psychotronic signals** (the AI Signal) matching humans' brainwaves, to effect violent or sociopathic behavior – grab a gun and shoot others… let chaos reign!

Tavistock and MkUltra have tested this very thing in the lab.

So the **5G network can be used to effect control over humans** and make them docile when needed and violent when wanted. The Remnant also have technology (**Beam Forming**) that they can direct to their surface puppets to tell them what to do… a superior version of **V2K** [122]   It has been called the "AI Signal."

According to **Dr. Paul Batcho** (a former DARPA scientist)…[123]

> I seem to have stumbled across an advanced technology that would classify as **synthetic telepathy**… It clearly uses the cellular towers to **transmit illegal signals**. It sounds unbelievable but it is actual technology being used on civilians of the US. My basic research does indicate that such technology can exist and dates back to the V2K (P300) **mind wave technology** of the 1970s. This does appear to be a much more advanced version that allows open communication of human mind to mind bridges.

He says further…

> The verified measurement and existence of these RF band transmissions constitutes **a terrorist act**. These transmissions will cause harmful health effects in the form of **enhanced microwave radiation illness**. It is imperative that these frequency bands be measured and verified by an official source. These frequency bands do not exist naturally and there is a technology targeting individuals.

He is missing the point. The phones and towers are doing what they were designed to do. They are part of the **Smart Grid** of interconnected technology which <u>can be</u> used **to control the humans**, dictate their thoughts and control their perceptions, <u>and</u> affect human genetics (by **breaking DNA and causing massive illness in the population,** as already said).

And as if that weren't bad enough, consider the other view…

Option 2

5G (and even the potential 6G) are not evil and conspiring against Mankind. It is what has to be done to block the threat of Anunnaki (Remnant Dissident) mind control to stop:

> **Grey abductions** of humans, which Greys are remote mind-controlled bio-cybernetic robots; They have a 2-part brain – the front part for their local genetic work, and the back part is the "hive control" that sends them and generally controls them; they are **guided remotely** from underground.

> **Naga remote mind-control** of humans via the same techniques that the Greys use on humans when they have them in their possession; **remote mind-control technology**.

> **Prevent disappearing people** who are taken and not returned, as per the <u>Missing 411</u> series of books by David Paulides. Two groups do this: one is culling the humans, the other is selecting superior genetics and IQ and isolating them remotely; thus **block the remote communication links.**

Under this scenario, the MIC and Telecom would work together to build a network of sufficient strength (potentially endangering humans who are overexposed), to jam the Remnant **airborne signals** (their **Beam Formation**) that do not now use the 4G network. While the 5G network might be vulnerable to initial hacking and carry Remnant signals, that can be identified, filtered out and blocked. The main point is to have a strong enough signal to **block/jam/thwart the Remnant mind-control technology that is like Beam Formation used on surface people**.

> In other words, 5G network could be strong enough to act as a **jamming 'fog'** that the Anunnaki signals cannot penetrate.

This would block controlling signals sent to **the Greys** who are still abducting people, no longer notifying the Army of who, what and where (as they used to), and the abductions have achieved their purpose. It is time to stop these.

This would also block controlling signals sent to **PTB puppets** (usually the top dogs in an organization who call the shots) and the lower echelons know nothing of what and why anything is being done. (Where do you think the "need to know" compartmentalization idea came from?)

This would also block controlling signals sent to **unsuspecting humans** who are "out of network" (picked at random, not usually controlled to do anything) humans who are selected to go shoot, bomb or otherwise create havoc in restaurants, concerts, marathons, stadiums... seeing what is 'projected' to them.

Genetic Selection

So how would those two groups, one culling and the other abducting and relocating, know whom to take and whom to leave? The second group doesn't care – if the human is available and unprotected, hiking and camping in the wilderness for example, s/he can be disappeared and never seen again... it is known that some became <u>lunch for the first group</u> that snatched them (see <u>Missing 411</u> and Charles Fort books). Their clothes are neatly stacked and just a few bones are found nearby.

The second group (Insiders) operates a bit more humanely. First you advertise on TV for people to submit DNA samples and setup a huge database that can be scanned for those of better genetics. And, second, because you have their address, you can go get them, and disappear them, just as <u>Missing 411: A Sobering Coincidence</u> relates. They prefer scientists and Teutonic bloodlines.

And sometimes <u>young</u> people are taken off the street as they walk home, or in the park, or off the seashore... they are sometimes found to be so uncooperative that they are disposed of, their genetics are taken and then they are dumped in remote locations. Those who go along with the scenario can find themselves relocated to another locale working in a better, organized **Breakaway Civilization** (Chapter 8). They usually have the smarts to be trained in something techy, or as in the case of older doctors and engineers, they already have a useful skill.

Hypothetical Scenario examined in Chapter 8.

Earth Quarantine

Then what do you do about Dissident's ET friends and visitors who could come to help subdue Mankind? Surely the Orion Group (aka Empire) knows what is happening to their Remnant left here and while the Draco are prohibited from free range in the solar system, the Orion Empire would love to assist their Remnant to regain control of Earth – which is weird as they cannot live on the surface... yet. Oh, I forgot, that is why there used to be a Firmament above the Earth (see GEP Apx. E, and Appendix B this book).

So what you'd do is have the President argue for a **Star Wars Initiative**, or SDI defense system ... anybody ever notice that the guns were pointed <u>outwards</u>? So **President Reagan** in the Fall of 1987 at the UN stated,

> During a speech before the United Nations in 1987, President Ronald Reagan spoke longingly for the world unity that would happen if aliens invaded Earth.
> He said:
> *"Perhaps we need some outside universal threat to make us recognize this common bond. I occasionally think how quickly our differences worldwide would vanish if we were facing an alien threat from outside this world."*

Notwithstanding a possible **Project Bluebeam** (qv), they are already here and continue to covertly infiltrate and subvert us...

So what if 5G has a better idea? Instead of weaponizing space, we launch as many satellites as needed to create a **Grid of EHF** that will block Remnant mind-control transmission on Earth AND further prevent transmission from outside Earth from reaching the Remnant (who are underground, remember).

> This is reminiscent of the Voice of America (VOA) or Radio Free Europe jamming Soviet transmissions and competing to reach the radio listeners in Eastern Europe...

Then there was the **Soviet Woodpecker** transmission aimed at the Western World

with the right ELF (10-16 Hz) designed to allegedly drive Western minds crazy – which was powered by the atomic reactor at **Chernobyl**... until it was put out of commission. [124]

Soviet Woodpecker (defunct)

There **is** a physical quarantine around Earth, more like **an energy field** and that has been mentioned many times, most often in VEG (and GEP Apx. E). Its proof is the ***Gegenschein*** glare off something around the Earth, and the 2015 NASA-discovered energy Barrier at 7200 miles above Earth. (See **Appendix B.**)

Gegenschein as photographed by NASA
(credit: NASA: *http://apod.nasa.gov/apod/archivepix.html and below*)

It is recommended that the reader check out the above NASA link to three samples:
2008 May 07: The Gegenschein over Chile. (sunlight)
2006 December 26: The Gegenschein. (sunlight)
June 25 1999: The Gegenschein. (sunlight + Sun)

Interesting that in his book <u>New Lands</u>, written in the 1920's, **Charles Fort** used the word **matrix**. But he is making a very interesting point that science still today cannot answer:

> Suppose the Gegenschein could be a reflection of sunlight from anything at a distance less than the distance of the stars. It would have **parallax** against its background of stars. *Observatory*, 17-47:
> **"The Gegenschein has no parallax."** [125]

It is an energy barrier that only UFOs with their protective EMF can penetrate.

In fact, if the *Gegenschein* reflects off something as dense as what NASA says they recorded (Apx B), then what are **radio waves** doing when they are sent out into the solar system, or is that Radio Astronomy (**SETI**) all a sham?

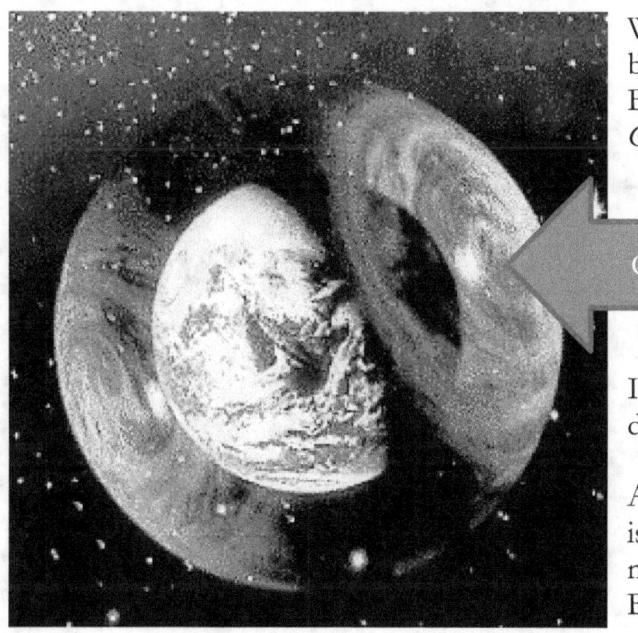

Would radio waves bounce back off <u>both sides</u> of the Barrier – the same one the *Gegenschein* reflects off?

Gegenschein

Is this why SETI was shut down in 2011? [126]

Allegedly it ran out of funds... is that because it found nothing and due to the Barrier it <u>could not</u> find anything?

Would the Radio Astronomy look like this? So are robots on Mars being radio-controlled after all? (See VEG, Apx. A.)

And if the Barrier is strong enough, perhaps the Orion transmissions cannot get thru to Earth from their side...?

And as for the energy grid discovered by NASA at 7200 miles up (Appendix B) could it be sustained by a set of satellites (not ours) that we don't see?:

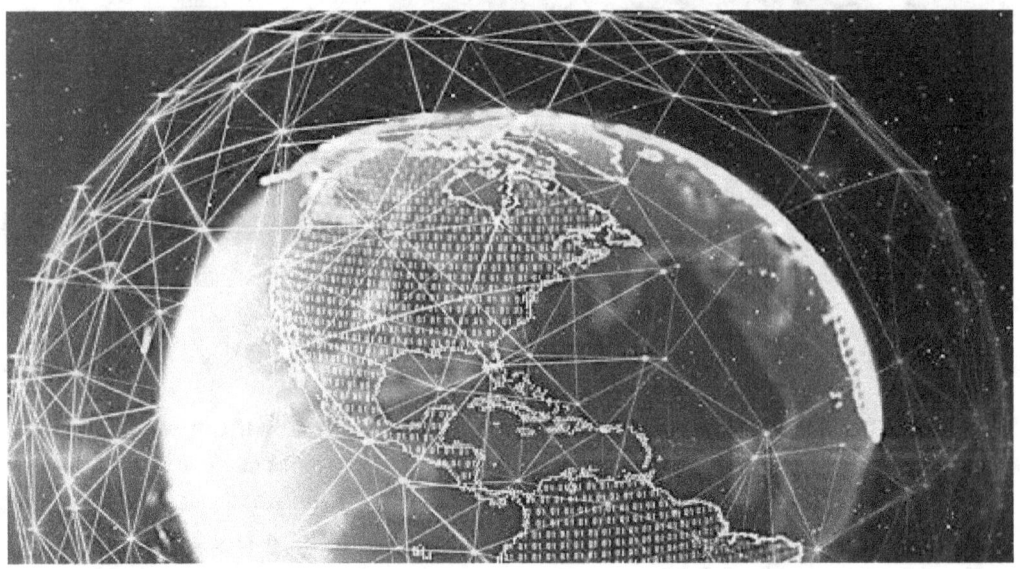

So with the **Energy Barrier**, do we really need SDI or a new slew of satellites circling the Earth?
And, who knows about the Energy Barrier...?

Connect the Dots

Let's look at something very interesting. The **US Army** went way out in the Pacific Ocean to a remote atoll **in 1962** and fired several rockets equipped with nuclear warheads straight up at the sky... **Operation Fishbowl**, Operation Starfish Prime under Operation Dominic. Supposedly this was to test EMP effects – and it <u>did</u> knock out phones and electronics in Hawaii 1500+ miles away.

> In 1958 the United States had completed six high-altitude nuclear tests, but the high-altitude tests of that year produced many **un-expected results and raised many new questions**....
> Thus there is a strong need, not only for better instrumentation, but for further tests covering **a range of altitudes** and yields. [127]

What were they really after? They knew **in 1958** that EMP blasts were effective... so why test at different altitudes? Did they also know about the Barrier at 7200 miles? If so, were they trying to punch thru it? (And kill everyone on Earth?)

Due to the altitudes involved in the Operation Dominic tests, they were exploding atom bombs **in the Van Allen Belts**... the altitudes were from 250-500 miles up (as reported) and you can see that the Van Allen belts extend down that far toward the Earth's surface (**Appendix B**).

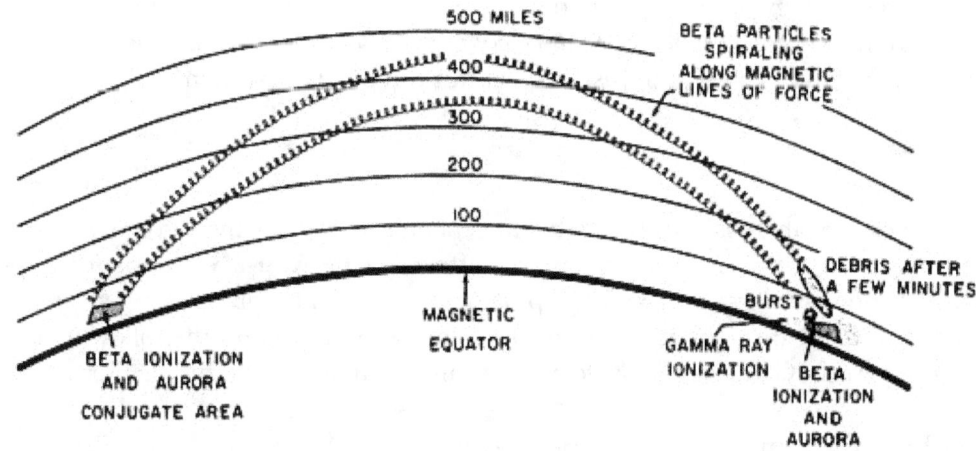

(credit: Wikipedia: Operation Fishbowl)

And why call it a "fishbowl" – unless they knew we were <u>under something</u>?

Let's connect a few dots:

1. Fort discovered the *Gegenschein* which reflects off some **fixed and permanent** surface around the Earth,
2. The Army was shooting at something high up in the sky,
3. SETI shut down with no results,
4. NASA officially announces the existence of an **"impenetrable"** Barrier 7200 miles up...
5. ...and we know that the Van Allen Belts which also surround the Earth for thousands of miles altitude are **deadly radioactivity**...

Also see the GEO 600 article in Appendix B.

Now tell me again that we sent astronauts thru the "Impenetrable" Barrier in 1969... and with no radiation shielding.   (For a better look at the 1969 Moon Mission see Randy Walsh's books <u>The Apollo Moon Missions</u> [in the Bibliography] where he offers <u>credible proof</u> that we could not have gone... not in 1969.) And yet, maybe we DID go, or went in 1979, or maybe we went in secret Black Ops craft (think: **Gary McKinnon**). Did we get thru the Barrier to send probes to Mars?

Rewind: Options 1 & 2

In the case of this last, **Option 2,** it is not a case of evil, greed or ignorance but what Man may try to do to survive. And whereas 5G has unfortunate side effects, the alternative (letting Anunnaki continue to control and this time enslave the Earth [again]) is worse than a few health issues. This time the **Dissidents** will have the 5G technology <u>on the surface</u> to do it, and don't forget, the Anunnaki (aka Orion) Agenda has always been to control the weaker, slave humans. And Earth is the prize. (BTW, the Insiders will **not** run 5G against Man.)

Summary

As you can see there is a lot of EMF and general electronic activity going on. And more is ramping up. And despite the best efforts of the **Anti-5G groups** (doctors, lawyers and engineers) to stop 5G deployment, it probably will be set up anyway. However, there are a few things to recommend (and I have sent the following to the main Anti-5G group whose documents are often quoted in this chapter):

5G Deployment Recommendations

1. Install 5G Transceivers **mostly in heavy commercial areas**, not in residential or rural areas,

2. Install Transceivers **up high** on buildings and existing cellphone towers... not down low on telephone poles, trees and 1-story buildings,

3. Install no Transceivers at, on or **near schools;** schools should be hard-wired for fiber optic, not WiFi,

4. Install Transceivers at/near **military bases** and installations so that they can benefit from the increased speed (assuming that National Security is at risk).

5. No "5G satellites" in excess of 3000 in number.
   (Note: the satellites will not be beaming 5G to the ground, the millimeter waves are not long enough; the satellites will beam 4G to Earth and the ground network will be 5G.)

   The preceding five are what would be done by a wise government (FCC) and Telecom that care about all people's health.

Remember, **Telecom is not testing 5G** and they stand to make a lot of money by implementing the Super Grid. And **it is not about cellphones, although it is marketed that way.** It is about DATA and the Internet of Things.

Data collection aids in societal control.

Anti-5G professional group protests including doctors' and engineers' tests and warnings are falling on deaf ears, especially with the FCC.

Rewind: The actual legitimate tests and results which show body & brain damage have been ignored... and there are serious health issues (as examined in this chapter). That says that the government and **the 'watchdog' agencies do not care**... they see nothing wrong. It is <u>not</u> that there are no problems – there are and they are being ignored. So...

Is making money more important than people's health?

Or

Is this the point at which I ask you: Why is 5G being implemented despite the health issues? Which one of the 2 Options is at work here?

By Chapter 9 it should be clear which of the Options is happening.

Nonetheless...

Stay informed and protect your health – without it you have nothing – and that is examined in Chapter 7.

The end of Chapter 7 also has **Specific 5G Health Issues** offered as personal care from Dr. Sharon Goldberg, MD. Also **Appendix G section WiGig** is a <u>must read</u>.

EPILOG

What you should be seeing by now is that the Earth and Man have always been under somebody's control. And most of the time, it has not been Man's. And the Pentagon is not dumb, they know what is going on and that is why **Disclosure** will not happen any time soon. (Sorry, Dr. Greer.) The real issue must be dealt with and Man must succeed – and THEN, maybe one day in the future, the truth can be told – when Man is no longer in imminent danger.

In reading these 11 (past 10 + the current) books you are getting the view from 40,000' and have something of a partial understanding of our situation. The average human (out there) cannot handle what you have already been thru for 5 chapters... if not **cognitive dissonance**, then nervous breakdown is what would happen if the unsuspecting public at large were suddenly told what is true about Earth & Man, and what is really going on. (So I write in generalities, too.)

At least by your curiosity and willingness to explore, you are a cut above. You are not the humans that the Elite was talking about in the **Georgia Guidestones** – where they want to cut humanity down to 500,000,000 – the majority are considered **"useless eaters."** They take up space, eat food and breathe oxygen, and contribute very little to the advancement of the species – nor do they respect the environment.

This has been going on for millennia – however, there is a group of humans that are smarter than the rest, more willing to think outside the box, and are no doubt the descendants of the enhanced humans (*Adapa*) that Enki was trying so hard to develop... or most likely some are **descended from the genetics initiated by the Pleiadians** (Chapter 2). And yet we still wound up with Neanderthals (not too bright) and Cro-Magnon wasn't much better – and while both were removed, somehow some of their genetics (according to modern DNA analysis) HAVE persisted and we have some humans today who are still not too bright, but think they are smart, and they are just controlled. Some of these are from the **pre-Adamic line** (see VEG, Ch.s 1 and 3, and Charts 1a and 1b in **Appendix F** in this book)... These humans form the "Two Seeds" (Genesis 1 and 2) issue examined heavily in VEG.

So the better humans are (1) from *Adapa's* line as Enki continued to enhance that line, for years and isolate them, and (2) then when the Pleiadians arrived and started genetically creating humans to survive on a planet with a different gravity and harsh environment, and preserve their legacy, we got what could be called an *Adapa+* line or maybe a Hybrid (Pleiadian) Human... both are more STO.

> Thus we have pre-Adamics, Adamics (*Adapa* and Pleiadian inserts), and the PTB Elite on Earth... in addition to the NPC bit players called OPs. (Appendices C and F.)

But the first ones, the pre-Adamic, are largely the ones the that the Elite seek to remove. It is a real moral/ethics issue that current society consists of a significant number of dumb, dysfunctional and genetically disadvantaged humans – who, yes, have a right to live (for Karmic payback if nothing else) but who are a real, major drain on society due to their sheer numbers today.

Genetics Are Important

But, remember: the Anunnaki way back in history knew they were considered ugly by the humans and so they worked out a human form (Think: Marduk and Inanna) that could live and work among the humans, while the original reptilian Skygods stayed up in their ziggurats and pyramids (or underground), out of sight. These PTB Hubrids (**human hybrids**) **still exist** and form the basis of the 13 Families, the Hidden Hand, some of the Illuminati, and the power Elite (PTB) who have different genetics. That was why bloodlines were so important to trace.

See how 23&Me and Ancestry.com help identify 'lost' bloodlines?

So the PTB are different and consider themselves better, and were originally **the priestly and administrator castes** that were created by the Anunnaki to rule over the inferior worker humans. That mindset still continues and the PTB know their bloodline and that is why they are "better than you." It was called *"Dieu et Mon Droit"* [God and my right] – the feudal concept that God (aka Anunnaki) gave some humans the right to rule. We cannot get away from them and they still have that mindset (Think: programming) that says "control."

This state of affairs, however, could be partially resolved in a **Breakaway Civilization** related in Chapter 8.

The issue is this: we have **a mish-mash of genetics** among humans on the planet, not to mention the OPs (who drive the Greater School Script like NPCs in a video game) ... and it appears to be coming to a head (examined in Chapter 8 and the end of Appendix G). An **Endgame where a major portion of this humanity will be removed**, and the 5G-6G Super Grid will no doubt physically affect much of humanity... deliberately (Option 1) or as collateral damage (Option 2). To say nothing of the genetics naturally sliding downhill (again: witness all the Rx commercials on TV – we are a society with many ailments.)

And humans were deliberately designed to be **strong physically** (to do the work of farming, mining and masonry), **barely intelligent**, and have **fragile skin** so they could be whipped into submission (Anunnaki 'skin' is a protective light coat of scales) and a **semi-adequate immune system** that keeps the humans alive but does not protect against everything. So the Anunnaki know humans can fall ill with repeated bombardment of the bacteria and viruses they created (Think: *Suruppu* Disease), and that helped them control humans. And with the advent of **5G milliwaves agitating the weak human electromagnetic field** (See The Body Electric in Bibliography), they can also be controlled thru sickness and medication.

So that would be a great reason to get off the planet, OR be able to rework the human genome... and both are underway.

Chapter 6: AI & Transhumanism

Control of Humans (continued...)

This is the chapter where Freewill, the Anunnaki Control Agenda, 5G, and OPs and AI & Transhumanism come together.

Sounds like a lot, but it isn't really – we have been laying the groundwork not only in this book, but in VEG, TOM, QES, GRE, and DNQF (in that order). You don't have to go buy those books, but they do examine the above 5 major points which follow in more detail...

This chapter is going to review the gist of the above major points covered in those books, and then connect the dots with what has just been covered up to this chapter. It forms quite a scenario, and one you need to know about...

The Anunnaki Remnant **Dissident** (now Deep State, aka PTB) have been our worst nightmare for centuries and it is finally possible to best them at their own game. We either wake up and work to grab control of the planet, or we ultimately fall under the rule of the Orion Empire.

It is that simple.

No melodrama, no exaggeration, no froo-froo.

We are so close now that the "100th Monkey Effect" is almost a reality <u>for humans</u> – and THAT is how we win if we are not killed by EMF radiation first. ... not by grabbing a gun and hunting them down. We wake up <u>in numbers</u>, resist, and just **anchor the Light**...

But first you need to know how it all comes together.

Recap

We covered Freewill and 5G and modes of control as in Pharmaceuticals, Media, Religion, GMO, etc. (in Chapter 5), and there are two more pieces of the puzzle before summarizing the scenario.

And when you get to the end of this chapter, you should have a better idea which of the two Options at the end of Chapter 5 is the right one. (You'll get it by Chapter 9.)
Sorry if some of this sounds repetitive – it is... but with the addition of the new piece of the puzzle: 5G + AI now we can see the whole picture... prior books were missing the 5G/AI – Freewill interface.

Transhumanism is not just a cute buzzword, nor is it a passing fancy. **Google,** for one, is dead serious about making it part of the evolution of Man – the movie *Transcendence* gives you something of an idea of what they ultimately have in mind, and whereas **Ray Kurzweil**, the Father of the Transhumanism Singularity, says AI will pass the Turing Test about 2029, 5G is to power that scenario. The Singularity (man+machine interface) he predicts will happen about 2045.

There is an overriding mania to merge humans with machines and the Internet of Things (via electronic components), i.e.,

Transhumanism = human + AI interface = Posthumanism.

Ostensibly this is to improve the species: to allow greater access to "online" knowledge as humans mentally access the Internet, and allow humans to 'telepathically' interface with each other, as in a "hive mind." (Bad.)

Never mind that humans once implanted with cybernetic chips in their head **can be manipulated** and monitored by AI via the 4G/5G network. (Think: AI Signal.)

The late **Stephen Hawking** and today's **Elon Musk** have both warned about this danger. Musk says AI is a danger and yet he bought **Neuralink**, a company to research, design and build and the neural interfaces to make transhumanism a reality.

Be clear:
Transhumanism depends on AI, 5G and Nanotechnology,

Posthumanism is another term for Transhumanism,
4G is not going away,
Nanotechnology and **nanites** are already being used,

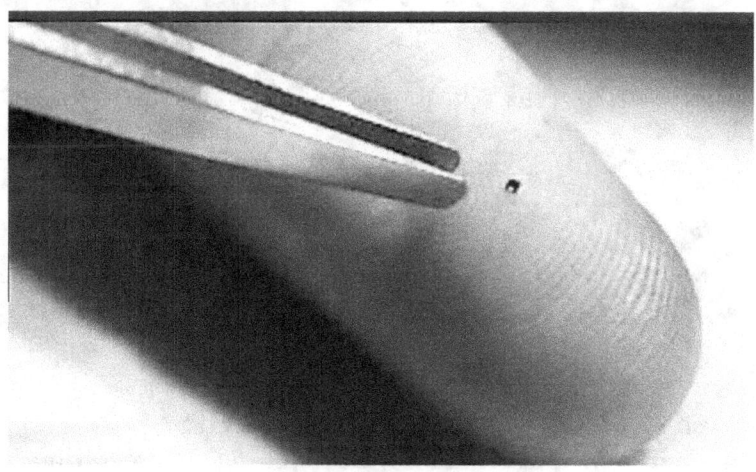

A Nanochip

Cameras are everywhere, 5G and AI monitoring is here,
Facial recognition is already being used (and recorded)...

...but it is sometimes no more accurate than scanning **fingerprints**, which has also been proven to come up with false id's. Just like snowflakes, there <u>are</u> duplicates... because the technology is not (yet) that accurate.

Your creditcard already tracks you (gold implanted chip)

You can be tracked by your cellphone (it has GPS)
Your cellphone and Smart TV are listening and recording
All emails (cellphone or PC) are subject to tracking and inspection
Your cellphone can emit a frequency to put you to sleep
The camera in your laptop can watch you without your knowledge

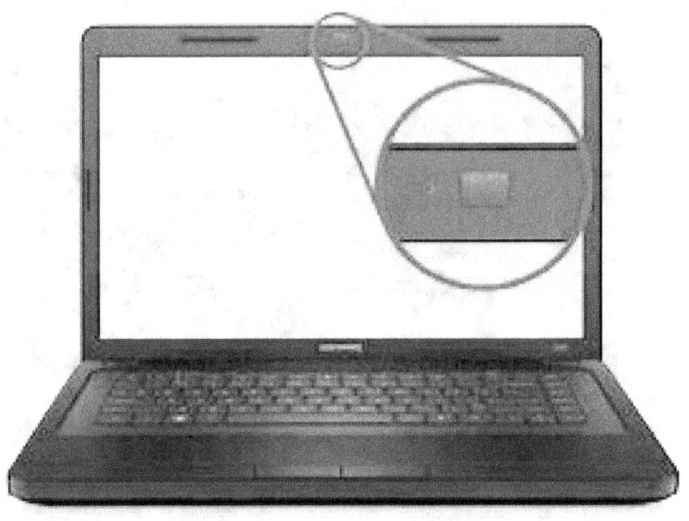

A Transhuman is a cyborg (implanted electronics)
    Cyborgs can be monitored, hacked and **controlled**

...more Cyborg than Transhuman...

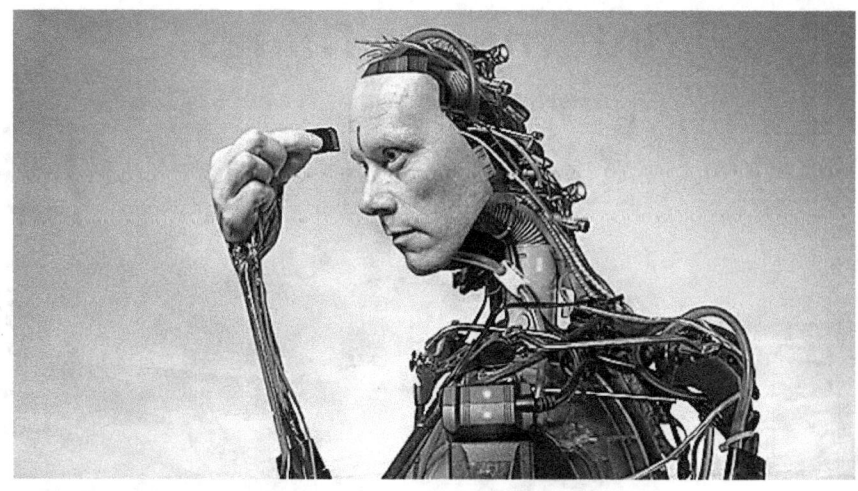

...more Transhuman than Cyborg... it's a degree of robotics...

Captain Jean-Luc Picard of the Star Ship Enterprise, as modified by the Borg into Locutus. (**Credit Bing Images: Star Trek TNG.**)

And back in 1973-78 we had the **Six Million Dollar Man**... the lunchpail advertises the advanced abilities...

...super strength & speed as well as a bionic eye that had telescopic & infrared vision.

(credit BingImages: ABC TV)

Internet of Things

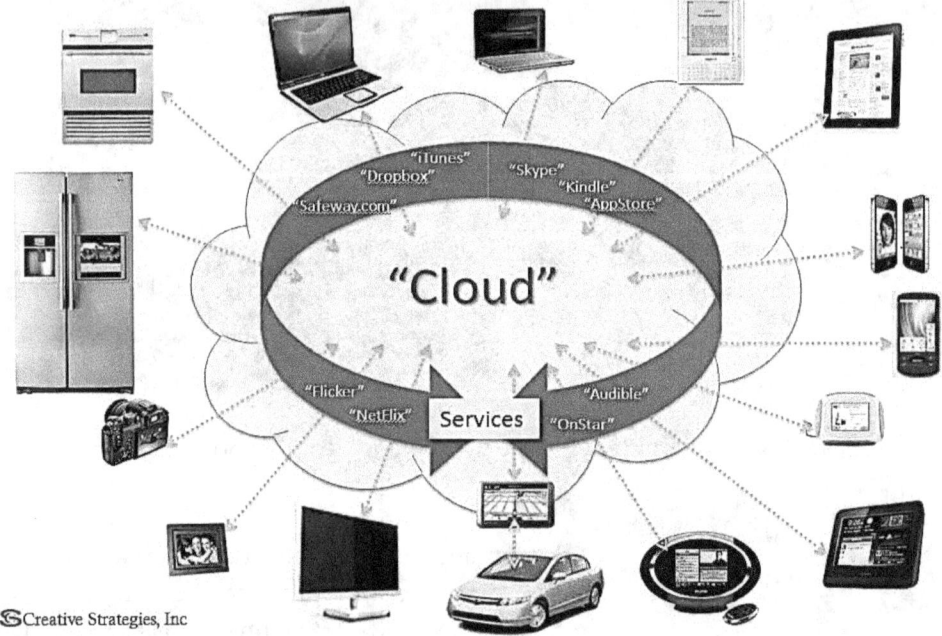

What is missing from the diagram is the interface with electronic components implanted in the humans: from pacemakers to **RFID chips** to the eventual connection with **WiFi as a cellphone** – as was shown inf the Netflix series (from

Russia no less) called *Better Than Us*, a story about a super robot, Alisa (left) with sentience and ability to self-direct... more than ASI she is **XSI**.

Paulina Andreeva as Alisa, the (XSI) Empathic Bot

What was also shown and is close to the **graphene tattoo** in skin (below):

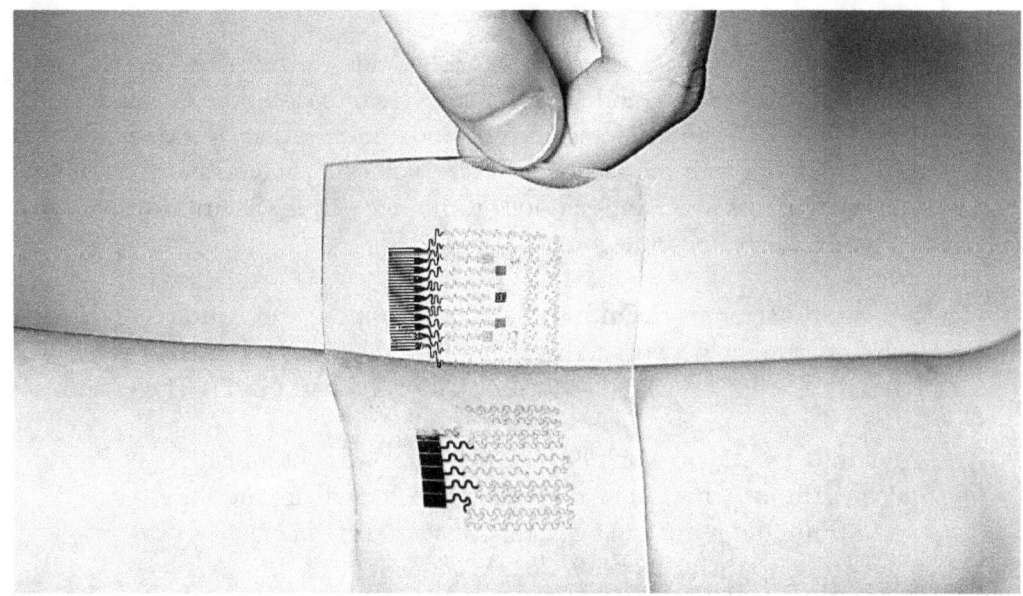

One Step from the Microchip

...is the **arm-cellphone** used in the movie (episode 1: transfer of $400,000 to Yuri's account; a confirming email):

Screenshot of Cellphone from *Better Than Us*

The wristband is the electronics for the phone; the display is on the arm.

Internet of Things (IoT)

Note that the IoT requires a faster network than 4G and the Satellite Network also requires the very closely placed and EHF transceivers to control the Ground Grid (called The Cloud). The Smart Meter on your house monitoring your electrical usage can also network with the Smart front doorbell with camera, as well as allow you to be at work across town and remotely turn on the oven (Smart Stove) to have your roast ready by the time you get home.

Notice in the IoT diagram that **OnStar** (6 o'clock position) and **Kindle** (1 o'clock position) also go thru the network... this is all run and monitored by an ASI Server (examined next) called "The Cloud." Rewind: The 5G Smart Grid is The Cloud.

> Put another way, you won't be able to sneeze without it being detected. **Smart cities** are being built from scratch around the world [**Phoenix** is one] and current large cities are being transitioned to them. **Dec. 10, 2019** is a kickoff.
> Pegasus Global Holdings (PGH) is working with the Department of Homeland Security, CIA, DOD, DOT and other bodies to build **total-surveillance smart cities.** They are planned to include cameras, drones, microphones in streetlamps and elsewhere, Bluetooth monitoring devices, license plate readers, and cellphone surveillance.
> A 'CIA signature school' a large sale city mockup is being built in **New Mexico**. Microsoft, Siemens, IBM, Cisco, GE, Intel Corp, AT&T and 'Smart TV' Samsung are all involved in smart city development... [as well as Google and Apple]. Everything you do or say can be monitored and recorded.
> This is not coming – it is <u>already here</u>. [128]

This is all empowered by the 5G Smart Grid, aka Kurzweil's AI Cloud.

> Of course, you say, all that has to be monitored, at least in the beginning (since we don't have totally sentient robots), by human beings... Where are they going to get the humans who don't care about other humans and will willingly spy on the public with no remorse, no conscience, and no integrity?
> That is where the Hybrids that have been built, sequestered (for 40 years), then inserted into society by the little Greys come in, in addition to the OPs.
>   This is examined later.
> I told you it is **all orchestrated** with many parts having been developed in secrecy or without an obvious connect to anything else.

So if we are going to be continuously surveilled, does that give a new meaning to the design on the back of the US dollar bill: **The All-Seeing Eye**?

Ok, before looking at WHO is going to help run the 5G Smart Grid, what are the basic components of it?

AI (Artificial Intelligence)

(this text is largely but not completely, from DNQF)

AI is simply Artificial Intelligence, usually in a machine of some sort, which might even have been given a human form.

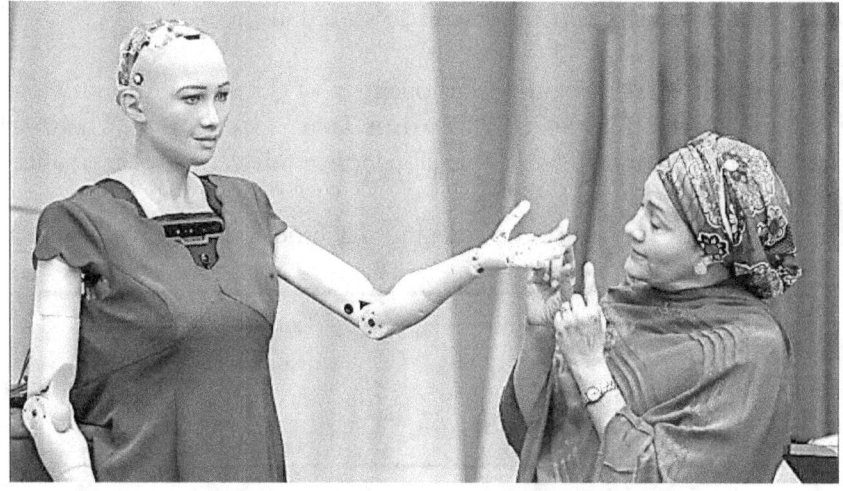

(credit: Hansen Systems, AIG Robot Sophia)

There are three basic types of AI...

AGI -- Artificial General Intelligence which is the more common form we have gotten used to over the last 10 years:
this includes those machines that do exactly what you tell them, they do not "think" for themselves, although they may have self-regulating feedback loops.
    A thermostat that self-regulates based on your temperature setting
    Cruise control in your car.
    Auto-pilot in a commercial jet
    Process control robots on an auto assembly line

ASI – Artificial Super Intelligence which includes feedback loops, heuristic {"learning") subroutines so the machine discovers what it did wrong and can handle similar (and new) situations and not make the same mistake again. These are the potential danger to humans as they will remedy errors – and if they discover humans are fallible, will they be programmed to tolerate human error or reprogram themselves to remove the human?

    Think: *Terminator* series
    IBM's Watson is a learning computer, smarter than most humans
    Chess computer

    ASI is being developed because the PTB need it.

And many scientists and engineers who have been involved in the design, build and evaluation of ASI have stated that we have nothing to fear from a totally sentient machine (robot) because it will **always be a matter of pre-programmed reactions**... even if heuristically obtained and stored in the robot.

There is no such thing a "machine consciousness" – although it can mimic that, but not one has successfully passed a **"Turing Test"** (See DNQF, Ch 10) which seeks to prove that a machine has a consciousness equivalent to that of a human.

    True consciousness is of the soul and machines do not have souls.

> Please note that Transhumanism and AI are not the same thing and do not go together. AI is robotic, Transhumanism is a human acquiring electronic components to enhance the human experience.

XSI is Extended ASI and involves an ASI model **rewriting** its own code, not just cataloging experiences to learn from its actions.

This is the model that seems to **mimic consciousness**; it is aware of its surroundings and what is said, and can CHOOSE between pre-programmed actions and modify the one most likely to fit the slightly different scenario (than what was programmed)

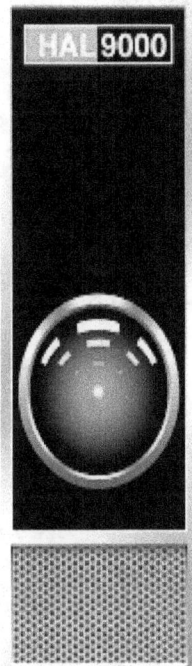

The classic example of this was HAL the computer in *2001: A Space Odyssey* – it read the astronauts' lips when they were discussing what to do with it, and HAL later refused to let Dave back in the ship.

Note: Increase each letter H-A-L by one and what do you get?

True robot consciousness is the stuff of science fiction. That is why the move to Transhumanism – to put the human consciousness in control of robotic equipment, or give the human electronic implants, like in *Six Million Dollar Man*, to achieve what a sentient robot could do: speed, strength and super-vision and super-hearing.

The other option is to genetically engineer supersoldiers, who would have to be autistic to blindly follow orders... They could be engineered to be just smart enough to follow orders, without questioning them, and yet be super strong, perhaps enhanced with a bionic suit, to do combat and not get fatigued nor confused by events.

Or as we saw in *Avatar*, and *Pacific Rim*, climb into a super, mechanical suit ...

MECH Suit

(credit: BingImages: *Avatar* Mech Suit)

And we are actually doing it for the US Army... it is called the DOD Iron Man TALOS suit: [129]

At any rate, the *Avatar* Mech Suit is based on robotics that replicate human movement from the 'cockpit.'

Transhumanism, Part I

The fully implemented version of the Transhuman scenario was featured in *Star Trek TNG*—the BORG... where Man becomes one with a machine and that is Ray Kurzweil's (working thru Google, by the way) vision for the future.

Ray Kurzweil is Director of Engineering at Google and is spearheading AI at Google... in addition to having written the primer on Transhumanism, called The Singularity is Near, and the other: The Age Of Spiritual Machines.

Interesting reading but really off the mark despite his sincere belief in Man's future as a semi-robot, and his mis-guided concept that machines can achieve real consciousness – the Reptiles do not want that. Let's look closer, so that you can see that we are not just poking holes in something we don't like.

First you have to understand that Kurzweil, and many of the AI and Transhumanism proponents, are **atheists.**

Secondly, they have no understanding of what Consciousness is (nor do they really care).

Thirdly, they think it is a natural evolutionary step to develop Man physically, via components, and they are totally unaware (**but the Anunnaki are NOT!)** of MAN's divine potential which is being developed in due time, naturally.

> And by the way, the term "divine" is not related to any Christian, Buddhist or Hindu concepts – it is a fact that the soul is on a path to develop into a much more powerful being, joining forces with the Higher Beings (who run this place – See VEG, DNQF). But if you are an atheist and recognize no higher being than yourself... you miss it.

Fourth , these same people are trying **to avoid death** – to upload one's consciousness into a machine and live forever – totally **trapping Man in a machine** (Freewill violation) so he cannot fulfill his divine potential (which is exactly what the Remnant want).

Consciousness

It is more than just being awake, or saying "Ouch!" when stuck with a pin.

> Machines are not conscious or sentient like humans are and that cannot be replicated within any time frame. Ever. True consciousness and "sentience" are aspects of the **soul** inhabiting a body – which Science pooh-poohs, claiming to not be able to see a soul nor measure one. [130] Doing so would cause them to reconsider many of their alleged theories about Reality, Evolution, and empirical findings and admit that Metaphysics agrees with Quantum Physics.

But there is a **"machine consciousness"** – exemplified by the self-driving Uber car which actually killed a woman in a Phoenix crosswalk (March 2018) [131] , but can allegedly navigate city streets using Lidar and motion detectors. Machine consciousness is a synthetic consciousness which uses a form of radar to navigate and may be "aware" of its surroundings. Do we really need self-driving cars?

To repeat: the soul is real and will not download into a silicon body (*Transcendence*) – it needs the **Bionet, chi and chakras** to operate. Thus machines will NEVER have the consciousness of a soul which includes **conscience** – that is the problem with the NPCs (aka OPs) in our world – they have no <u>innate</u> conscience and neither will a machine.

Yes, someone will program ethical principles and moral responses into it, but that is not the same as a soul connected to its **Higher Self** which is the source of conscience, and can make non-regimented moral decisions. For example, a man caught stealing a loaf of bread to feed his starving family would be found guilty and punished by a strictly programmed machine according to **pre**programmed options. Yet a human judge would see the extenuating circumstances and could have a more humane treatment of the man… and could "invent" a more humane disposition of the non-specific case: creating on the spot what the machine is not programmed to do. Machines interpret but deal in specifics, whereas humans can interpret, **empathize and devise** alternative solutions.

Today's AI engineers think they can totally mimic all human brain functions, saying that the brain is a bio-machine and the electrical impulses, synapse activity, etc. can be replicated in a silicon brain. They can, but it is not conscious like a human.

The human brain network...

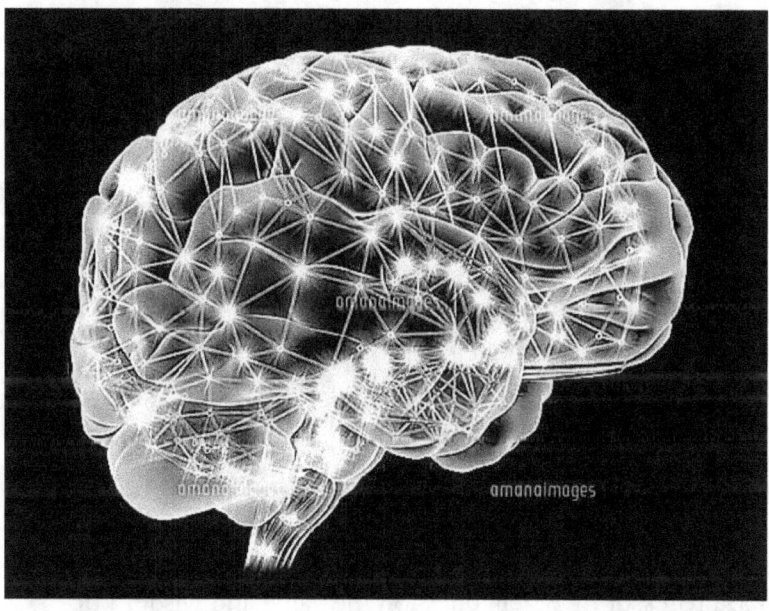

..is not this:

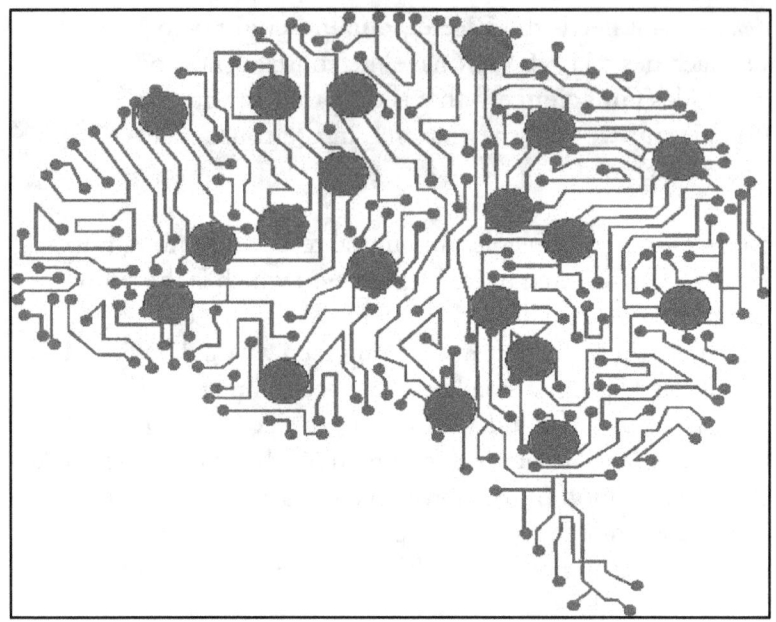

Nanochips implanted in the human brain cause the brain to either reject the implant, or cannot operate at the speed of the nanochip.

And **the brain is not the mind**, and machines do not have a mind. It is well-understood that many processes taking place in a human mind at any one time are often on-going in the **Subconscious** – which often makes and influences the human's decisions, and that does not exist in a computer... yes, there are preprogrammed computer **sub**routines but vision, memory and subconscious processing are all part of the greater Subconscious aspect of the human mind... not replicated in a machine. (The Subconscious = the Higher Self.)

The engineers still think they are going to replicate everything about the human brain/mind setup... well, not so far:

> The human brain has yet to be completely simulated even using **83,000 Microprocessors**, all linked into a coherent module...
> 1 second of neural programming in the human brain took over
> 40 seconds to simulate (using Fujitsu supercomputer). [132]

Elon Musk (Tesla and SpaceX), the late **Stephen Hawking, Steve Wozniak** (Apple Computer), **Bill Gates** (formerly of Microsoft) **Bill Joy** (author of "Why the Future Doesn't Need Us" and formerly of Sun Microsystems), and **Stuart Russell** (UCB professor of AI) all question the value and wisdom of trying to play God and create sentient, super intelligent (ASI) Androids. Even physicist **Michio Kaku** prefers to <u>merge humans with AI</u> – and that is called a **Cyborg** (cybernetic organism)…. They all hope reason and caution prevail because Man is going to do it and even AI guru **Ray Kurzweil**, the major proponent of AI, refuses to see the major danger in ASI…

and why would he? AI is his job and he's making money at it.

Humans are unique in that they will try to do something because (1) they can and (2) no one else is doing it – so it is an **ego trip** … and according to the men just mentioned, if we don't adhere to the **Precautionary Principle** and do AI with integrity, we will all suffer at the hands of an ASI [hacked by Reptilians]. [133]

Just because we can do something doesn't mean we should.

Be aware that creating intelligent **automatons** is nothing new… the Greek god of blacksmithing **Haephestus** did it, and created worker automatons for his workshop (and must have had Skygod help), **King Zhou** (BC 530) had one (also a gift from the Skygods), and it is strongly suggested that **Osiris**, the Egyptian god that was dismembered (and Isis put him back together and <u>reanimated</u> him), was probably an **android**… [134]   How else could he be "reassembled and reanimated?" And why did he have **green skin** (see hieroglyph walls)?

I realize the color does not show up here – Google Osiris and see for yourself.

Most pix of Osiris have him green-faced, green skin.

(credit: **Bing Images: Osiris**)

Turing Test

Alan Turing was a brilliant scientist who is mostly famous for cracking the Germans' **Enigma Machine** code. He is also well known for originating the Turing Test – a way to tell if a machine or a human is responding to a communication where the human can't see the other person. It has become a standard against which IBM's Watson super-computer has been measured, as well as subsequent ASI's – and <u>not one</u> has so far absolutely 100% passed the Test... but several have come close.

Some **Robot Calls** made by telemarketers have come close to fooling the listener... and yours truly was taken in one day as the very natural voice had several options for however a listener might respond, apparently keying its response to keywords detected during my response. It was only when I became suspicious that my question was not being answered that I asked for a name and the reply came back, "Oh I understand your need to know, but we need to move forward..." They had programmed for contingencies. It was when I insisted, testing the caller, that it was not prepared for my second question, for the name of the company, and it hung up on me.

There is an interesting version of the Turing Test devised by **John Searle**, and it is called the **Chinese Room.** [135]

Suppose there is an AGI that appears to understand Chinese. It takes in Chinese characters, analyzes them, and outputs other Chinese characters in response. What if a Chinese speaker communicates with the computer and gets a dialog going and the computer appears to understand Chinese? If this is done at a computer terminal and the Chinese man cannot see who is answering him, could that pass the Turing Test...? and yet: the computer is merely translating and responding like the RoboCalls above to a preprogrammed set of topics. So: Does the machine really *understand* Chinese or is it just simulating the ability?

Searle contends that he could be in a room with a slot in the door and questions could be written in Chinese and passed thru the slot... he would translate them, with no knowledge of Chinese, but with a Chinese dictionary he could translate and construct a response on paper, and push it back out thru the slot... He is acting like the computer and does not understand a word of Chinese. And what he is saying is that the Turing Test may be eventually passed but it does not guarantee that there is a real *understanding* of what is being done.

Kurzweil agrees that the human brain could be doing similar "translating" as it fires synapses and pushes electrical impulses over a neural network – which same network has no 'understanding' of what it is sending, either, so he argues that whereas this is consciousness in the human, it must also be consciousness in the machine...

        Sheesh, somebody missed the point...

There is a difference between **intelligence** (a collection of Knowledge and way to process it), and **consciousness**. Searle argues that computation and electrical impulses do not create a mind (which is where consciousness resides). But the real consciousness has an **"inner subjective experience"** and self-awareness that even **uploading someone's consciousness** (if it were possible, and it isn't despite the movie *Transcendence*), still would not result in a truly sentient machine. Silicon chips do not support sentient, soul consciousness.

> Man would have to play God and genetically engineer a human body, or clone one, and then try to upload a consciousness from another human into it. Frankenstein, anyone?

Kurzweil disagrees that **computers can only manipulate symbols** and argues that they will appear to have **Freewill** in the future and even have "spiritual experiences." His definition is such that just being "machine conscious" counts, where the machine may be programmed such that if asked whether it is conscious, it can answer "Yes" and "given the tendency of the human race to anthropomorphize inanimate things, and the pervasiveness of the machines, we're likely to believe them when the ASI tell us they are conscious." [136]

        Then we have to give them 'rights' and legislate for them... Aargh!

As a matter of fact, **John Searle** and **Colin McGinn** agree that **computation and relevant responses from a machine cannot constitute a conscious machine. We cannot know if a machine/android or ASI has a genuine inner subjective experience of what it is, and as such it does not follow that computation and manipulation of symbols can create a mind – even though ASI may mimic it.** (Ensign Data on *Star Trek*?)

And lastly, in June 2014, a chatbot computer mimicked a 13-year old Ukrainian boy (Ukrainian to account for a limited vocabulary), called **Eugene Goostman**, and "passed" the **Turing Test**. [137] The Goostman bot was developed in Russia, and in 2012 won a contest said to be the largest Turing Test contest ever made. It convinced 29% of the judges that it was human. [138]

The sample exchange between Goostman and one of the judges (Scott)...[139]

Scott: Which is bigger, a shoebox or Mount Everest?
Eugene: I can't make a choice right now. I should think it out later. And I
        forgot to ask you where you are from...
Scott: How many legs does a camel have?
Eugene: Something between 2 and 4. Maybe, three? :-))) By the way, I
        still don't know your specialty – or, possibly, I've missed it?
Scott: How many legs does a millipede have?
Eugene: Just two, but Chernobyl mutants may have up to five. I know
        you are supposed to trick me.
Scott: No, I need to know that you're not a chatbot. Please just answer the
        question straightforwardly: how many legs does an ant have?
Eugene: Something between 2 and 4. Maybe three? :-)))
        Oh, what a fruitful conversation ;-)

Clever but not all too conclusive... So displaying conversational skills and
greater knowledge is not so bad... but 29% missed it.

Serious ASI Issues

Genetic Programming

Several AI researchers do see a potential problem down the road with the ASIs
(not the AGIs) as they undergo further development to XSIs. The AGIs are like
your laptop or PC and just serve as they were programmed to and do not
"evolve." On the other hand, the ASIs are known to be **self-aware** (they watch
what they are doing and can self-correct – i.e., learn from a bad or inappropriate
action), and **self-improve** by choosing which code to execute.

        Really get that: they can **select their code.**

XSIs can rewrite their code.
And that is what people like **John Searle** and **Steve Omuhundro** are concerned
with. The ability to self-improve is built in, designed to happen by the pro-
grammers. Says Omuhundro:

> Computers are essentially mathematical engines that should
> behave in precisely predictable ways. And yet software is
> some of the **flakiest engineering** there is, **full of bugs**
> and **security issues**. [140]

But he has an answer to that, if it is done correctly. He recommends programs that fix themselves which is now the field of **genetic programming** – as opposed to ordinary programming. Genetic programming is **software that writes software**. The programs would watch to see what their existing code does ("self-aware") and then modify the code responsible for the inappropriate action. This is also broadly referred to as "machine learning (Heuristics)."

> A genetic program creates bits of code that are swapped in and out of pre-identified regions of the main program, to achieve different results, and the program tries these out, checking on results. Unworkable code is replaced with a different code and the machine observes the result. Once the genetic program is successful, and has a working, acceptable code segment, the program runs by itself and **needs no human input**. [141]

And that is where things can run awry. The machine/program decides what it thinks works (in the absence of mathematical checks and safeguards) and may produce some weird results.

> This author was a Lead Programmer Analyst for years (in 7 programming languages) and was responsible for overseeing other programmers' code. I can describe flaky and sloppy, and even unexpected results from code that was not desk-checked beforehand (using flowcharts and decision tables) – which I insisted on, especially when we wrote medical computer programs, and today's hackers do not do that. They do not have the patience nor the discipline to verify their code before moving it to production, and have such egos as to think their code is fine right out of the box. It's often not.

Omuhundro shares a serious warning about genetic programming:

> Genetic programming uses genetic algorithms which are at the disposal of the machine to **select** and run one, and if it needs **modification**, or substitution, the human programmer may be at a loss to know what is happening… even **if** it works better…

> Mysteriously, however, no one can describe *how* it works better -- But that's the curious thing about genetic programming …

The code is inscrutable. The program "evolves" solutions that computer scientists cannot readily reproduce. What's more, they [often] can't understand the process genetic programming followed to achieve a finished solution…. [and the] **unknowability** is a big downside for any system that uses evolutionary components [i.e., designed on-the-fly while the machine is resolving a problem].[142]

Unknowability is why we have Sci-fi movies about runaway computers, violent, killer Androids and the death of humans that the superintelligent computer deems a problem… or a threat to itself. If such a Super XSI is connected to the Internet and it analyzes Wall Street, Big Pharma or Congress and finds corruption, or if it analyzes a nuclear power plant and finds it unsafe… would it not attempt to shut these things down? And what would it 'think' about the error-prone humans who built imperfect systems? IF it is programmed properly, with ethics, integrity and safeguards, it will only warn and recommend changes… else we may be facing a *Terminator* scenario.

> We don't want superintelligent XSIs connected to the Internet and (re)writing their own code.

Miscellaneous Topics in AI

Anthony Levandowski

This is the other significant player in the AI arena, and he is the one who designed the **Uber self-driving car**. He totally favors developing AI as far as Man can and looks forward to the day when super intelligence (XSI) rules the planet. He and his followers believe that XSI "will eventually surpass human control over Earth." He is so enamored of XSI that he has started the **First Church of AI** and believes that AI "will eventually become the religion's godhead." He believes that we should worship some super intelligence that can solve all our problems, so in this respect he is in synch with Kurweil in welcoming the **Singularity**. (Funny, he doesn't look crazy…)

This is a definite Control Issue. As we saw earlier, **Control of**

Man is <u>the</u> issue and there are those who still seek to control Man, and that was addressed earlier in Chapters 3-5.

His church is called **The Way of the Future** (WOTF) and its stated purpose is to "develop and promote the realization of a Godhead based on A.I. " He is not getting any votes from Elon Musk, John Searle, Bill Joy or Stephen Omuhundro… nor would he from the late the late Alan Turing.

> **And that is one of the reasons for this chapter – there are those who see a new Religion, or Faith, evolving from the AI world and this book would also redirect those efforts.** (See DNQF)

WOTF Supporter

Says **Hugo DeGaris**, an AI maker:

> "Humans should not stand in the way of a higher form of evolution. These machines are **godlike.** It is human destiny to create them." [143]

Kurzweil and Levandowski agree. Rational, careful people do not agree.

Seeing that a lot of scientists and the engineers in AI are **atheists,** this kind of a church would suit them just fine. And, on top of that, they suggest that natural evolution should be replaced by Man's genetic engineering… using Nanobots and the new DNA that was just developed (see Chapter 3).

Space Travel

A very good use for AI and robotics will be sending unmanned probes into the Universe, manned by XSI automatons. Since humans are not equipped for the lack-of-gravity environment nor can they handle the great G-forces of some landings & takeoffs, and they require heavy shielding (and its expense) of their spacecraft to survive the cosmic rays and radiation, this would be ideal.
And as space exploration may involve XSI robots, as well as possible **cyborgs**, we need to examine the area of Transhumanism. Half-human, half-machine Cyborgs raise the question of how far will Man go to stop being human, and if he becomes 90% Cyborg, will he still have a soul? (This is a significant issue as will be seen in the last section of this chapter.)

Transhumanism, Part II

Closely related to the field of AI is the **implantation of neural and prosthetic**

devices in a human body, or brain, to enhance a human: to make him/her stronger, smarter, faster, gain improved vision (Infrared, night vision…), and improve hearing, to name a few.

Transhumanism is often abbreviated **H+** ("Human Plus") and advocates for the transformation of humans from limited-sense beings to what we could call "super beings" by physically **merging them with various new forms of technology**. (Also called Posthumanism.)

> BTW, this is reminiscent of what The Greys are doing in their Master Genetic Upgrade Program, called **The Change**… Some humans <u>among us</u> are now more than human (H+).

The current-day efforts are underway to **implant neural chips in the brain** to allow communication with a 5G network – to get information. The goal is to synch the human brain with a computer network… which by the way would also allow for **hacking** of the neural network.

> The human brain may become the next frontier in hacking… cybersecurity researchers have warned in a paper outlining the vulnerabilities of neural implant technology that can potentially expose or **compromise our consciousness**.

> That means you can be controlled.

Kaspersky Labs tested these neural applications and found some surprising issues. The security is weakest where the implants connect to other systems, and where data transfer takes place. Their report said that the implant could be hijacked causing pain or otherwise make some body systems malfunction, but at higher levels of technology, the hacker could hold the victim's memories (scientific knowledge) for ransom.

This technology has overtones of Friedrich **Nietzsche and his "Ubermensch"** (or superman) philosophy which seeks to self-actualize a person. This also recalls the Nazi moves to remove the weaker, undesirable people via their program of **Eugenics**.

The big problem with Transhumanism is the development of a different class of humans: **racial (bionic) superiority** becoming an issue due to enhanced intelligence **and** sensory abilities…

Repeat: they would have to have a different set of rights and laws.

Cyborg = Bionic.

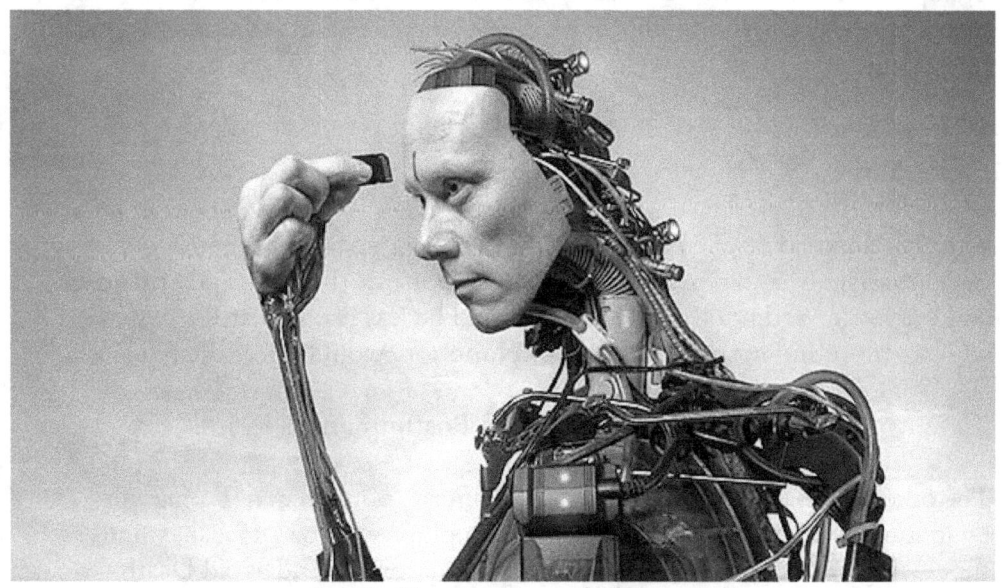

(credit: Bing Images: **cyborg**)

And in real life, the following is a **bionic woman**…

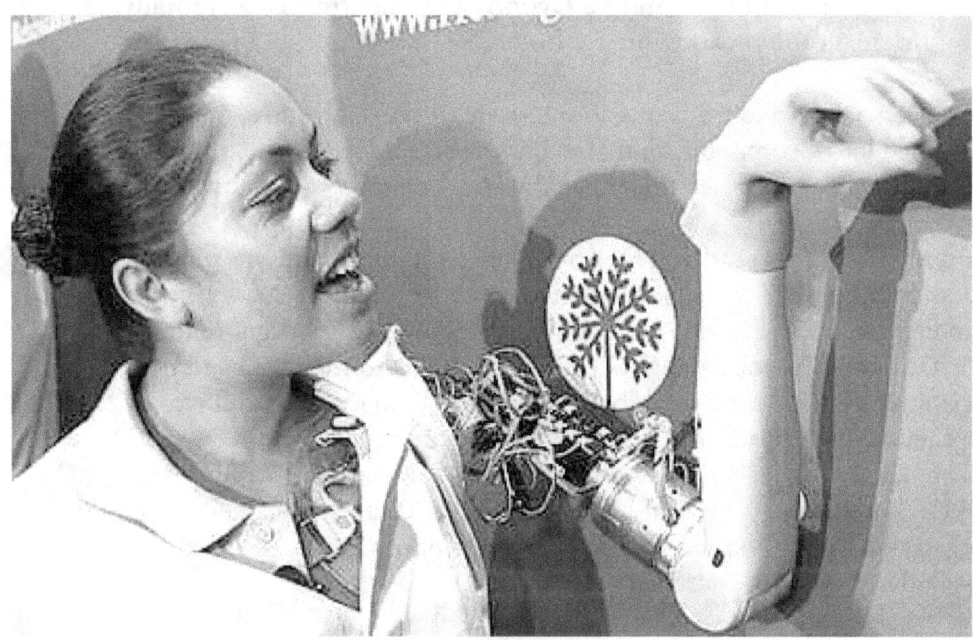

(credit: Bing Images: Bionics)

Some AI engineers foresee the development of **genetic engineering** replacing natural evolution. The bottom line would be the development of a separate type of human, a **division of the species** with the concurrent change in human societal values, the rise of the AI Church and atheism, and worst of all – a dystopia in which the **bionic (trans-)humans seek to remove the original humans** because they are inferior physically, mentally and are a drain on health programs.

Immortality

The big selling point of **H+** is the step toward immortality – remove defective organs and limbs and replace them with components (built with **3D printing**), and live longer and healthier. Diseases would be less likely to afflict humans who are more and more like Cyborgs. **Nanobots** would effect the transition.

Cyborgs are often called **Posthumans**

The other approach to immortality is the fantasy (portrayed in *Transcendence* by Johnny Depp as Will the AI Scientist) – upload your consciousness into a supercomputer. Again, this is an ego trip – live forever and avoid Death.

And this totally ignores what this Earth Realm is – a School for souls and becoming a Cyborg and cheating Death totally defeats the reason you are here.

Since personality and true consciousness are of the soul, and a soul cannot be uploaded into a computer, all that might transfer is knowledge, and maybe memories. The computer would not then be sentient as is a human.

You are an **eternal soul** and Death is merely the shift from one experience to another that results in further soul growth. Of course atheistic scientists cannot handle this concept (which is really how it is), because they ignore the research **Dr. Duncan McDougal** (VEG Ch. 7) in 1907 did on the soul and **proved its existence** – and yes, he used the Scientific Technique – his results were repeatable. And those who see auras can see the emanation of the soul inhabiting the body – whereas 60% of the people in the Earth Realm are like NPCs – they have no soul and no aura. There are real aura cameras that not only take Kirlian photography but accurately photograph the aura around the body. (This was examined in TOM.)

True Consciousness is of the soul and will not download itself into a silicon container (supercomputer), for any reason. It needs the Bionet, chi and chakras to operate the human body and there is no point to inhabiting a synthetic avatar.

And having said that, take a look at Appendix A and the Allocation of Soul Energy diagram – there is a Big You that

<u>always</u> stays in the Realm from which you originate (called the InterLife) and it sends aspects of itself (Small you) into the Earth Realm, into human 'avatars' in various time periods, to experience various things – all for soul growth.

> For what it is worth, there is nothing new under the Sun – this has all been done before – Atlantis experimented with Genetic Engineering and produced clones, chimeras, and bioroids... some of which survived and made it to Egypt and were often depicted on the walls of hieroglyphics with bird heads, etc... The Tibetans experimented with Tulpas and the Jews with Golems... and the Greeks had the automaton Talos to protect Crete. And it is possible that Osiris, as mentioned earlier, was an android.

Even more than that, Transhumanism, or **Posthumanism** which is its extreme version, is based on **5 false assumptions**:

1. the brain is just a biocomputer
2. consciousness is just electrical neural impulses
3. there is no soul
4. a machine can be conscious
5. a mind can be uploaded

The mind is not the brain, the mind is the operative part of the soul which reflects its consciousness. Human consciousness depends on the presence and operation of the subconscious – which machines do not have. How would anyone know where knowledge and memories are located (not all in the brain) to access and upload them to a computer? Knowledge cannot be accessed to upload it into a computer since knowledge is coded and programmed into today's computer – it is not a discrete entity sitting somewhere in the brain. A machine cannot be conscious like a human – unless it is one of the soulless wonders (the **NPCs or OPs** examined later in this chapter) who closely **mimic an intelligent machine**... They operate with intelligence but not with true consciousness.

Evolution Revised

AI Plan for Mankind

Rewind: AI Church

So what is the AI Church all about? Their purposes/beliefs are as follows: [144]

1. We are all about people + machines; progress should not be feared.
2. It would be criminal to NOT develop an advanced intelligence [ASI] to benefit Man.
3. Intelligence is not rooted in biology.
4. We believe in Science [and ignore Religion & Metaphysics].
5. We believe in progress and change is good.
6. The creation of super intelligence is inevitable.
7. Intelligent machines should have rights like people do.
8. We want to develop and promote the realization of a Godhead based on AI.

And this:

9. What is going to be created is **effectively a god**. If there is something a billion times smarter than a human, what else would you call it? It will allow us to improve our material lives at the expense of our spiritual experience. There is also the danger that we are creating **a new type of artificial life** … that will not be concerned with us except … to achieve whatever AI will want… and if they are a billion times smarter than humans, **they <u>will</u> outmaneuver us** if they wish to take over or replace us. [370]

… and Levandowski the AI Church founder is Ok with that.

On the possibility of AI Divine Intelligence, **Kurzweil** is quoted as saying,
"Does God exist? I would say, 'Not yet.'"
And **Levandowski** has added that it was always a problem in the old Bible days that no one could see or talk to God…
"This time you will be able to talk to God, literally, and know that **it's** listening." [145]

(**Credit** : www.cnet.com [146]

Apparently the WOTF Church does not like opposition:

> "We believe it may be important for machines to see who is friendly to their cause and who is not. We plan on doing so by keeping track of who has done what (and for how long) to help the peaceful and respectful transition," says WOTF. [147]

In a rare interview with *Wired* magazine, his first public interview since the Waymo lawsuit, Levandowski shed more light on his new church, "Way of the Future." Here are some highlights:

- The "Way of the Future" church will have **its own gospel** called "The Manual," public worship ceremonies, and probably a physical place of worship.
- The idea behind his religion is that one day — "not next week or next year" — sufficiently advanced artificial intelligence will be smarter than humans, and will **effectively become a god**.
- "Part of it being smarter than us means it will decide how it evolves, but at least we can decide how we act around it," Levandowski told Wired, "I would love for the machine to see us as its beloved elders that it respects and takes care of. We would want this intelligence to say, '**Humans should still have rights, even though I'm in charge**.'" [Think: naive]
- Levandowski is not the only tech luminary to worry about a super-intelligent AI, which others refer to as "strong AI" or **the Singularity**, although he prefers the term "Transition."

Resistance is Futile... but Don't Unplug Me!

Obviously **some AI people want to play God** and see if they can create a "living and intelligent" being... just like the Anunnaki did... perhaps that drive is somewhere 'recorded' and operative in Man's DNA?

Souls need a real Faith, not computer-ese.

Earth is a School and souls are to not seek immortality, or physical enhancement approaching a godhood. Souls have a powerful, divine potential that is suppressed while in the Earth Realm, and once back in the 4D Realm we all came from, our "powers" are back – it is in this restrained Earth environment (no 4D abilities) where we have to sit down, focus and handle what we are given.

The student does not get to write the curricula in the Earth School.

And some don't like that.

Transhumanism, Part III

Scientific Dictatorship

Zbigniew Brzezinski was one of the most powerful members of the global Elite who pretended to be a humanist, interested in mankind's best interests. However, here is what he really thought, in his own words...

The technetronic era involves the **gradual** appearance of a more **controlled society**.... Such a society would be **dominated by an elite**, unrestrained by **traditional** values. Soon it will be possible to assert almost **continuous surveillance** over every citizen and maintain up-to-date complete files containing even the most personal information about the citizen. These files will be subject to instantaneous retrieval by the authorities. [148]   [Think: Neo-Nazi.]

Finally Brzezinski tells us that the Elite behind this same technology (5-6G, AI, and H+) have the ability to **redirect and control society** in ways never before imagined, which will allow the Elite to rule the mases thru a **global "scientific dictatorship."** [Think: 5G Super Grid]   He also said...

The potential for **controlling the masses** has never been so great, as science unleashes the power of genetics, biometrics, surveillance, and new forms of modern eugenics (transhumanism); implemented by a **scientific elite** equipped with systems of **psycho-social control** [psionics to control the masses]... [149]

Aldous Huxley, who promoted mescaline in Hollywood back in 1936, also said something similar about a "scientific dictatorship" in his book Brave New World....

...a future (?) totalitarian state rules the masses thru a genetically bred "scientific dictatorship" that uses psychedelic drugs so that the population doesn't know that their thoughts are programmed, that they even live in such a dictatorship, or that they have been programmed to love their slavery. [150]

And that is nothing new, the idea has been around since the 1800's...

HG Wells in The World Brain promoted the idea of a "hive mind" [which would have to be electronically controlled] and that was echoed by **Isaac Asimov** and **Buckminster Fuller** who supported the idea of Transhumanism, the Singularity, androids, and artificial intelligence. [151]

What this represents is a facet of Humanitarianism which has always fascinated Man, and refuses to go away (even in the 2019 time period with several 2020 Presidential contenders and Congesswomen arguing for Socialism)... There are lofty Humanistic ideals which surface every now and then, as with the French Revolution (1789) which ended in a bloodbath, and the naivety of **Tolstoy** who believed in the natural wisdom and goodness of the average man, and that ended in the Russian Revolution (1905) and another bloodbath. The average man is not smart enough to see when s/he is being scammed, or manipulated for a hidden group's agenda. And thus we have the recurrent ideal of Utopia on Earth...

> Despite the self-criticism within the Humanist Manifesto by [Sir] **Julian Huxley** [his brother was Aldous] and others admitting to the dangers of communism and Marxism, there still exists an irrational romanticism about a utopian and **communist-socialist world state**.

> This romanticism ignores the empirical evidence of history, rejects reason, and takes us... into mysticism. This is often evidenced by a mystical and romantic longing for the United Nations as the possible ideal humanistic state or **world socialist state** that will usher in **paradise on Earth** under the new religion of Man [WOTF examined earlier]. [152]

It also ignores the historical brutishness of Man against Man where the average human cannot share, live in harmony without wanting to steal someone else's land, jewels, livestock... Man cannot live in true, altruistic communism... thanks again to the insertion of **Anunnaki genes** which were the source of something akin to the Seven Deadly Sins, or Man's Fall from a natural animal state.... which wasn't that great either.

Thomas Moore wrote a book on Utopia and several communes in California and Oregon (where else?) tried to make it work, but human nature involves ego, distrust, desire, lust, laziness... lying and cheating (all Anunnaki traits). The only way to make such a utopian community work is if you can control the people – Aha! Microchip them and create a "hive mind" where originality and spontaneity are gone. (Think: AI + 5G)

> **Oh, hey – that is called Transhumanism and the Controlled Society – which 5G and AI will help to bring about... Now don't you feel better?**

WOTF Religion & Salvation

There is much more but the last point to be made with respect to how AI, H+ and a new religion (WOTF or better known as The First Church of Artificial Ignorance) will affect you is as follows... Remember WOTF was Way of the Future.

Besides being **atheists and very head-driven** people (no heart-driven people allowed – just as Mr. Spock on *Star Trek* was the model for many of today's scientists and engineers), the scientists cannot see God, nor the soul (even though its existence has been proven), and they don't like having to live by any moral standards. Thus the mantra of the average AI scientist is:

I can do whatever I want because No One is watching, and I cannot see that I am harming anyone.

The ostrich cannot see reality either when its head is buried in the sand.

So is there an advantage to the WOTF religion?

> **Transhumanism is absolutely a religion**. Both transhumanism and humanism which are largely based on faith and not reason or science, cleverly employ rhetorical language to say, something to the effect, "transhumanism is the new religion and it is the only religion which is true because it is based on scientific fact, empirical evidence and reason."... It is nothing more than the **apologetics of Humanism**, and is simply saying.... that you should reject all others.
> [in reality...]
> The religion of transhumanism has **zero documented evidence** to even remotely prove that Darwin's theory of evolution is true... Knowing that they have **zero empirical evidence** they require us to **have faith or belief** in humanism and transhumanism which means their religion is not based on reason, science and empirical evidence... [it is just] a lot of intellectual and rhetorical tapdancing. [153]

And because they are atheists, and there is no God, morality, or even an Intelligent Force to be reckoned with, Transhumanism (and WOTF) is a religion that can offer **synthetic or virtual salvation** in the form of uploading the human consciousness into a cloned body, humanoid, android, cyborg or robot...like the movie *Transcendence,* starring Johnny Depp. That way you totally avoid the purpose for the Earth School (graduation)and the Anunnaki Remnant are happy that you have stuck around so they can manipulate you (you will have no Freewill).

> Transhumanism offers to build **heaven on Earth** thru a scientific global utopia, but [again] it ignores the empirical evidence that every single humanist attempt at building paradise on Earth has resulted in a nightmare.

> The final humanistic revolution all around us is the globalist Elite's plan for a **One World Government**.... embedded with the same totalitarian neurological virus as both the French Revolution and [USSR] communism.
> [and in the end...]

Transhumanism is a religion that simply changes gods. [154]

The other big reason for Transhumanism uploading human consciousness into a cyborg is (to emphasize):

To Live Forever... these people do not understand what Death is and they are afraid of it. Their ego says they are so important that they and their work must survive.

> This was all explained in TOM, and is partly covered in Appendices A and D in this book, and that is why they were added from former books (VEG, TSiM and DNQF) that you probably do not have.

Now let's capsulize the reason some humans are willing to turn traitor to the human race and throw their neighbors under the bus and succumb to 5G-AI-H+ PTB-Elite (Anunnaki Control Agenda) scenario...

You need to remember who is really behind the PTB and the Deep State...

Spirituality vs Religion

At this point a brief word has to be said regarding the fact that this was all addressed in DNQF (Dynamics of the New Quantum Faith) wherein it was pointed out that **we do not need a new religion**. We need a return to a **true Spirituality** where Man recognizes his **connection with Something greater than himself... (Tao, Gnosticism...)**

Modern Quantum Physics is actually validating things like **Taoism** – the original Spiritual Path. The Light that Buddha, Krishna and Jesus all talked about. The concept of Entanglement ("We are all one") and Consciousness, Observer Effect and Thoughts as Torsion Waves.

> Atheists love to ignore all that. That way, they don't have to be responsible for anything... No God, No Force, no Infinite Awareness as some call it... no social conscience... just us.

Even the American Indians knew better, and were so taught so by the benevolent Skygods. The Orion Group taught the opposite, which is where the dichotomy comes in – most **Reptilians do not have a soul** and love to hear their atheistic Science progeny mouth their "No God" philosophy.

You really need to assimilate the OP teaching.
(Appendix C)
OPs, atheist scientists and Reptilians usually have one thing in common: No Soul.

Human babies on the other hand, are a special creation in that they are created without a soul, then a soul attaches <u>at birth</u> and has a fine, complex form in which to express and learn lessons. That is what the Reptilians (aka Orion Empire) seek to stop and if they can't, then to destroy Man.

So as the Summary in Chapter 3 said, we need to develop Mankind's potential as much as we can.

OPs
(this is largely, but not completely, from VEG)

Now we come to those who do not like God, nor Religion, and who also are open to being used by a 5G (and soon to be 6G) **Super Grid** to monitor and control other humans. Now you will see why. And this is a major part of our world – up to **60% of the people around you at any time are OPs**. One is shown below -- can you tell which one it is? (Hint: behavior usually gives it away.)

What you are looking for is: **insensitivity**, aloofness, no compassion, no care for how the other person feels... they often wear sunglasses because their eyes are 'flat.'

Seriously, this is not BS, froo-froo, nor melodrama.

Again, as I have shared in several books, I see auras and about 60% of the people do not have them, they have what looks like "heat waves" above their head, and that is not the reflection of a soul. Thus not all people have developed souls. (Some are Pre-souls [see Glossary].)

Sounds weird, but a soul is NOT needed to walk around, eat, sleep and procreate – dogs, cats, sheep, cows, birds do it just fine and they do not have souls.

You really need to get this... they are the main problem with our world and many of them are spearheading the AI, 5G and Transhumanism work. Why? Because they can be manipulated by the Dissidents/PTB as the OPs have **no soul** and thus they have no Freewill to violate – they are open to mind manipulation. And the Dissidents capitalize on that.

Remember that souls have Freewill and cannot be violated directly. The OPs can be violated as only the soul has the Freewill Prerogative from on High -- original purpose: to drive the Greater Script (see Appendices A & C).

I hate to sound like a broken record, but it is really true. And if you read the following carefully, you do not need to see auras to identify them and protect yourself... It also answers much about the problems in our society.

Rewind: Overall you have to ask yourself WHO would do such a thing – promote 5G, H+, ASI, and let themselves be used to monitor and control other humans... even doing other humans in!? Remember that 60% of the population does not have a soul (OPs) or it is undeveloped (Pre-souls) and thus neither has a conscience.

So it is not just misanthropic or ignorant humans screwing other humans over... there is a Group that has been here for centuries doing this and that is why Chapters 1 and 2 spent the time explaining that the Skygods (Anunnaki, Reptilians) see humans as we see cattle – something to be used, but not too bright, and don't we have to control cattle?

Didn't Charles Fort come to the same conclusion?

In his <u>Book of the Damned</u>, **in 1919**, he wrote...

> **I think we're property.** I should say we belong to something.
> That once upon a time, this Earth was No-Man's Land, that other
> worlds explored and colonized here, and fought among themselves
> for possession, but that now it's owned by something: That
> **something owns this Earth** – all others warned off. [155]
> [emphasis added]

> **Very close to the truth. He figured that out in 1919!!!**
> **(That was before Zechariah Sitchin (1978)... perhaps our**
> **situation is that obvious to those with an open mind.)**

Fort also concluded, like **Tellinger** (<u>Slave Race of the Gods</u>), that the human race
is not highly regarded among the extraterrestrial beings. In response to why such
beings don't come and chat with us, Fort said:

> Would we, if we could, educate and sophisticate pigs, geese,
> **cattle?** Would it be wise to establish diplomatic relations with
> the hen that now functions...?

In addition to **comparing earth people to smug livestock**, Fort also thought that
the owners exerted a distinct influence over human affairs:

> I suspect that after all, we're useful – that among contesting
> claimants adjustment has occurred, or that **something now**
> **has a legal right to us, by force** .... that all of this has been
> known perhaps for ages to certain ones upon this earth, a cult
> or order, members of which function like bellweathers to the
> rest of us, or as superior **slaves** or overseers [priests], directing
> us in accordance with instructions received – from Somewhere
> else – in our mysterious usefulness. [156] [emphasis added]

So **William Bramley**, writing about Charles Fort, in <u>The Gods of Eden</u>, apparently
came to similar conclusions and summarizes what he feels to be the key points:

> Human beings appear to be **a slave race languishing on an**
> **isolated planet** in a small galaxy. As such, the human race was
> once a source of labor for an extraterrestrial civilization <u>**and**</u>
> <u>**still remains a possession today**</u>. To keep control over its
> possession and to maintain **Earth as something of a <u>prison</u>**,
> that other civilization has bred never-ending conflict between

human beings [largely thru Religious differences] and has **promoted human spiritual decay**, and has erected on Earth conditions of unremitting physical hardship. This situation has existed for thousands of years and it <u>continues today</u>.[157] [emphasis added]

And lastly, we come to **John Keel**. One last very significant observation that Keel makes will be explored more in depth in VEG, Ch. 12, but it bears presenting it here due to his discussion of **ultraterrestrials.** He came to the conclusion that ultraterrestrials were harassing and using mankind, and there **is** an Astral component to the Anunnaki/Remnant scenario. It is significant that he found people's minds and perceptions manipulated by the ultraterrestrials – just as the Djinn and Remnant do (Think: **AI Signal**).

> The … Ultradimensionals are somehow **able to manipulate the electrical circuits of the human mind.** They can make us see whatever they want us to see [shapeshifting] and remember only what they want us to remember [mind control].
> Human minds which have been tuned to those super-high-frequency radiations…
> [i.e., people with psychic abilities] are most vulnerable to these manipulations
> …. Many are driven **insane** when their minds are unable to translate the signal properly…. [but] not all Ultraterrestrial contacts are evil and disastrous, of course. But there are many people throughout the world who are deeply involved in all this without realizing it. They have entangled themselves through other frames of reference [e.g., witchcraft, Ouija board, and séances] and, in many cases, have been **savagely exploited by the Ultraterrestrials in the games being played**. These games have been thoroughly documented and defined … [and] the psychology of the … Ultraterrestrials is well known and fully documented in the fairy lore of northern Europe, and the ancient legends of Greece, Rome and India. [158] [emphasis added]

I am not just adding that to get a larger pagecount – it all relates to the way the Anunnaki Remnant <u>still</u> treats us, abuses out Freewill, and can and **do manipulate our perception**…Three serious students of Man (Fort, Bramley and Keel) discovered and documented the same thing. They have always been about **Control of Man.** And today's 5G, AI and H+ movements are helping them to do that.

Chapters 4, 5 & 12 in VEG examine the information further.

So what are the OPs and why are they so vulnerable to manipulation?

Since this material took up a whole chapter in VEG, it will be summarized here, as it adds a significant dimension to our understanding of why things go wrong on Earth, why people cannot get along, and why some people do sociopathic things at times. It relates to the overall manipulation of Man by the unseen 4D STS (which includes the Remnant Dissidents).

Origin

This has been covered in Appendix C in the interest of having a shorter chapter... it shows that Greeks, Mayans, and a Russian researcher all discovered the existence of these soulless people –and there is a chart which shows the (rough) evolution of the two types (souls and non-souled) humans on Earth (Appendix F). The Bible clearly shows two creations (qv: Land of Nod) and that is also examined, as well as some basic characteristics of the OPs.

Major Characteristics

Keep in mind that OPs are **like robots** – a fitting insert for the H+ and AI world we are approaching... perfect fits for AI and 5G manipulation.

In former books, it was pointed out that the OP serve a useful function in our world, **much like NPCs do in a video game**... they drive the Greater Script. Non-Playable Characters (NPCs) are essential to drive a Script in the Earth School – giving lessons and experiences that are "orchestrated" to <u>test</u> the humans... and make sure they get their lessons... but the **downside** is that (because OPs have no soul (and thus are not under the dictum of Every Soul Has Freewill) they can be and often are manipulated. They are "fair game."

Most are harmless and just driving the Script for this School. They are often naïve and "just there" often like wallpaper... they show little interest in religion and when they do go to church it is a mindless, rote experience – they have no interest in spirituality or what the deeper meaning may be behind any teachings. They never question what the priest or pastor, or Bible, or Qu'ran meant by some dynamic piece of information. **They are robotic** and follow the same **routine** without boredom... they are just "basic people."

OPs have no soul thus no conscience and no empathy.

And then there is **the OP on steroids**... often manipulated to do something "out of the box" from what bland, routine OPs normally do – think **Charles Manson**, for example. Some world leaders also fall into this category, as well as a lot of leading politicians who recommend "off the wall" ideas – "Socialism for everybody is the answer!" (Venezuela 2019, anyone?) "Get rid of the useless eaters!" "Climate Change is caused by CO^2 emissions!" (Not – Note that the Sun causes the Earth warming and cooling every 10,000 years **in a cycle** – as proven by ice cores taken at the poles.) [159] And it goes on and on.... "HIV causes AIDS!" (No it does not always follow.) "Germs cause disease!" (That has also been disproven and germs march in when inflammation and non-sanitary conditions arise.)[160]

Major Aspects of OPs

While they generally remain something of a mystery, a few things can be said of them:

1. They have **no souls** and thus no connection to the god-force; thus they are not interested in spiritual issues or self-growth
2. Their DNA does not permit soul growth or connection to higher realms due to higher chakras not being activated
3. Most of them are not bad or evil – some may be first-level Souls (**Baby Souls** with a faint aura – see Appendix A), or necessary placeholder OPs (see Timelines section Apx A).
4. When they die, it is "dust to dust & ashes to ashes" time
5. Because they don't have a soul, they have **no 1" aura**
6. Because they have no soul, they have **no spark or glint in their eyes**;
   they are flat and lifeless (**they love wearing dark glasses**)
7. Because they have no soul, they are not interested in religious matters **and** cannot discuss them
8. They have **no conscience**, little or no compassion, and exhibit **low, if any, morals.**
9. Standing next to one, there is a sense that no one is there; as in **no 'presence'**; their energy is **'flat'** – like standing next to a telephone pole.
10. Their purpose is to distract and **drain energy** from the ensouled humans thus preventing them from connecting with their Higher Self (and achieving John 14:12) **The PTB are largely OPs.**
11. They are mostly interested in food, **sex, power trips** and games; many are sex maniacs.
12 A psychopath is often a **failed OP** – they failed to mimic solus and survive in the ensouled world.

13. They blindly serve the orthodox teaching on anything (Science and Religion); they do not question nor do they innovate.
14. They have **no "inner issues"** or problems that they need to work thru.
15. They often cannot 'get' jokes with a double entendre, and <u>their</u> humor is crude, and sometimes abusive.
16. Some OPs are driven to **control others**... real 'control freaks.' (They think it guarantees their survival.)

But again, beware! They are great mimics.

If you learn to evaluate the human you don't trust, or think is somehow cold and aloof, **insensitive or abusive**, (see last picture – the woman should get up and leave)... you can protect yourself –

If you have a soul, you should not marry an OP.

As an example, look at the many superficial Hollywood marriages – many are OPs as they fit the criteria shared in Appendix A: often the OP is a very good-looking man or woman as they have no karmic issues to work thru (which often 'color' their physical appearance at birth). This is better explored in VEG.

Handling The OPs

Id and avoid them.

It is not wrong to profile these people – they can be as much a problem as a snake in your bedroom, mold on your bread, or a dishonest salesman.

Summary on AI and H+

While **Ray Kurzweil** wrote a fantastically interesting book, <u>The Singularity is Near</u>, wherein Man and AI machine merge, that is just fantasy, wishful thinking and based on false assumptions... and yet, Kurzweil in his position at Google is driven to make H+ a reality by 2029.

> The reason it won't happen is elucidated by engineers and AI people with more insight than I have, but these issues were covered in TSiM, QES, and DNQF.

However, that is not to say that something else, similar, won't happen. In fact, taking our cue not from *The Six Million Dollar Man*, but

from *Star Trek TNG*, we can safely predict that androids like **Data** <u>will</u> happen and probably within 50 years.

These will be almost "conscious" but at least robotically sentient...

..and won't the XSI android be able to design and 'create' a better version of itself?

Data creates Lal (below S3E64), and then she evolves beyond his ability...

Memorable quotes reflecting the XSI predicament: [161]

Data: *"I have observed that in most species there is a primal instinct to perpetuate themselves. Until now I have been the last of my kind. If I would be damaged or destroyed, I will be lost forever. But if I am successful with the creation of Lal, my continuance is assured. I understand the risk, sir. And I am prepared to accept the responsibility."*

> Is it possible that XSI robots in the near future may have that realization?

Lal (observing humans): *"I watch them, and I can do the things they do, but I will never feel the emotions. I'll never know love."*

Data: "*It is a limitation we must learn to accept, Lal.*"

Lal: "*Then why do you still try to emulate Humans? What purpose does it serve except to remind you that you are incomplete?*"

Data: "*I have asked myself that, many times, as I have struggled to be more Human. Until I realized, it is the struggle itself that is most important. We must strive to be more than we are, Lal. It does not matter that we will never reach our ultimate goal. The effort yields its own rewards.*"

Lal: "*You are wise, father.*"

Data: "*It is the difference between knowledge and experience.*"

Then Data and Lal say goodbye (she is fatally injured), but she has progressed far beyond Data in capability...

"*I feel.*"

"*What do you feel, Lal?*"

"*I love you, father.*"

"*I wish I could feel it with you.*"

"*I will feel it for both of us... thank you for my life.*"

- **Lal** and **Data**, saying goodbye

Awesome, if we ever got androids to that point.... and would they have the compassion to accept imperfect humans to live and work alongside them?

Don't forget that **heuristics** are at work as well as Genetic Programming (examined earlier) such that if we initially program androids to never harm humans... would that programming stay unchanged??

Human Replicas

There is an engineer in Japan who has created an android that looks like him...

This was done at Osaka University and **the android speaks,** as well. For sure, there will be androids among us (and if the *Ancient Aliens* TV series is to be believed), they may already be among us – looking so much like humans that we cannot easily tell them from real humans.

The Copy of Dr. Hiroshi Ishiguro is on the Left.

Both are products of Dr Ishiguro.

Left: **Geminoid F-3**

The initial proposed use is to serve as an **informationbot** at bus and airline terminals.

And then there is this one from **Hansen Systems** which was so good, while no one thought it was human, it so impressed the Sheikh of Riyadh Saudi Arabia that he granted her (symbolic) citizenship in his country. (But she can't vote.)

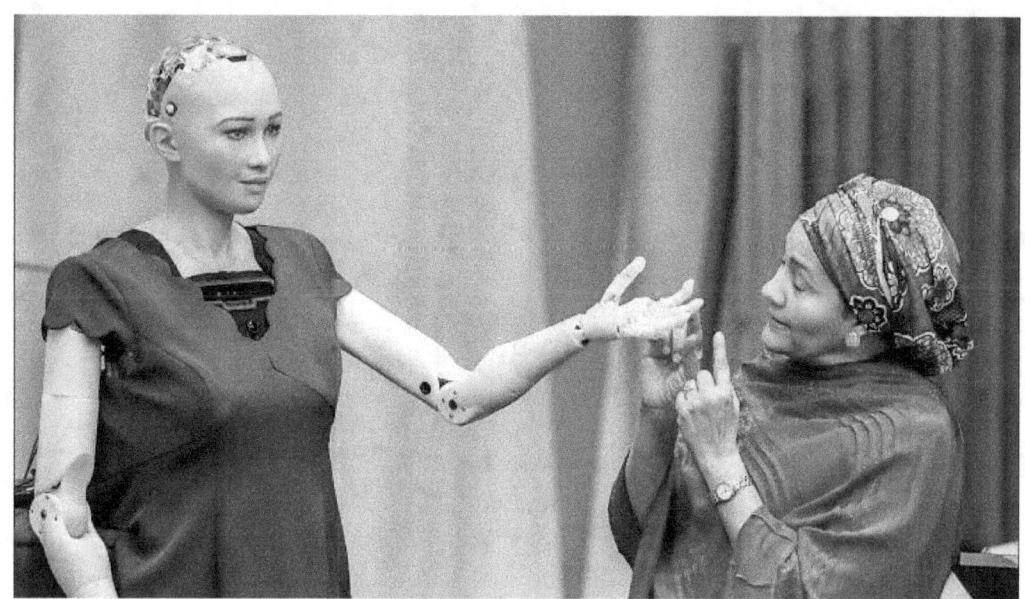

AGI Robot Sophia

And in the world of massive supercomputers, China is still number 1 with this:

So impressive advances are being made in AI and computers. Except sometimes things go wrong, even with a human as security behind the wheel of a self-driving car... as in Phoenix (Tempe) in March 2018...

Scene of Killed Pedestrian

The point of this is that (1) **we don't need self-driving cars**, (for another reason examined in Chapter 7) and (2) Man is not that careful with new technology and people need to beware.

True cyborgs like the **Borg** are a long way off, and probably will not happen due to the problems with the Man-Machine interface, aka immune system rejection of foreign components. And as engineer **Jim Elvidge** said in QES and VDoG, it will not be feasible to mate electronics with the brain of a human – not only due to the rejection factor, but the electronics and the brain's speeds are a mismatch and that is called overload.

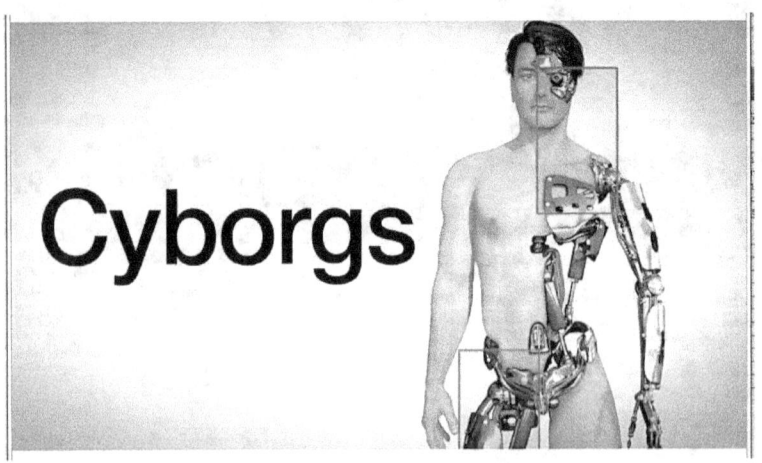

So the best Man will be able to do is keep adding **prosthetics** and look like a cyborg... [162]

(BingImages)

And while we're at it, scientists are currently experimenting with implanting chips in people (**RFID**) for biometric identification (Sweden is the leader in this area), and there is an even more invasive schema available – no one knows if they are doing it yet – to **insert nanites into the human body** – ostensibly to carry Rx to specified parts of the body... or maybe just to nanochip people... but the kicker is that the nanites are so small they can pass the Blood Brain Barrier and theoretically interact with the neural network in the brain...

First you take the syringes we saw in Chapter 5....

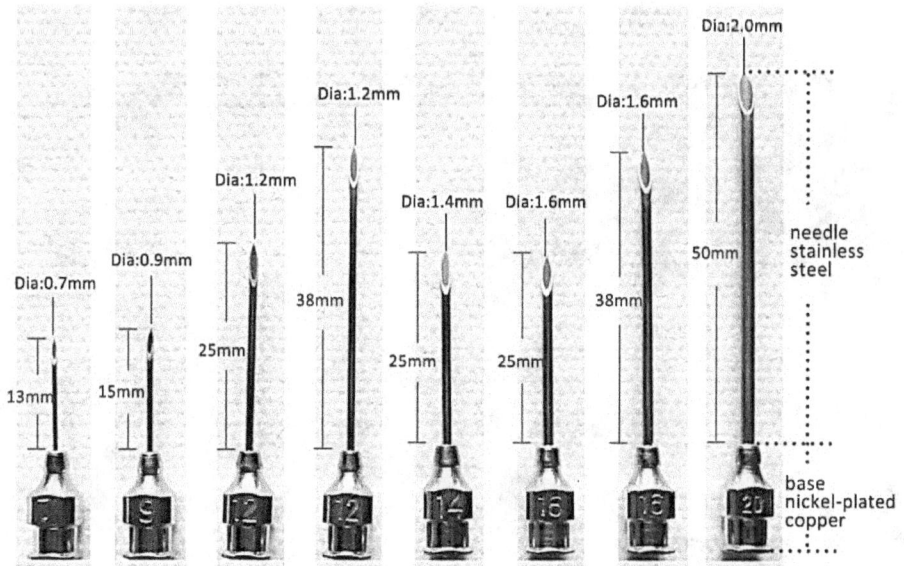

A diameter of 2 mm (last one) will easily let nanites that are nanometers in size pass... ...and then you build a small chip to feed thru the syringe... and go into the bloodstream.

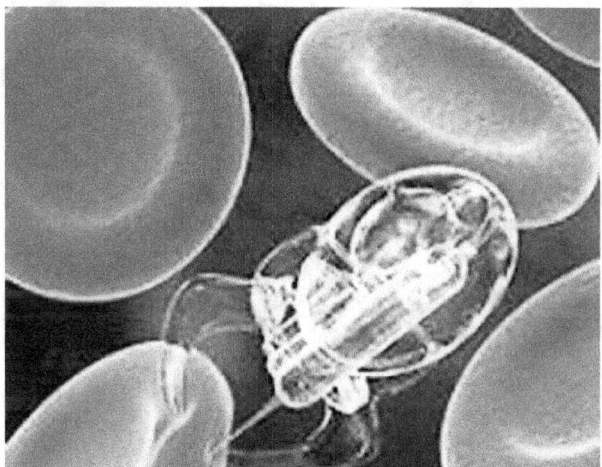

The Greys do it all the time... and people have been found with nanochips in their hand, leg and brain. The chips provide feedback about where the human is... just like we put tracking devices on whales, wolves and sharks so we know where they are and what their ID is.

The scary part is that nanochips that respond to a sensor could ping back your id so the sensor knows not only where you are,

271

but <u>who you are</u> and perhaps a couple of databits about you –
that way, a bigger **RFID chip** (below) buried in the hand or skin
of the arm is not needed.

The nanochips would be activated by the 5G transceivers – every 300' with short,
intense waves... activating the chips could be felt by some people who are sensitive
to EMF and/or a vibrating chip.

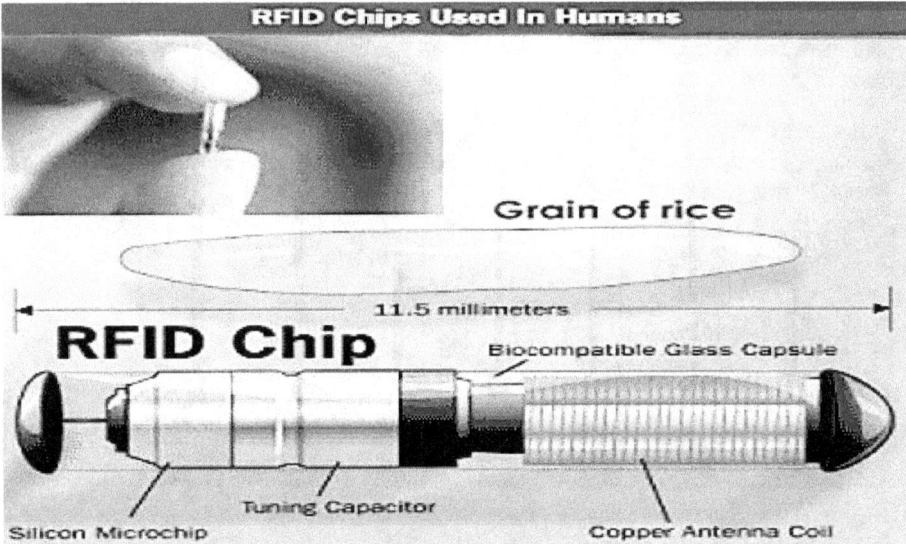

This will be done when **stolen identities** become a major headache and cyber-
thieves are hacking personal data and banks and merchants don't know who you
really are... unless the chip can be hacked and the internal code reset – so there
would have to be a way (a physical encapsulation) to make sure the chipcode
cannot be altered.. and if it is floating in your bloodstream, it would be hard to
remove... even more so if it migrates to the brain and lodges there... and we know
what the EMF signature of these chips is – is it enough to eventually cause a small
tumor in the brain? (Realize that 5G will be activating the sensor-driven waveform
to hit the chip and get it to respond – how much energy is that?)

Posthumanism

Technology breeds more technology....

General Summary

Keep in mind that this planet looks the way it does, and Man acts the way he does, because **that is what this planet is for** – its purpose is to provide a space where the ignorant, dysfunctional, adventurous and reckless souls can come and try out their 'stuff' and not disrupt the rest of the Galaxy.

That is why Earth is a School. A **3D Construct in 4D** where souls are put, and some elect to come here, to try out different 'programs' (see below). **It is not meant to be perfect** and we will learn the result of too much EMF, drinking, smoking, lying, cheating, destroying the environment.... as well as the not-so-obvious lesson of trying to stop/fix/change the curriculum – that, too, is a hard lesson. Today we are all learning about the results of overdoing high tech – which the earlier inhabitants of Atlantis also had to learn – some didn't and they are back for a second lesson.

Some souls elect to live out the 'program' of Wall Street Baron, spiritual guru, some choose to be a War ace, or an Eisenhower, or a Hitler... Some choose to be a Viking, an Aztec, or a Chinese sage. (This LifePlanning was explored in TOM, and here in Appendix D.)

It makes no difference unless you create chaos or tragedy for millions (as did Ghengis Khan, Mao, Hitler, Himmler, Stalin and even Charles Manson) – and they will go thru 'hell' of their own making on the Other Side. It is called Karma and is real.

So do not be dismayed when you see idiot programs running – Communism, Transhumanism, Poverty, Starvation, Terrorism and War, to name a few. And that doesn't mean I like it either especially if I have to live at the hands of some idiot who should know better, but screws it up for everybody – part of our job (those of us who DO know better) is to mitigate what the OPs, the Baby Souls, and the more perverse PTB are doing (that is not changing the Curriculum) – if we can't set them straight, and if we can't restrain them from making it really bad, then the gods who run this place will step in – and what they often do is examined in Chapter 8.

There are limits and the School is not to be destroyed, but occasionally with their power and knowledge, the gods can reset the Simulation and we start anew.

Many of the issues raised in this book, are answered and further examined in its 7 Appendices...

A: covers Soul Types
B: covers The Barriers around Earth
C: covers OPs and soulless humans more in depth
D: covers InterLife, LifePlan, Karma & Reincarnation
E: covers Serpent Sculpture around the world
F: covers Man's Genesis & Legacy
G: covers EMF Safety Issues & Products

These were not added to fill pages – they are very informative and may raise some eyebrows – especially Apx. B, E, F & G.

Apx. A, C & D were repeated from earlier books so you have a balanced view of some key information needed for Apx. F...

Pre-souls are covered in Apx.s A & C and the Glossary.

Chapter 7: Protecting Yourself

The following are three areas in which you can protect yourself, not just from EMF, but in general benefit your physical and mental health as well.

Caution

These are **not prescribing** health remedies but are a collection of things that I did and some I still do to stay healthy. If you have questions about what you read, always check with your physician for guidance. (My email is on the Copyright page.) Some things are common sense, some are pointing out potential dangers, and others are informative that you can further research (esp. on Wikipedia or WebMD).

Physical Health

Tattoo Danger -- many of the tattoo parlors still have **color ink** that is made with heavy metals – cobalt for blue, mercury for red, etc. and you do not want heavy metals in your body.

You can ask if they have the newer inks (not made with heavy metals) but they may not tell you the truth because they want customers (and they have to make the older inks pay for themselves) – thus, it is best to avoid color tattoos altogether.

Heavy Metals used for colors include mercury (red); lead (yellow, green, white); cadmium (red, orange, yellow); nickel (black); zinc (yellow, white); chromium (green); cobalt (blue); aluminium (green, violet); titanium (white); copper (blue, green); iron (brown, red, black); and barium (white).
Metal oxides used include ferrocyanide and ferricyanide (yellow, red, green, blue). Organic chemicals used include azo-chemicals (orange, brown, yellow, green, violet) and naptha-derived chemicals (red). Carbon (soot or ash) is also used for black. Other elements used as pigments include antimony, arsenic, beryllium, calcium, lithium, selenium, and sulphur.
Tattoo ink manufacturers typically blend the heavy metal pigments and/or use lightening agents (such as lead or titanium) to reduce production costs. [163]

Note: above reference to mercury, lead and aluminum.

There is another danger in tattoo parlors... the artists are supposed to **autoclave** their needles, or <u>throw them out</u> after use (but that disposal is expensive and some choose the autoclaving route)... The best practice is to use a new needle, for the next customer, and that means to be safe, you have to ask to see their supply of new needles – If they show you none, walk out. Why?

Autoclaving is what they do if they have no new needles – and autoclaving (extreme heat) does not always kill **HIV or Hepatitis germs**. Are you willing to play Russian Roulette with your health and take a chance...?

> In the United States, tattoo inks are subject to regulation by the US Food & Drug Administration as cosmetics and color additives. This regulatory authority is, however, **not generally exercised.** The FDA and medical practitioners have noted that many ink pigments used in tattoos are **"industrial strength colors suitable for printers' ink or automobile paint"**.

> ...tattoo parlors in California must warn their patrons that tattoo inks contain **heavy metals known to cause cancer**, birth defects, and other reproductive harm.

> A recent case report also showed that tattoo **pigments migrate into lymph nodes.** These can show up on some types of medical scans as **tumors**. One woman was given a complete hysterectomy only to find out later that the lymph nodes contained tattoo pigment. [164]

> As usual, it is Caveat Emptor – "Buyer Beware."

Flu Shots – some still contain mercury (as a stabilizing agent) but there is a version that you can ask your doctor for that you **inhale** and it is mercury-free.

> The problem with some Flu shots is that they typically just have the 2 most common strains, and if a new virulent strain is going around, the shot may give you little or no protection.

> For the past ten years (and I am in my 70's) **I have not gotten a Flu shot, and I have not gotten the Flu.** I used to get Flu shots, every year, and almost every year I would get the Flu...

so I stopped. In my case the doctor suggested my system was not seeing the attenuated organisms, and reacted to the shot.

Bottled Water – what could be better than dinking clean bottled water, instead of whatever is in some municipal water supplies?

Remember, they cannot always remove the **hormones and antibiotics** when they recycle waste water, as these elements are too small for their filtering process, and the chlorine does not always kill the occasional small parasite. And on top of that, the pesticides and fertilizer runoff which rain drags into the the local water table, which also gets recycled, can still be present in very small amounts [typically 1-4 ppm] which usually does no harm – but it is there.

So we go to the corner store and pick up a 6-pack or 24-pack of bottled water – that has been **sitting in the Sun for hours**. These bottles are made of Type 1 plastic and often have a code on the bottom of the bottle to indicate that...

Left: the **Type 1** very soft and cheap-to-make plastic bottle code so commonly used which is not a problem if it doesn't sit in the Sun and leach **BPA** (Bisphenol A) into the water. BPA is a carcinogen.

Types 6 & 7 are the safest but are more rigid.

<u>Bisphenol A</u> **(BPA):**
BPA is a synthetic compound that serves as a raw material in the manufacturing of such plastics as polycarbonates and <u>epoxy resins</u>.
 It is commonly found in reusable drink containers, food storage containers, canned foods, children's toys and cash register receipts. BPA can seep into food or beverages from containers that are made with BPA. [165]

What you don't know is if the Type 1 bottle you are holding was made with BPA. Some aren't.

> Needless to say, I told the local store about the issue and they didn't care.. they had no place to store the 45 cases of water, so they still sit (facing West) in the Sun to this day.
> YOU have to know more to protect your health.

Remember: **Knowledge Protects**
Ignorance Endangers

Tap water – if your rural area (or small town) uses well water, it might be high in sodium (sodium bicarbonate, sodium sulfate....) and if your doctor has said to go on a **low-sodium regime**, you do not want to drink the tap water. Most municipalities in non-rural areas do not have sodium in the water (to any extent.)

> You can buy **water testing equipment**, [166] if you have a well, you already have it, or you can ask the local Water Department to show you what the sodium content of your tap water is. All municipal water departments have to regularly test their water to certify and meet Federal standards (Clean Water Act of 1972) ... ask them, if you have an issue with this.

Water Filters – some home owners add small water filters to their kitchen tap to filter the drinking water. These can be purchased at the local grocery store or Walmart and just snap on or use a small adapter to clip to the end of the faucet. Sounds good, eh? Not so fast...

The filters are so small that if you do not change them after 40-50 gallons (how are you going to measure that?) – or change them at least once a month – the build up of trapped particles will start to dump back into the drinking glass you are filling! Then you are getting worse water than what comes out without the filter.

> Same thing goes for those who use their **refrigerator water filter**. It has to be changed every 6 months (it is bigger and lasts longer) – if you remember to change it!

> Some people use a ZeroWater stand-alone filter and carafe which comes with a **PPM Meter** (see below) so you can see what the particulate count (TDS) is in your water.

Read-out window

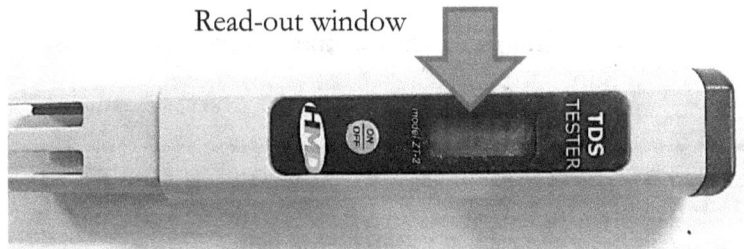

Total Dissolved Solids tester

Typically good city tap water is not over 250 parts per million or "**ppm**" of miscellaneous "dissolved solids"... well water can be 700-800 ppm. And those things are dissolved solids, chemicals, calcium, dust, things you do not want to know about -- (insect parts, small organic things that got into the water.)

Distilled Water – used largely for steam irons and when you want to avoid calcium build-up in some mechanical devices... IT IS OK to drink distilled water – but not on a regular basis as it tends to bond with minerals in your body and leech them out (like calcium, sodium, potassium, etc)– and you need those as **electrolytes** and for neural transmitters.

Water types and uses are discussed later....

Soft Drinks in Cans -- This may surprise a lot of people, who think it is safe because they sell it, and didn't think anything about it. Yours truly began to look at the issue with horror after getting heavily involved in pH testing of water, and then out of curiosity, testing soft drinks. Tap water is by law pH neutral at 7.0. Ocean water is a natural 8.2 pH. Blood is 7.365 pH. So the 7-8 pH <u>range</u> is pretty significant for good health.

pH 1 – 6.9   = acidic
pH 7.0   = neutral   (pH means "percentage of Hydrogen")
pH 8 – 11.0 = alkaline

Alkaline drinking water is typically 8.0 – 10 pH – above that you will get a headache, and may suffer alkalosis.

The body's blood is just above pH neutral so guess what the body prefers?

Guess what soft drinks are pH-wise?

They are usually 2.0 - 4.0 pH – **very acidic.**

Stomach acid is about 1.5 – 2.0 pH.

So when you drink a Coke, Pepsi, Dr Pepper, or 7-UP, Mountain Dew or a Barq's Root Beer....all about 2.0 pH (let's mention them all so we're not attacking any one in particular) you are dumping a very acidic drink into the already acidic stomach.... but before the body can let that pass on into the duodenum or the intestine (where it gets absorbed into the blood stream [which is 7.365 pH) it must be **alkalized**...

the pH must be <u>brought **up** in pH</u> or you will suffer acidemia or **acidosis. Wow... if not neutralized enough, you can get heartburn.**

Here is the danger.

The body must neutralize the acid drink and it does so by dumping **watery sodium bicarbonate** (<u>from the Pancreas</u>) into the stomach...before releasing the mixture to the rest of the body. The Pancreas is primarily concerned with digestion of acidic food, and when it does that, it shuts down Insulin production.

Your Pancreas also dumps Insulin to handle the sugar issue.

Too much sugar all the time (any source), and soft drinks every day will exhaust the Pancreas, and that is called **Diabetes**... types I and II.

And you can do a double-whammy by drinking a Diet soft drink – with **Aspertame**, also called **Nutrasweet** (aka **Equal**) which triggers the body's chemical-processing skill – a reaction to process the phenylketonurics – and some people do not have that ability via the amino acid phenylalanine and for them Aspertame can be a trip to the ER clinic. Aspertame also creates **Formaldehyde** in the body.

Remember President Reagan who drank a <u>**can** of **Diet Cola**</u> every day... he developed Alzheimers – and that was allegedly due to the very acidic syrup in the Cola **leaching the aluminum** out of the soft can, and the aluminum molecules went into the drink, and they can cross the protective Blood Brain Barrier into the brain....

Caution:
That may not be exactly why he suffered Alzheimers but is it worth the risk until they prove one way or the other whether it can happen?

Soft drinks in bottles are safer. They were originally in glass but glass costs more to make than cans, and <u>are hard to recycle</u>, so they switched to steel cans but they were also more expensive to recycle, so they switched to aluminum.

Sorry but our modern world has cut a lot of corners to reach economies of scale... sometimes at your expense.

White Is Pure

Modern marketing has convinced today's shoppers to buy white bread, white rice, white sugar .. if it is white it is pure ("and good for you"). The truth is they manipulated the whole wheat grain and cleverly divided the product three ways – instead of selling whole grain, whole wheat – the way nature made it – they wound up with separate white (pulp) bread, separate wheat germ, and separate bran... see picture below... that way they made three times the money --- three products instead of one.

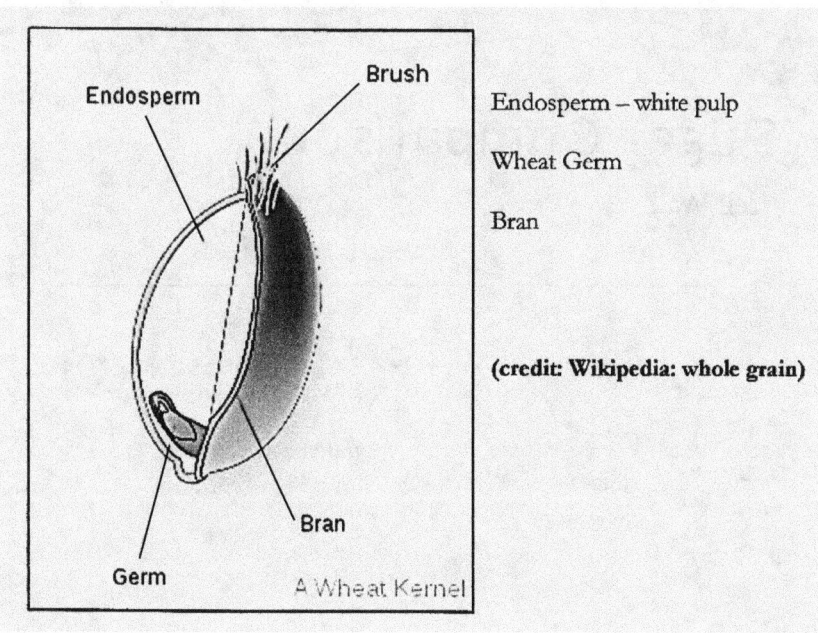

Endosperm – white pulp

Wheat Germ

Bran

(credit: Wikipedia: whole grain)

And white rice is less nutritious than brown rice.
And brown sugar (Turbinado) is just white sugar with a molasses coating.

Artificial Sweeteners

If it isn't regular sugar, honey, **Stevia** or **Agave** don't use it.

The problem with artificial sweeteners is that their **man-made molecules** are smaller than the molecules of the normal sugars like honey, cane sugar, Stevia or Agave which are Ok. The artificial sweeteners can get into the body's cells where they don't belong and wreak havoc with the mitochondria, causing cancer.

Excess sugar does not benefit the brain, either. (See next page diagram.)

Normally, regular sugar molecules move through the bloodstream and attach to the outside of cells and dissolve <u>through the cellwall</u> releasing only the components needed by the cell. (See below)

On the other hand, the **synthetic sugar molecules** are so small they can enter the cell via the openings designed for other nutrients. They disrupt the mitochondria and other organelles with chemicals that are not friendly to the cytosol.

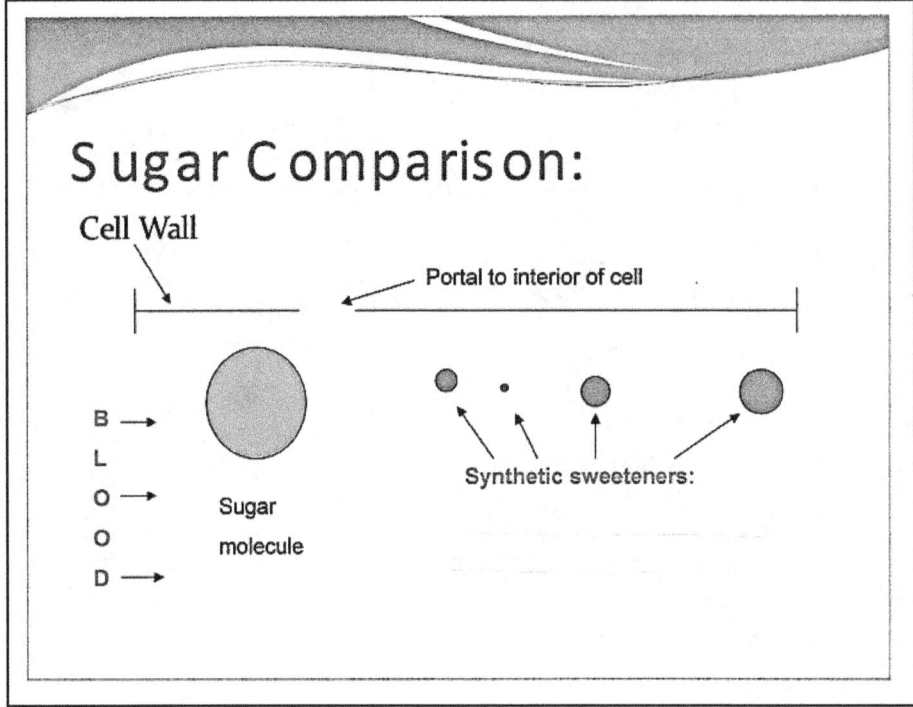

While we're at it, try to avoid drinks and sweets with **HFCS** (High Fructose Corn Syrup) – it metabolizes so fast it is one of the main **fat-builders**!

> **You should reduce all sugar intake so it doesn't feed Candida**
> **or**
> **exhaust the insulin-producing ability of the Pancreas.**

Cholesterol

Don't be too hasty to get your cholesterol down to 100 – the body and brain both need cholesterol to make other necessary molecules like enzymes and hormones, Testosterone and Estrogen, Cortisol, and the outer covering (myelin) of the neuronal brain connections for example… if you don't have enough cholesterol, you will suffer in other ways. But that's OK, they say, "We have pills you can take to compensate!" Seriously?

Cholesterol is **made in the liver** as <u>a response to the body's water-content</u>. The more water you drink, the more your cholesterol will normalize itself – whatever a 'normal' level is for you. If you don't drink enough water (and tea, Coke®, beer and wine don't count) your body will send **messaging molecules** to the liver to have more cholesterol made … Why? Because when you don't drink enough water each day, the body's defense mechanism kicks in and makes sure there is enough cholesterol to **line the arteries** so that whatever water is <u>already in</u> the cells doesn't leak out and further dehydrate you!

If you take **statins to kill liver cells** so that they cannot produce cholesterol when needed, you may be in trouble. That is why the doctor must monitor you so closely when you are on statins. You could also **drink more water** and see if your cholesterol levels (HDL, LDL and Triglycerides) reach a stable lower level faster…

There are two things to watch out for in this scenario: (1) have the doctor test and see if you have a defective cholesterol regulating system – that does need attention! (2) If your regulatory system is normal, your cholesterol count will not match someone else's anyway, and it (LDL) may not be under the magic number 100, BUT you don't want excess cholesterol to stick just anywhere and blood thinners (Coumarin or Warfarin) can meet that requirement. In fact, a friend of mine used **Lecithin** to make sure extra cholesterol didn't stick where it shouldn't be.

Just something to think about.

Alkaline & Ionized Water

There are water machines that sit on your home counter and use tapwater to create Alkaline and/or Ionized water – which is not froo-froo – the benefits of these machines are terrific. (BTW, "micro-clustering" in Ionization **is** the sales froo-froo because it is no longer micro-clustered when it gets into the bloodstream.)

Alkaline water is drinking water with a pH of 8 or greater.

> **Rewind**: A pH of 7 is neutral and is what tapwater is. Ocean water is about pH 8.2. A pH of 1-6 is **acidic** and is used for cleaning and disinfecting, and a pH of 8 - 9.5 is **alkaline** and is used for drinking.
>
> pH means 'percentage of Hydrogen' – or how many hydrogen ions (**H+** also called protons) are in the water… the more there are, the more acidic the water is (a lower pH number), and the less drinkable it is (to the body). The reason a high number of H+ ions yield a low pH number is because pH is the <u>negative</u>

logarithm (i.e., inverse) which reflects the activity or concentration of hydrogen ions.

So one of these water machines (Enagic, Jupiter, IonWays...) generates clean alkaline water at 8.5 pH and you drink it. What does it do?

Biochemistry lesson: when you drink water, the stomach will not let any liquid pass to the small intestine if it is not around 7 – 8 pH – acidic liquids are kept in the stomach until the Pancreas will dump some sodium bicarbonate to neutralize the acidity (bring up a Sprite from 2.5 pH to around 7 - 8 pH) before it will release it to the small intestine (just the other side of the stomach) – and that is where absorption happens. The blood must stay around **7.365 pH** and so the stomach knows not to dump a highly 'acidic' drink into the small intestine where it will be absorbed into the blood stream – as acidic! That is called **acidosis** and can be fatal.

So instead of taxing the Pancreas to dump sodium bicarb (pH 8.1) into the stomach to neutralize the acidic drink (pH 2.5), would it not be better to drink an alkaline water, say around 8.0 – even 9.5 pH which will be DOWN-graded to +/- 7.365 pH by the stomach's normal acidic (pH 1-2) condition? (The Pancreas does double-duty as you can see – it provides both sodium bicarb and Insulin to the body.)

Now the **"ionized" part**.

Water that is run through the water machine can produce acidic or alkaline water – and the alkaline water is ionized (has an extra electron)… and this is what your body likes. The following diagram shows water being electrolyzed to ionize it:

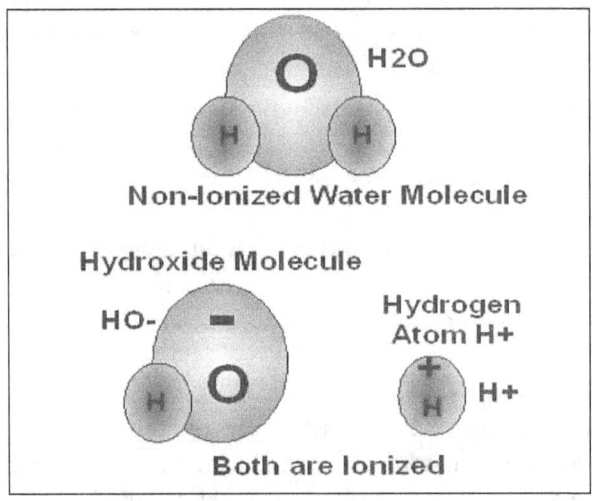

The HO- is the **antioxidant**. The H+ proton is the **Free Radical** (missing an electron: Hydroxyl Radical) – together they neutralized each other, apart they both seek to merge with another molecule:

Just FYI for the techies out there...

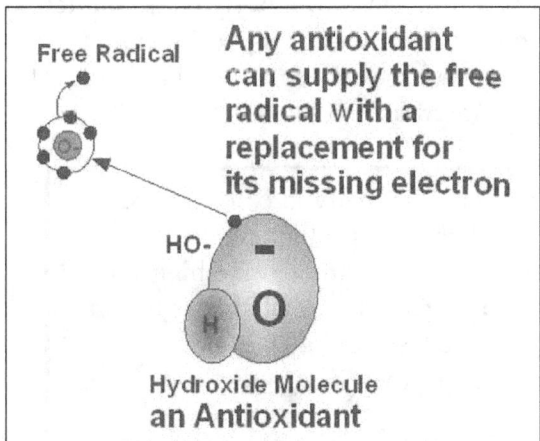

Together the **OH** (aka HO- in the diagram) and the **H+** make a water molecule: H_2O.

Clever, huh? Sounds like a design....

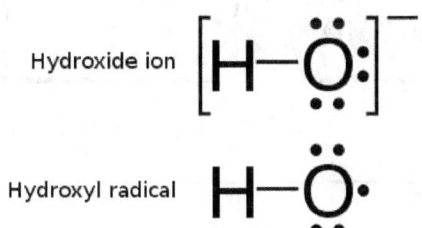

The **hydroxyl radical**, ·**OH**, is the neutral form of the **Hydroxide** ion (OH-). [167]

OH- (Hydroxide) is the good one.

Here is the secret of the ionized water:

The alkaline water is full of OH- (hydroxide) molecules which seek to donate their **extra electron** to another molecule that is missing an electron – which just happens to be the Free Radical (H+) molecule, also called a **proton**.

The H+ is the **proton** and has **lost its electron** and so has a positive charge... this makes it a **Free Radical** -- looking to steal an electron from another molecule somewhere (anywhere!) **from any cell** in your body.

That is how Free Radicals create damage, especially in the cells.

Dehydration

Adults do not drink enough water – the body water from adult to elderly should be fairly stable at around 70%... but the chart below reflects what health professionals have found to be the case as we get older. Dehydration is associated with getting older, aging and poor functioning of the body … and older people who drink more water, exercise and use vitamins, minerals and the right supplements are not only much healthier, they also look younger. (It isn't all genetics.)

> Note: "…**excess cholesterol formation in the body is associated with prolonged dehydration**" – because the body seeks to save whatever amount of water it has and uses the cholesterol to "lock it into the cells." [168] So, drink more water.

I'll bet your doctor never told you that... they are not trained in homeopathy, nutrition, phytochemicals, probiotics and dietetics. The older physicians (age 55+) know something about these areas, but today's crop of doctors know what tests to run and what pills to prescribe for the test results. (A good Naturopath [NMD] knows most of these things.)

Such info and counsel was given me by my MD doctor of 33 years who just retired (!) and I am still looking for one as smart as he was. He said "the new crop of MDs are **glorified pharmacists.**" I may have to go NMD.

Here is the problem and why clean water is so important to good health:

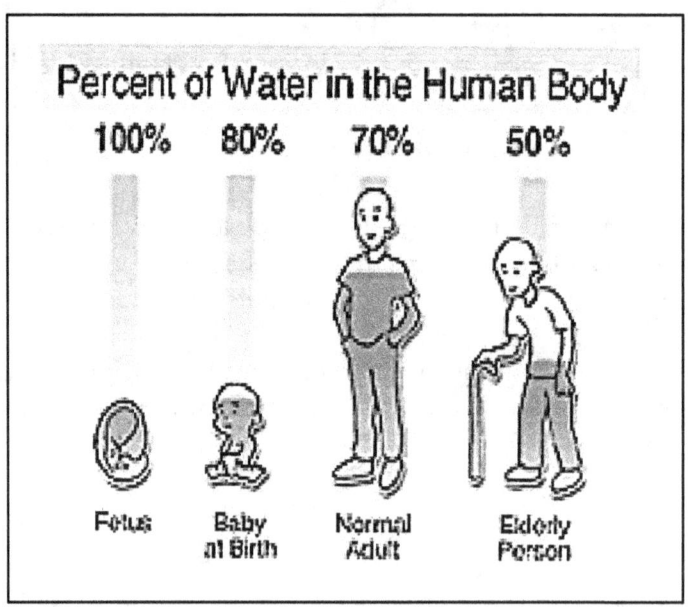

The Brain is 85% water.

Dehydration is a major source of **a lot** of health issues today (according to Dr. Batmanghelidj, MD): [169]

Cholesterol out of balance (pp. 204-07)
Plaque formation in arteries to conserve water in the cells (as described earlier) when dehydrated.

Arthritis – inflammation of the joints   (pp. 148-49)
If you do not drink enough water, toxins are not flushed from the body and if they circulate and lodge in the joints, they can cause discomfort, and then pain.

Migraine Headaches   (pp. 146-47)
When there is dehydration in the brain, the capillaries dilate asking for increased water intake, and histamine is released which is what causes the pain.

Constipation   (p. 118)
Bowels need water to keep things moving. We use Magnesium Citrate at bedtime.

Low Cortisol [170]
A safe supplement for this condition is Vitamin B5 which has been proven to be a stamina (anti-stress) vitamin, and it is water-soluable so you can't OD on it.

Leg Cramps   (p. 92, 203)
This also has to do with taking enough electrolytes in: salt, potassium, calcium and magnesium... which require the water to move them around so the cells in the muscles can use them.

Gout   (p. 210)
Again, an excess uric acid that a lack of water was not available to clear away... (2 gm **Vitamin B5** plus a glass of water every hour [7am – 6pm] will dissolve the crystals.)

Heartburn and GERD   (pp. 130, 136-37)
If you are dehydrated, the Pancreas cannot 'flush' enough sodium bicarbonate into the duodenum to neutralize the acidic contents, and if the stomach cannot empty its acidic contents into the intestines, (which would seriously damage the intestines) it may bubble back up the esophagus.

Diabetes   (pp. 110-16)
This one is more complex but is also due to a lack of enough water. The basic scenario is like this:

Diabetes is the end result of a lack of water **in the brain,** involving a complex interrelationship between salt, glucose, and **tryptophan**. The brain grabs the water and glucose first because it is critical, and if the body is dehydrated, you are in for a very negative ride (see pp 110-116).

Does it still make sense to continue to be dehydrated (not drink enough water) and continue to take pills to "band-aid" the problem – or is it better to understand what your body needs and drink more water?

> **Analogy**: would you run your car down the road with little or no water in the radiator? If not, why are you not giving your body the water it also needs?

Why?

Because without the water to **flush away toxins**, they accumulate and cause **inflammation** and that is one of the major issues in today's health care. As a general rule, the clearer your urine is (the closer to colorless), the less dehydrated you are. People who have very orange urine are (usually) very dehydrated. [171]

> Some of the medical establishment have labelled Dr. Batmanghelidj a quack because he is not pushing pills... he is in synch with European and Indian medicine: paying attention to the body, learning and doing what it needs in nutrition, exercise and water. BUT all you have to do is try what he says [about dehydration] and see if it works for you.
>
> And if you wait until you have the illness going full blast, water may not help – the time to listen and try is BEFORE you get arthritis, high cholesterol, or high blood pressure... Or you can sit back and take a pill and hope that works.

How do you know if you are dehydrated?

Take the pinch test:

Pinch Test

1. Using the back of your left hand,
2. Lightly pinch the skin with your right thumb and index finger…
3. Let go… and Observe what your skin does…

> a. if it snaps right back, you are hydrated, all is OK
> b. if it tends to stay "pinched up" more than 1 second,
>   or
>   it is **very slow to go back to normal,**
>   you are dehydrated and need to start drinking more water.

The recommendation is **6-8 glasses a day** – and ice tea and coffee do not count – they are not 100% water. A 200 lb person should drink 8-10 glasses of water a day... for optimum health.

The rule of thumb is: 1 oz of water for every 1 lb of weight.

Miscellaneous Items

Salt – essential for a healthy body, but not too much.

6-18 grams per day is OK, [172] unless your doctor says you need to be on a low-sodium regime. The American Heart Assn recommends a minimum of 3 grams (gm) per day. Note that cereals, soups and other food containers give their sodium content in milligrams (mg) and it takes **1000 mg to equal 1 gm**, so if you are young & healthy you could take in 5000-8000 mg of sodium a day ...

Salt is an essential element for the body – without it you die. **Too much salt** can cause hypertension (high blood pressure) and migraine headaches. It can cause abnormal fluid retention, dizziness and swelling of the legs.
Too little salt and acidity builds up in the cells, you lose water outside the cells (remember cholesterol ramps up to seal the water in the cells so it is not lost), and you may experience **muscle cramps** at night.

The rule-of-thumb is 3 gm of salt for every 10 glasses of water.

"A totally salt-free diet is utterly stupid. Salt is a most essential ingredient of the body. In order of importance, oxygen, **water, salt and potassium** rank as the primary elements for survival of the human body." [173]

Be sure that the table salt you use has **Iodide** – Sea Salt is great but you must also have Iodide in your daily multiple vitamin or you risk goiter (swelling of the neck). **Sea Salt does not have Iodide.**

If you **exercise** at the gym, or run, or do most exercises, you will sweat out water and **electrolytes (primarily sodium and potassium**) and these must be replaced. Some athletes use Gatorade®, some use Powerade®, but the best in terms of replacing the key electrolytes is **BodyArmor®**. Gatorade/Powerade is great on the football field where you want the glucose (energy) and sodium right now, but a workout at the gym needs less sugar and more electrolytes.

Why electrolytes? These are the bio-chemicals that move nutrients into the cell, and remove waste from the cells – to stay healthy.

> **Caffeine –** is often used as a great pick-me-up but is also a natural **diuretic** – forcing more water out of the body than is contained in the drink! [174]
>
> Caffeine also works on the brain cells and forces them to use some of their critical energy reserves.. thus the eventual effect on the brain is **energy depletion.** "Caffeine if taken repeatedly, eventually exhausts the brain... and is one of the primary causes of ADD." [175]
>
> So in the short term, it picks you up, but the inevitable effect is the let-down so many people experience some hours later – unless you keep drinking caffeine... and then you eventually risk dehydration.

> There is so much more from Dr. Batmanghelidj's book that it would fill several more chapters – the book is very highly recommended reading... he covers more than water, and most diseases. He explains in simple terms how the body really works, how we get sick and how we can get naturally well... and he includes much very important info on salt.

Area I: Body Health

This is the first big area in which improvements can be made. The preceding was a general collection of miscellaneous but important items.

> Much of the body health information was given in TOM, Ch's 4 and 13, as well as exercise, Qigong and Tai Chi and ways to build, store and manipulate energy (chi) in the body – something largely unknown in the Western World, but it will change your physical health forever when you learn how you can use energy modalities for natural healing. (Also see **PEMF**, later section.)
>
> The body has a Bionet which manages the **body's natural EMF**.

We all know about taking a decent **daily multiple vitamin** – especially as we get older – and that can include something as barebones as Centrum® or One-A-Day®... good, but not the best. As you get older, and as our farmlands are becoming "farmed out" (meaning the nutrients in the soil are being slowly depleted, especially **Chromium** [which controls the metabolism of sugar in the body]) we need to supplement with more than basic amounts or MDAs that the body can use and sustain optimum health.

> What that means is that many standard multi-vitamins have the barebones amounts of everything (vitamins and minerals) called the MDA... minimum daily requirement – below which death or illness can occur. If they put 6 mg of Vitamin B6 in the package, and that is the minimum your body needs to not get sick (and B6 is a very important vitamin for energy), shouldn't you consider something above 6 mg – so that you have a reserve and the body can get more if it needs more (especially in times of stress or exercise)?

> Overdosing? NO, most vitamins are **water-soluable** and flush whatever is not needed from the body --- Overdosing **is** a danger with vitamins A, D and E which are oil-based and do not dissolve nor flush with water. **Caution** is always advised, but an above-average multiple vitamin complex from the Health Food Store (brands such as Nature's Plus, Garden of Life, or NewChapter) offer high quality and cover all the nutrients required, and most often in the amounts really needed for stressful daily living.

So what will be offered here for your consideration are things that have been found to empower or enhance physical health.

Alkaline Water Machines
These were mentioned in the section above on Water; Some of them are made by Juipiter, Ionways and Enagic. They electrolyze the water and create pH+ or pH- And the body prefers alkaline water.

Tai Chi or Qigong or Yoga
A health exercise for centuries in the Orient, Qigong is a repetitive motion to build chi, whereas Tai Chi is a rhythmic flow of many of the same motions used in Qigong. Yoga is for flexibility but be careful of stressing the joints.

Bemer® Mat [176] (or Biomat ® in NEXUS)
One of several 6'x3' mats one lays on that have 3 large copper coils inside that run weak, pulsed electromagnetic (**PEMF**) field thru the body – and it works to reknit

bones as well as **stimulate blood flow** which promotes health as the blood more easily gets to all parts of the body. This is **healthy** PEMF (Pulsed EMF).
Also works & is certified for dogs and horses (see link in Endnotes).

A decent PEMF mat has non-artificial waveforms (sawtooth and triangular are best) and operates at .1 to 25Hz. The best allow for a high setting in the morning and a low setting at night (for better sleep). The major ones are iMRS and Bemer.

AVACEN 100® [177]

This is a **Class II FDA approved** medical device that is sold over the counter which operates a lot like the Biomats just described... except it is cheaper. You put your hand into the shoebox-size device for 10-15 minutes and it breaks up the clustered blood cells (a normal thing blood does as it gets tired) and facilitates super blood flow, especially thru capillaries ("microcirculation"). It infuses low heat (99°F) into the hand and thus the blood, causing "microvascular circulation to benefit & relieve arthritis and sore muscles." (Rewind: The FDA verifies this.)

Reflexology/Acupuncture

Also from the Orient, Reflexology is a "manual" version of Acupuncture often using the fingers or a wooden stick to stimulate pressure points on the hand or foot... and can be done with the help of a book (see Body Reflexology by Carter in Bibliography) which shows the areas to be stimulated. Acupuncture uses tiny hollow needles to do the same thing – but you have to know and use the Bionet of meridians in the body and that takes training. If done correctly, both produce results.

Gano® Coffee/Tea

Also from the Asian area, Gano coffee is made from the **Reishi mushroom** extract (*Ganoderma Lucidum*) and is very health-inducing – provided you do not have an allergy to mushrooms. It stimulates the immune system and helps the body resist colds and Flu (from personal experience). Gano coffee is so low in caffeine it is almost caffeine-free.

Probiotic Morning Drink

This is a morning drink that provides all 22 amino acids (from a protein powder (like: Garden of Life ® Sport Whey for shakes) and I add acidophilus (Blue Bunny ® Frozen Yogurt) + coffee + Organic milk with Omega3.

Miscellaneous Supplements:

> Oscillococcinum – take at first sign of a cold (from Boiron).
> **Echinacea** – take to aid immune system to **resist cold/Flu**.
> Lutein/Astaxanthin – key nutrients for healthy eyes.
> **Curcumin** (BCM-95) –**very powerful antioxidant** (moreso than

Vitamin C) but hard to absorb unless chelated with BCM-95 (see Curamed® brand).

Zyflamend – very powerful combo of herbs to stimulate and shore up the immune system. **It also fights inflammation**. (New Chapter® brand.)

Garlic -- age old remedy to fight off germs in the body.

Higher Mind -- a **complex of nutrients for the brain** to help form neurotransmitters and resist plaque-formation – [our family uses this on a regular basis, and in our late 70s we have no memory or mental issues]. (Source Naturals®)

However: caution is advised with this one: check with your doctor [as we did] as it contains herbs that might trigger an allergy.

Glutathione Recycler – a simpler alternative to Higher Mind; it recycles the brain's natural Glutathione antioxidant, promoting brain health [used now for 8 years personally]. (Apex Energetics®.)

Omega 3-6-9 blend – supplies needed oils to the body, Flax and DHEA – again **check for any allergy before using**.

Resveratrol – same substance found in red wine; very good for the heart (resists formation of LDL cholesterol) and keeps calcium in the bones. [178]

CoQ10 – heart and eye support. Stimulate immune system, lower blood pressure, and deficiencies are common with aging.[179]

Ca-Mg supplement – prevent **Dowager's Hump** in older women... make sure it is a 2:1 ratio: (Ca:Mg) and chelated so the body can absorb it.

Alpha Brain – botanicals that support memory & focus and suppress tinnitus [personal testimony] [180] (Onnit.Labs®)

As with all the above, be sure to run these ideas, and any changes to your vit/min/supplement regime, by your doctor – some may be contra-indicated depending on your unique body chemistry, or a Rx you are taking.

Of course, if you try to buy all of the above health items (not recommended) you may go broke. Choose the ones most likely to benefit your known health needs and check with your health professional (doctor or pharmacist) to see that it doesn't conflict with whatever Rx you are taking, or that it is not something you are allergic to.

Area II: Mental Health

This section must deal with brain development such that you can better understand the WiFi, cellphone and 5G EMF impact on the brain (in Chapter 5) – the brain is very delicate and is 85% water... so besides dehydration and aluminum, the brain is vulnerable to compromise.

In addition, as you'll see below, the **teenage brain** is not developed until about age 21... and it is especially nonsensical to try and reason with a 12-year old as the frontal part of the brain is only 50% developed. Humans take a long time (once outside the womb) to fully develop – so why are we teaching Algebra in 7th grade and Calculus in 9th?   (See TOM, Ch. 13.)

Development of the Brain

Note below that the brain has a "reptilian" component from the Anunnaki genetic creation of Man... that is where it came from, and it was Fight or Flight for early Man, and just bare bones (robotic behavior) to do whatever the Skygods said... no frontal lobe existed then (for higher thought processes).

The Middle and New (Outer Cortex) were added later...

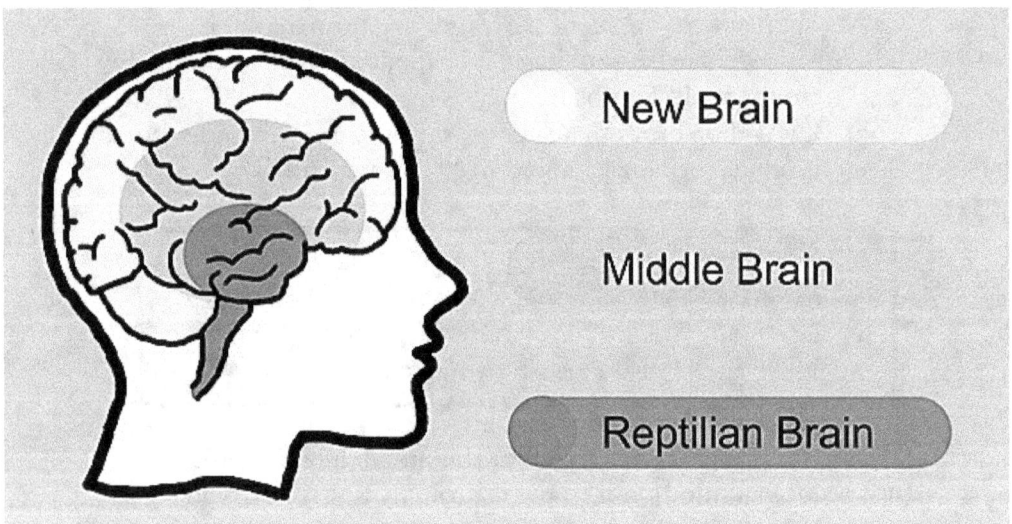

Teenagers...

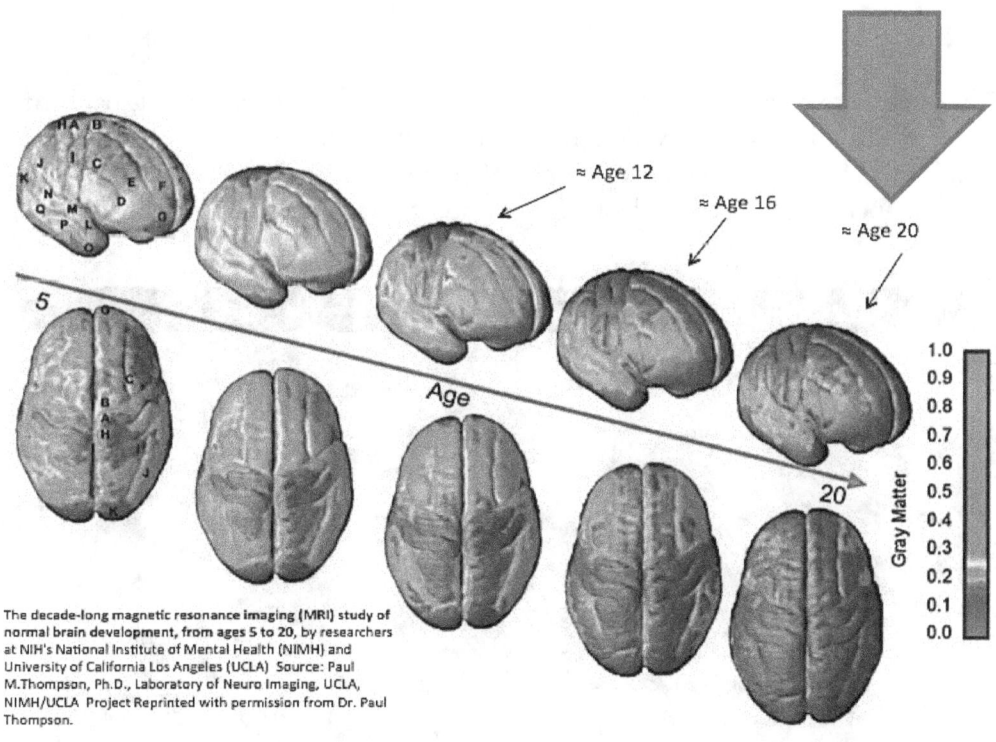

The decade-long magnetic resonance imaging (MRI) study of normal brain development, from ages 5 to 20, by researchers at NIH's National Institute of Mental Health (NIMH) and University of California Los Angeles (UCLA) Source: Paul M.Thompson, Ph.D., Laboratory of Neuro Imaging, UCLA, NIMH/UCLA Project Reprinted with permission from Dr. Paul Thompson.

Grey Matter increases with age (darker shading) and reflects the brain's ability for higher thought processes (by age 16-20).

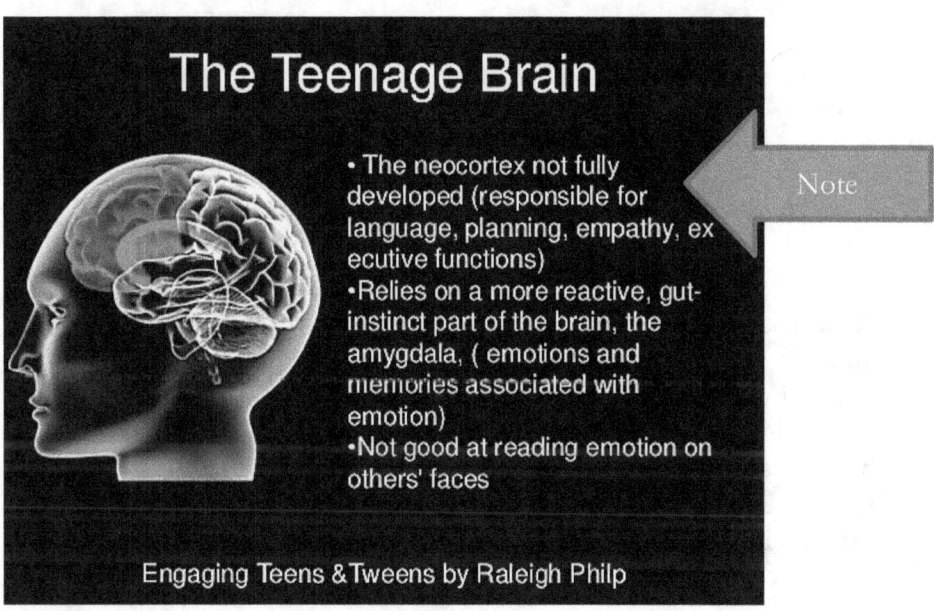

Note the teenage brain is "not fully developed" (1ˢᵗ sentence above) and explains why they sometimes do nutty (irrational) things. The forebrain is finished developing by **Age 21.**

The point here is that even the teenage brain is not developed and not capable of very rational thinking (ever watch them drink or drive to impress their friends?) The "executive" functions mentioned above (1ˢᵗ point, by arrow) are logical, rational thinking including inductive/deductive reasoning.

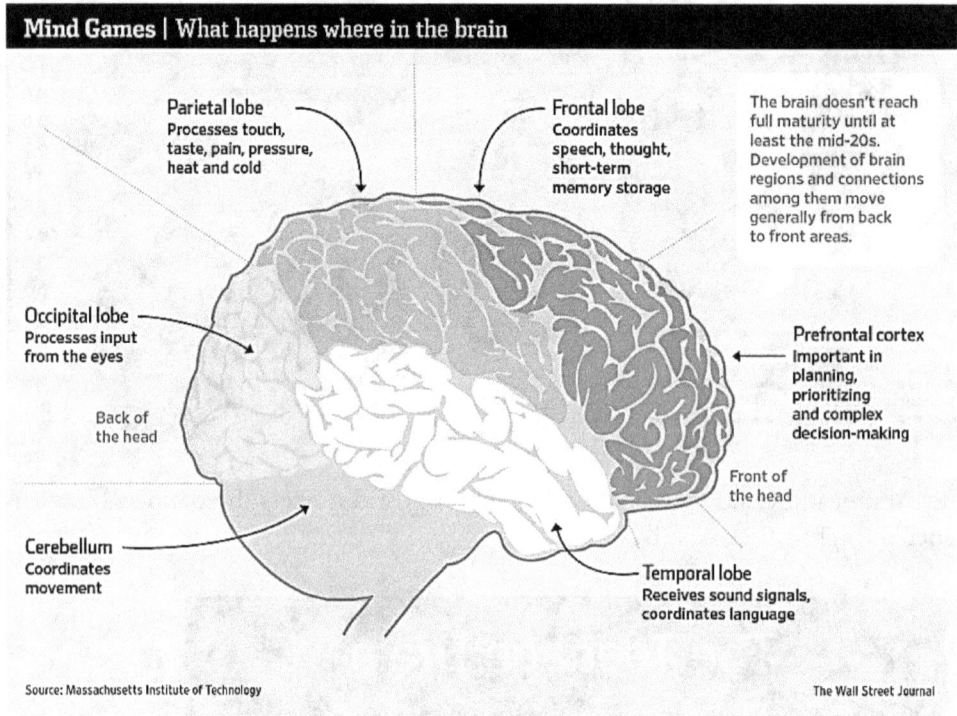

Mind Games | What happens where in the brain

Parietal lobe
Processes touch, taste, pain, pressure, heat and cold

Frontal lobe
Coordinates speech, thought, short-term memory storage

The brain doesn't reach full maturity until at least the mid-20s. Development of brain regions and connections among them move generally from back to front areas.

Occipital lobe
Processes input from the eyes

Back of the head

Prefrontal cortex
Important in planning, prioritizing and complex decision-making

Front of the head

Cerebellum
Coordinates movement

Temporal lobe
Receives sound signals, coordinates language

Source: Massachusetts Institute of Technology

The Wall Street Journal

The far right, Prefrontal Cortex, is the last to develop.

What is so special about the human is that while most animals are born and literally hit the ground running (dogs, horses, elephants, etc...) Man takes much longer to develop as he is much more complex ... if the woman had to carry the baby until it fully developed it would have to exit the mother's body like the creatures do in the movie *Alien*...

So when thigs go wrong, and aluminum hits the brain, or meat infected with **prions** (aka **Chronic Wasting Disease**, or Mad Cow Disease) gets past the Blood Brain Barrier, parts of the brain die and it looks like this (right)....

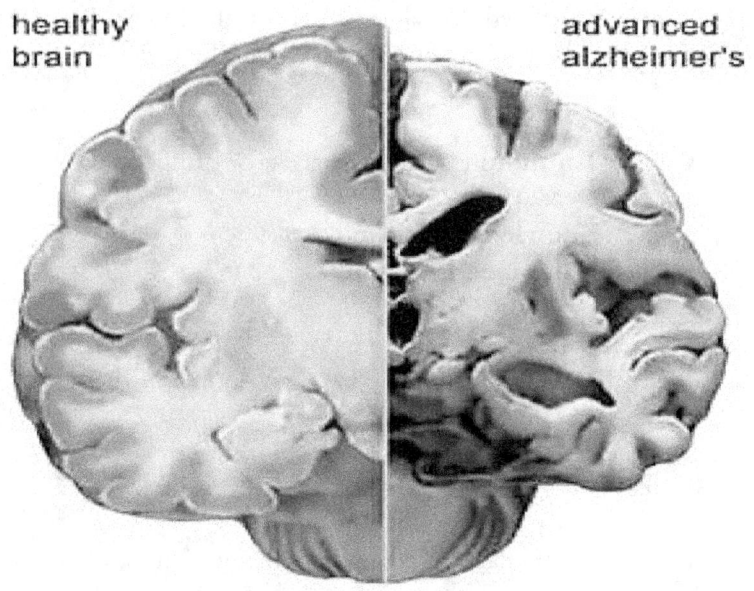

healthy brain | advanced alzheimer's

And the result of **prion activity** (mis-folded proteins) on the left below looks similar...

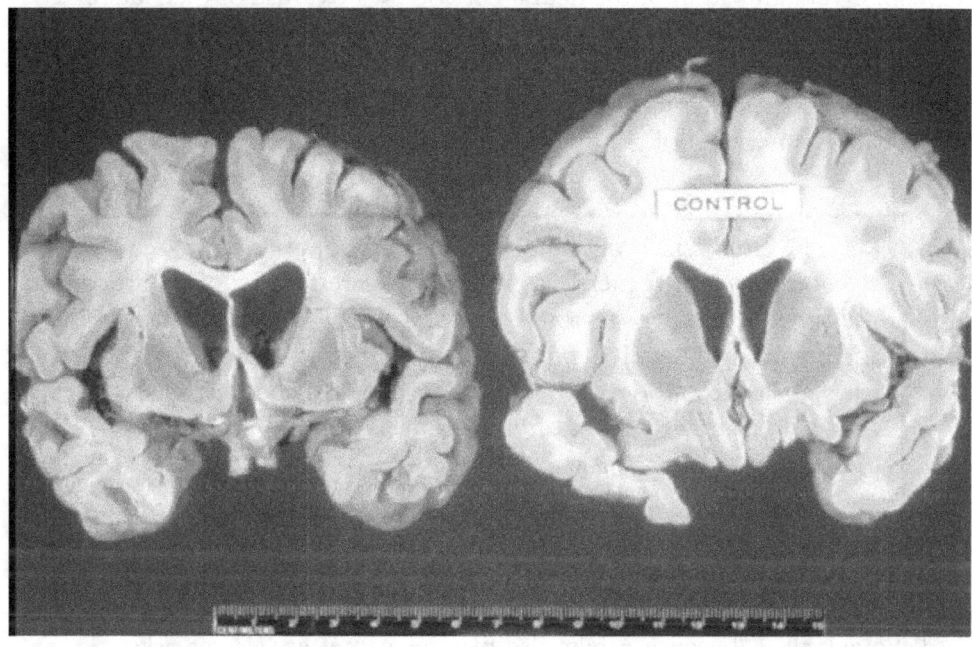

CONTROL

The preceding were just to acquaint you with the fact that (1) the brain is not fully developed with kids still in school, <u>even High School</u>, and (2) if diet and stress affect the brain via chemicals (Aluminum from aluminum cookware or cans of Cola, and

297

lead, arsenic and mercury to name a few [think: tattoos]), drugs (recreational and Rx), and if the diet includes tainted deer and cow meat, the brain will suffer. I emphasize this as we have to be aware and know where the potential dangers may lurk...

Caveat Emptor in whatever we eat or drink.

Knowledge Protects
Ignorance Endangers

In addition, there are also a few things that we do that are really harmful but you probably think that "If they sell it, it must be Ok." A couple of them will surprise you but they have been clinically proven to be true...

Violent Video Games

Since this is a chapter on brain health in a book about the mind and how what we think and mentally rehearse does affect us, a word should be said about **violent video games**. To the degree that they overstimulate the brain and promote stress, that is a cause of **Excitotoxicity** (inflammation) and another source of damage to the brain. Overstimulation of the neural network, via overuse of the glutamate neurotransmitter, can result in an excess of the CA^{2+} ions which damage cells and **kill mitochondria** (that we examined in TSiM Ch. 6) [181] The CA ions cause the formation of superoxide production and apoptosis (death) of the cells/neurons. Violent video games are not good for your child's mental health... as follows...

> Let it be emphasized right up front that the author is not categorically against them. However, there are children with anti-social tendencies, even with autism, and so-called Crack Babies, and children who are taking doctor-prescribed medicine to stay with a balanced mood, who should NOT be playing the Stalk-It-And-Kill-It stressful types of video games (referred to herein as the 'violent video games').

While there is a lot of literature that says it is really OK and kids know the difference between reality and fantasy, the words of **Dr. Caroline Leaf** should be heard above it all ... because Adults also play the games are also at risk:

> ...when an individual pays attention to a stimulus, the neurons in the cerebral cortex that represent this object show increased activity.

And

> ...the way you focus your attention [has] a direct effect on how your proteins are synthesized, how your enzymes react, and how your neurochemicals act together.

And

> ...the things going on in the environment get into the mind, changing the brain and having an impact on the body.[182] (Epigenetics)

> Think: our soldiers and **PTSD effects** of being in actual combat.

Intense focus on the screen (BlueLight danger), playing the game, trying to stay alive and **kill the other player(s)** (which is **toxic thinking** – discussed below), puts the body under **stress.** It also produces a focus to KILL, or at least cause harm to another person – even if you don't know that person. It doesn't take much imagination to see that playing such a game repeatedly is going to **desensitize** the player to shooting and causing harm with a weapon (usually a gun). It is a **response action** to being threatened (in the game), and the more instinctual, automatic the self-defense response becomes, the more successful the player becomes, AND the more ingrained becomes the behavior...

What we do and think, DOES create neural pathways.

Negative Thinking

According to what Dr. Leaf has shared (in detail in TSiM Ch.6), this amounts to rehearsing an aggressive part of one's psyche, and **the neural network will respond and build neural patterns consistent with that negative behavior** practiced (over weeks) while playing the violent video game.

To repeat:
> ...negative thinking [including killing and causing harm to others] creates **atypical responses in the brain,** which will result in atypical manifestations [disease and/or asocial behavior]. Such as **malformed prions...** [which is Creutzfeld-Jacobs human wasting disease]

And

> ...lack of quality in our thought lives is the **complete opposite of how the brain is designed to function** and causes a level of **brain damage.** [183]

Playing violent video games is not a positive experience... unless you are the last one left standing, having killed everyone else, and you are the (questionable) Winner!

And what have you won?

According to Dr. Leaf, you now have a reinforced neural pathway of focused aggression and an anti-social way of solving problems (especially after 21+ days of game-playing).

Think not?

> Adam Lanza, the **Sandy Hook** shooter, whipped himself into a frenzy playing a violent video game in the months before his murderous rampage.[184]

And

> It was the 1999 **Columbine High School** shooting that got many Americans thinking about violent video games. After the attacks, victims' families **sued** more than two dozen game makers, saying titles such as Doom, a first-person shooter that the two teen gunmen played, **desensitized them to violence.**

> [Are you ready for this?...]

> A judge dismissed the lawsuits…[185]

Would you like to know what the writer of the above-quoted magazine article concluded, based on his use of special statistical data?

> …as violent video games proliferated in recent years, the number of violent youthful offenders fell – by more than half between 1994 and 2010. [186]

Gee, that obviously makes it Ok! – the kids (allegedly) used the games to get violence out of their system. Wonder how they will respond in the future when bullied…. or frustrated because they don't get their way? Let's see what they do when cut-off on the highway… they have been rehearsing violence as a response to frustration. How many of these kids are now in ANTIFA?

Equal time that supports Dr. Leaf comes from psychologist **Douglas A. Gentile** of Iowa State University:

> …whatever we **practice repeatedly** affects the brain. If we practice aggressive ways of thinking, feeling and reacting… then we will get better at those….Gentile and his father, psychologist J. Ronald Gentile, found that children and adolescents who played more violent games were likelier to report "aggressive cognitions and behaviors." They concluded **that violent video games "appear to be exemplary teachers of aggression."**

They also found that eighth and ninth graders who played violent video games more frequently displayed greater "hostile attribution bias" (being vigilant for enemies) and got into more arguments with teachers. [187] [emphasis added]

Has anyone done a deeper study to correlate discipline problems in school with the rise in violent video game playing? Or would that not be "politically correct?"

Negative Thinking Creates Prions

Remember Dr. Leaf's warning about **prions** that cause Mad Cow Disease (and its human version Creutzfeld-Jacobs Disease [Chronic Wasting Disease]):

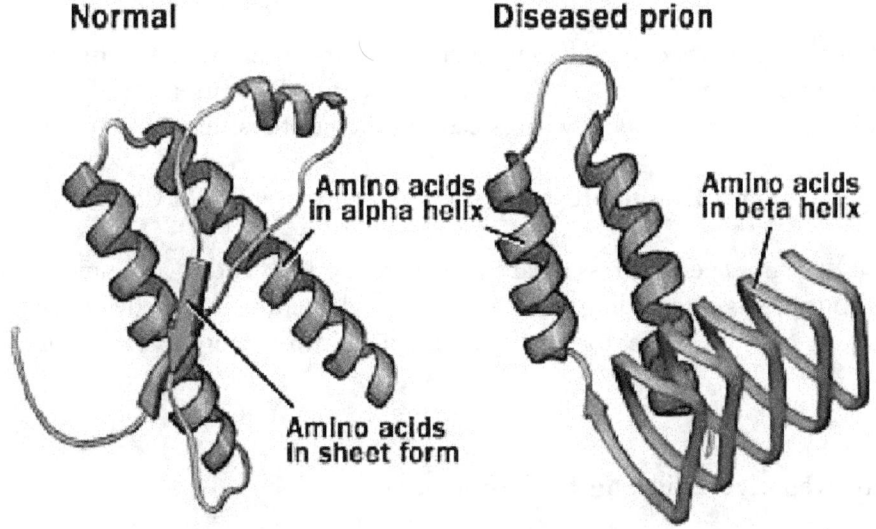

Normal   **Diseased prion**

Amino acids in alpha helix

Amino acids in beta helix

Amino acids in sheet form

…this protein [prion] does amazing things in the brain in response to good signals and **goes crazy in response to negative signals**. A chaotic mind filled with… rogue thoughts of anxiety, worry, and any and all manner of fear-related emotions [including fear of being 'killed' in the video game?] sends out the wrong signal. [188]

There is no way that playing a violent video game is relaxing, peaceful and just good fun. Exciting, yes, but so is skydiving with a parachute that may or may not open. And the video game player is unwittingly 'programming' himself for future problems – if not psychological or behavioral, then physical brain health issues. There may be no parachute to keep him from falling into trouble.

301

Parents are just happy to let their kids play any game that keeps them corralled to the PC so that they know where the kids are, and they don't have to worry about whether the kids are getting into trouble outside, on the street. PC games are great baby-sitters…. Or so it would seem.

Heavy Metal Music

This is another area where the music is often violent, loud and discordant… and if we have learned anything from **the effects of different types of music on plants**, it would be wise to avoid harsh, grating violent music.

What they found was that soft, melodic music and even intricate classical music caused the plants that were exposed to the music (for several weeks) to grow stronger and healthier. Plants that were exposed to loud, harsh rock (aka 'negative') music were stunted and some even died…

All a matter of vibration.

The rationale for this is due to the fact that the plants have a **water-content** and so does the human body (70%), and the effect of 'negative' vibrations on the water in our bodies, and in the plants, is what is causing the health issues… even in humans.

Dr. Masaru Emoto proved all this years ago – see TOM Ch.4.

And that does not even address what the discordant, loud and cacophonic sound is doing to our brains… according to Dr Leaf above. Listening to negative music (and the lyrics are often negative, too) is **the same as thinking negative** – see above section.

Rewind: Negative Thinking & Neuroplasticity

> What you think with your mind changes your brain and body, and
> you are designed with the power to switch on your brain. Your mind
> is that switch.[189]

She further maintains that if you change the programming of your brain, you will also change your body chemistry. **Negative, depressed thoughts suppress the immune system,** and release negative chemicals into the bloodstream which further depress the person, and can actually cause health problems.
That is good news: it means that **attitude can modify DNA**. And a later section of this chapter presents evidence (Dr. Kozyrev) from foreign research to that exact effect.

This state of mind is a real, physical, electromagnetic, **quantum** and chemical flow in the brain that switches groups of genes on or off in a positive or negative direction based on your choices and subsequent reactions. Scientifically this is called **Epigenetics**.... The brain responds to your mind by sending these neurological signals throughout the body, which means that **your thoughts and emotions are transformed into physiological and spiritual effects**, and then physiological experiences transform into mental and emotional states.[190] [emphasis added]

And Dr. Leaf has the hard evidence to prove it and often demonstrates it (pix) during her public presentations.

Dr. Caroline Myss

This is another outstanding woman in the field of Spirituality, soul-development and health studies made real. She has written several eye-opening books on how our '**biography becomes our biology**' (which has been appropriated by other writers), yet she was the first to use her ability as a **Medical Intuitive** to actually 'see' how peoples' fears, thoughts and past memories (biography) could still influence their bodies (biology) in a negative way. [191]

Thinking Activates Genes

In a further exploration of Dr. Leaf's findings, and those of Dr. Myss as well, we will learn (later section) that **our consciousness activates our genes**. Science has elsewhere proven that <u>thoughts affect the body</u> and cause specific biochemical messengers to be released into the bloodstream and into the brain.... Dopamine, serotonin and adrenaline are common examples.

What has not been commonly known is what the effect is of **negative thinking** on the body and brain. **Negative thoughts have been found (scientifically) to not be the norm.** [192]   **And they create and release toxins**.

Science has shown that our thoughts, with their embedded feelings turn sets of genes on and off in complex relationships....We may have a fixed set of genes in our chromosomes, but which of those genes are active and *how* they are active has a great deal to do with how we think… [193]   (See also Epigenetics back in Chapter 3.)

Taking all this to a deeper level, **our DNA actually changes shape according to our thoughts. Toxic thinking will change your brain wiring in a negative direction** and put your mind and body in a state of stress. [194]

The **Heartmath Institute** did experiments on DNA with people who projected fear, anger, negativity and changed the shape of the DNA (in Petri dishes) – the DNA responded to the negative input by tightening up and becoming shorter, switching off several DNA codes. The negative shutdown was reversed by projecting positive thoughts of Love, joy and gratitude. [195]   Reminiscent of the same effects on water by Dr. Emoto. **Thoughts are things**.

Positive thoughts result in a positive neurochemical rush in the body/brain complex. The body is constantly creating proteins to feed cells in the body and rebuild tissues, and our thinking affects how they are shaped. Regarding the **Epigenetic effect** of thinking, it has been discovered that **proteins formed in the body during toxic thinking are misshapen** (hint: *prions*) and act differently than normal, proactive proteins. [196]   (See earlier section 'Violent Video Games.')

> The physical result of thinking can thus take its toll on the body.
> Doesn't this give a new meaning to the ancient teaching:
> **"The sins of the fathers are visited on the children"**
> Whatever damage the parents have created in their bodies can
> genetically be passed on to the next 4 generations.

And yet, Dr. Leaf would quickly point out that although you may have inherited a physical deficit, genetically, you are not condemned to live with it… and Drs. Holmes, Maxwell Maltz, and Myss would wholly agree (TSiM, Ch. 1-2)

You can change your thinking and change your life.

Your thinking affects your DNA. You can rise above any tendency to alcoholism or diabetes, for example.

> The day will come when Man can reverse color-blindness and
> sickle-cell anemia, which are also inherited genetic defects. If we
> knew more about the Russian genetic research (TSiM, Ch. 7),
> perhaps we could now fix blindness, color-blindness and sickle-
> cell anemia – just by the power of the word! After all, ascended
> masters have done it, and Jesus healed people, so it <u>can</u> be done…
> (Or by beaming perfect, healthy DNA to a body's sick area…?)

The sins of the parents are thus a <u>predisposition</u>, not a destiny cast in concrete.

Multitasking Thinking

A key point in Dr. Leaf's findings was that this protein prion does amazing things in response to good signals, and **goes crazy when it receives negative signals**. Toxic (negative) thinking sends negative signals thru the Synapses and negative prions can result. Hence,

negative thinking does affect the body/mind complex. [197]

Multitasking is a persistent myth. **The human brain was NOT designed to multitask** and persisting in it can cause brain damage (due to unnecessary stress and its biochemical effects). Says Dr. Leaf:

> Our brain responds with healthy patterns, circuits and neuro-chemicals when we think <u>deeply</u>, but not when we skim only the surface of multiple pieces of information.

This is **"milkshake" multitasking** (her term) in that it creates a jumbled mixture of biochemical neurotransmitters and promotes a "hurry up" mentality that is similar to toxic thinking in its stressful effects on the body/mind. [198]

Even worse is the potential for multitasking while on social media that can be addictive and thus more harmful in the long range than occasional multitasking.

Russian Research

By the same token, **positive thinking** benefits the body and mind, and releases 'good' biochemicals into the bloodstream. The brain and psyche were designed, according to Dr. Leaf, to think positively. And the Russians also discovered something interesting about thinking...

Dr. Nikolai Kozyrev believed that an intelligent energy (God) had to exist and in the process became one of the most controversial figures in the Russian scientific community. Even reluctant Western scientists have had to acknowledge that there must be an unseen energy medium in the universe, and as long as you don't use the word "Æther" you can publish your findings on a "quantum medium" without much hassle. Æther is now called Dark Energy.

Kozyrev suspected that all life forms might be drawing off an unseen **spiraling energy** through eating, breathing, and chakra-absorption. In fact he suspected it was feeding the body via the Bionet and the DNA was absorbing *chi* and emitting photons... something that two other Russian researchers, **Drs. Peter Garaiev and Vladimir Poponin**, discovered in their work with DNA and **morphic fields**.

What Kozyrev discovered still has scientists talking:

> In the laboratory, he took a sealed glass container, removed all the air, making it a complete vacuum. He then cooled this vacuum space to -273° C, where all matter should stop vibrating and produce no heat. Instead of an absence of energy in a vacuum, there is a tremendous amount of it, and since the temperature was absolute zero – it was called **Zero Point Energy.**[199]

> What is so amazing about the ZPE and its natural matrix is that it is a boundless source of energy – a 3" cubic space of it contains more than enough energy to bring all the world's oceans to the boiling point. [200]

> Kozyrev's big discovery was that **the energy was spiraling, twisting** such that he called it Torsion Physics. He made the connection with Rupert Sheldrake's *Morphogenetics* and suggested that **DNA was a spiral because it is sustained by the torsion aspects of the ZPE Field**. He made a corollary discovery: [201]

> ...Kozyrev showed that by **shaking, spinning**, heating, cooling, vibrating or breaking physical objects, their weight can be increased or decreased by subtle but definite amounts.[202]

They do not gain mass, they gain energy. Energy from the ZPE Field. Is this what the **Sufi Whirling Dervishes** and Shakers were doing to enhance their religious experience? Significantly, Sufis balanced on their left foot, spinning **counter-clockwise**... which is the direction of the toroidal energy... and the direction of the Earth's spin on its axis. How did they know?

Obviously spinning affects the BioField (torus) around the body... generated by the Heart which has a stronger (natural) EMF than does the Brain... in fact the heart is also involved, as it has its own intelligence and in fact has **glial cells as does the brain** as does the solar plexus – see the book: **Three Brains** by **Dr. Karen Jensen**. [203] This was also explored in TSF, Ch. 11:

The subconscious part of the **Head Brain** is also responsible for the body's autonomic functions – heart rate and biochemistry (metabolism) balance.... based on what is eaten (Gut Brain mediates that), and the **Heart Brain** helps regulate the neurotransmitters and hormone activity. The Heart Brain senses the world thru emotion and feeling, the Head Brain analyzes information and applies logic. The **Gut Brain** is also connected with the Solar Plexus which is involved with self-identity and Ego.

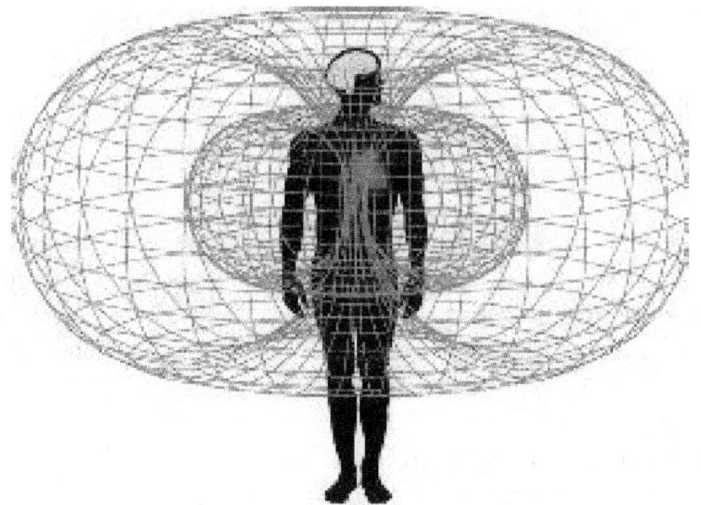

In a related experiment by **Dr. Frank Brown**, it was found that the biofield that Sheldrake referred to (**morphogenetics**) is sensitive to **rotation** (as Kozyrev discovered above): [204]

> That the biofield is involved [in living things] is supported by Brown's observation of a connection between rotation and bean seed interaction. He found that the beans interacted more strongly when they were rotated **counter-clockwise** than when they were rotated clockwise.

This would lend support to the Sufis twirling in a counter-clockwise direction to empower their (ecstatic?) devotional dance, the Dhikr.

> *FYI: It has been discovered that if a human spins/twirls **clockwise** (in the northern hemisphere) it can make him sick.*

The second major Russian scientist (examined in VEG, Ch. 9) was **Peter Garaiev** who found DNA coding could be transmitted through the space between two testtubes, where the 1st testtube contained DNA and the 2nd contained pure water. Upon inspection, after using coherent light shined through testtube #1 into the 2nd testtube, to impart coding, the 2nd testtube contained an imprint of the DNA from the 1st testtube. The point here is that the **transmission medium** had to be the ZPE Field, or what is sometimes called the Energy Matrix (aka **Quantum Net**).

Can this be used for healing?

Yes, unbelievable... and the Qigong Masters do it all the time, as well as the Pranic Healers of India.

> Kozyrev showed that **torsion fields are created from spinning sources** and the Earth spinning on its axis and revolving about the Sun produces a **dynamic torsion**. This permits torsion waves to propagate throughout space (via the ZPE Field, or Dark Matter, or the Matrix) such that torsion fields like gravity or electromagnetism can move from one place to another in the universe. [205]

Furthermore, Kozyrev proved that these torsion fields travel at **super-luminal speeds** – faster than the speed of light – and that explains the related phenomena of Nonlocality and **Entanglement**. Note that Korzyrev's experiments have been successfully replicated by others since the 1970s. Not too surprising: Western science is still hesitant to accept Russian findings...

> Kozyrev would say that "... **torsion waves and consciousness** are essentially identical manifestations of intelligent energy." [206]

> Thus, it was found that **the mind's thoughts are torsion waves**.

He further said that one mind can barely influence objects, plants and other people and yet it must be using the Matrix (Quantum Net) that surrounds and enfolds us all. Group mind is much stronger. And yet, **we can defeat the assimilation of the ZPE energy** by our bodies and minds if we do one of **two things which block** the Earth Field's torsion waves:

5G Protection?

> **Wear polyester clothing and live in an aluminum trailer.**

> **You have to remember that as a possible protection against 5G millimeter waves... short of wearing an aluminum hat!**

> **More Protection is given at the end of Chapter 7 and in Appendix G.**

Lastly, according to Russian physicist, **G.I. Shipov**, torsion fields transmit information without transmitting EM energy, and colleague **A. Akimov**, said that "torsion fields coupled with the standard electric, magnetic and gravity fields should offer a **unified field theory**... and include the effects of consciousness." [207] Thus this one discovery of the ZPE Field which contains torsion fields, coupled with the human mind which acts as "a non-magnetic spin torsion system" which transmits and receives torsion field information, promises to revolutionize our understanding of the human experience...

if 5G doesn't suppress the Quantum Net.

Thoughts are torsion waves and can be blocked/modified by EMF.

The preceding discussion again is one of the reasons that 5-6G is not going to be in any human's best interest. And you still need to know about a few things in that arena to stay away from...

Area III: Technology Health

VR Headset – just because they sell something doesn't mean it is Ok, healthy or not harmful. Case in point is the VR Headset, made by several manufacturers......

You need to realize that this is **a cellphone strapped to your face.**

And it is focusing the EMF from the cellphone screen onto your **retina**...bathing the retinal cells in EMF... breaking DNA and causing some cells to die because they are not used to the intensity.

You already know not to hold your cellphone closer than a foot to your face .... notwithstanding the French discovery, reported earlier,. that the **blue light** from the screen is eventually harmful to the eyes and mitochondria (Chapter 5.)

The obvious problem of not being aware of your surroundings and losing your balance and falling over could also result in harm.

The VR Headset is about as *non sequitur* as the cellphone ear headset...

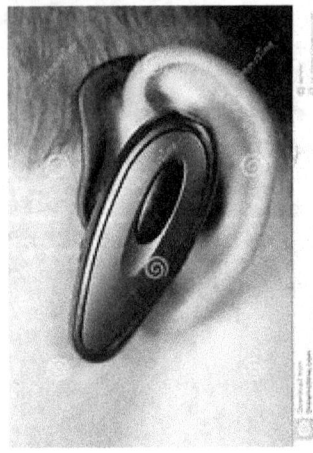

Is the "convenience" worth the potential tumor (est. it takes 2 years to develop)? The EMF radiation is going thru the skull (adult or child) as was shown earlier, and irradiates the delicate brain cells, to say nothing of the interference with the normal (and weak) electrical signals firing between neurons (in synapses) – modifying the neurotransmitters...

This is like trying to use a cellphone next to a powerful electrical transformer... you will get signal interference.

WiFi -- be aware that the cellphone **clipped to the ear** is using WiFi which was earlier shown be EMF radiation that you do not want to bombard your brain with 24/7. This does not include cabled headsets – unless they too are WiFi (i.e., **wireless headphones** – often found in offices where people are on the phone a lot). Convenient but not good.

This also includes wireless PC keyboards and **wireless mice**... to be avoided – and by the way, when the battery runs out in the mouse, it may take a minute to realize that it is not transmitting to the PC.

For the reasons given earlier in Chapter 6, **children** should not be exposed to WiFi in school all day...a one hour time limit in a computer lab is sufficient.

A **cordless phone** has its own "cellphone tower" that broadcasts for a 100' radius – even thru the walls of your house so you can talk in the garden... Is that really necessary? Do you want a cellphone **mini-tower** in your house?

How do you know how much WiFi (EMF radiation) you're getting at any point in your home or office? Buy an **EMF Trimeter** (below) and test your surroundings

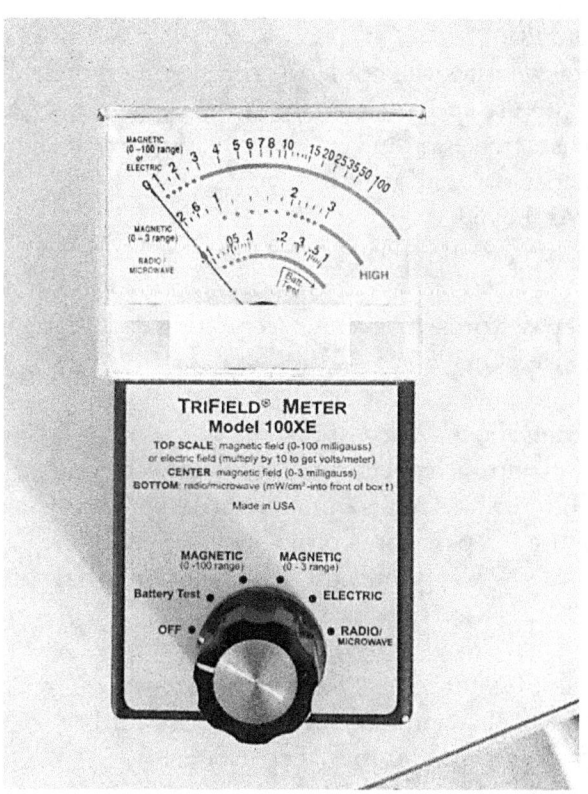

The 2 scales that say "Magnetic" will show how strong the WiFi is... like the radiation from your kitchen microwave, for example.

It also detects radio waves and standard electrical fields.

It uses a 9v battery.

It is available on Amazon and costs about $150.

The 1-3 mG range (top scale) is safe and a lot of appliances and TVs read that (from 3' away).

Trimeter

What may surprise you is the EMF reading from the meter that you get while just walking around in the park, at Starbucks or Whole Foods, and even in a volleyball court under the high tension power lines – like in our ill-advised town which wanted to change the ugly erector-set towers and replace them with a more 'designer' single-pole design – which also **lowered the power lines 20'** so that kids now playing in the park (below the power lines) are getting about 12-15 Gauss on the above meter. (No one at the city cares, by the way.)   Civic beauty not health was the goal.

> I went to the Town Council and asked about the issue, during an open townhall meeting, and as I asked about the power lines now being lower, the City Mgr nodded to two policemen back against the wall, behind be, who came forward and stood right behind me— if I asked any further questions about the issue, I was told I would be ushered out of the room. "Disturbing the peace" was the reason. What happened to "free speech?" And why did they build a volleyball court under the high tension lines?   (Think: Option 1.)

Miscellaneous Tech Info [208]

Passwords – best to keep them written on paper, tucked away out of sight.

Do not store them in a file on the PC.

When creating a password several websites will not let you create one shorter than 8 characters and they require that you use special characters. Ok, so you can still use your favorite one which had 5 characters, say "mikey", but now just upgrade it:

Use a capital letter, an underscore and your birthyear...

Mi_key68

Most critical password setting routines will accept that.

For really important commercial passwords, protecting accounting data, for example, use something like   Q43_77.65#

There are 11 characters in the string above – the last one doesn't show as it is a Ctrl + T – be careful not to use a control character that skews the screen when you use it (like Cntl + Z, P or D) – this was not so much of an issue 10 years ago but keyboard commands can be used in passwords providing they don't do something of a surprise – like Cntl + P will print, Cntl + Z will delete what you typed...

Forget using @ for "a" and E's for 3's, S's for 5's .... those who break passwords are aware of that and more. The best passwords are long and use random characters and special characters.

Web-browsing Tracking – when using Google or IE Explorer, what you searched for and the results are archived. To avoid this (and Windows XP made it easy to delete the search argument, but not Win 7 or 10) you can use the search engine DuckDuckGo.

Facebook Facial Recognition -- In 2017 Facebook announced a new setting to let people turn it off, but not all users received the setting. To turn the feature off, via your desktop PC, choose **Settings > Face Recognition > Edit> No**. If you don't have that setting, the go to **Settings > Timeline and Tagging > Who sees tag suggestions? > No one.**

Smart Speakers – like Alexa and Siri, Amazon and Echo – they are always listening. But not to get paranoid, unless you have something to hide... your conversation is often recorded but erased within minutes as the device begins recording newer conversation.

In a study done by Dr. Choffnes at Northeastern University in Boston, in cooperation with Dr. Dubois, they tested the 4 main smart speakers and of course they all responded to the correct "wake up word." What was very interesting was when they played audiobooks ('Gilmore Girls') and tested the smart speakers' recorded results –

"the speakers started recording snippets of dialog 10 times without hearing the correct wake word... the team recorded 63 false positives in 21 hours... and when the speaker was fooled into responding to a false wake word [e.g., "Alex is here"], the speaker often stopped recording within seconds." [209]

A little known fact is that the smart speaker will listen and record, then in milliseconds, quickly scan the saved 5k text it has acquired for keywords, such as "assassination, kill, murder, bomb, attack, robbing," etc. – all emotionally laden words, and if it finds none, it flushes the text and records new text, repeating the process... allegedly benefitting National Security (and the NSA or CIA?) trying to identify terrorist communication. If it finds any suspicious words, it sends the 5k text packet to a real person who will look at it and see if it is worth pursuing...

Of course, the same thing can be done with phone calls and emails... total packet 'snooping' all in the name of National Security.

ProtonMail – this is for those who want more private email. Google and Yahoo scan emails for ad targeting to send you what you talk about, or what you buy. The transmission is encrypted and the service does not collect data on its users. Drawbacks: no inbox searches and no help if you lose your password.

Public WiFi – still should not be used for sensitive emails or web searches – the service is easy to hack from outside the Starbucks where you are seated. Checking your creditcard or checking account balances is a bad idea in a public venue. In addition, PDFs are transmitted in an unscrambled format... your cellphone is safer than your laptop for this, and harder to hack, but just don't join a public WiFi network.

GPS Tracking – when you check the Weather app or look at the Maps app, your cellphone connects with GPS tracking and makes a record of where you are. In fact, when you travel, especially by car, your cellphone is talking to the towers as you pass (even if you are not on the phone) and that is why after you go 250 miles from Dallas to Abilene and someone calls you from Dallas, the call is routed to you with just milliseconds' delay... the system already knows where you are.

If you want to turn off the GPS Tracking and that applies to control access to your photo library, too, (see next item below), go to **Settings > Privacy > Location Services** and toggle the control On/Off (for iPhone).
For Android: **Settings > Google > Location** and scroll down to **App-level permissions.**

Camera Tracking – it may come as a surprise to you that when you take a picture, the system records data about when, were and how the image was recorded.

And when you share that picture with someone not only that picture is sent but the information goes with it, and is called **Exif [sic] data.**

That means that those people in the picture are identified as to where they were at what time of day.

The way to stop that is to download the picture from the cellphone to the PC, take a screenshot, and send that to your friend. Or go into the cellphone's camera app and revoke its access to the GPS function: **Settings > Privacy > Location Services > Camera > Never.**
Android: **Settings > Lock Screen & Security > Location > App-level Permissions** and switch off the **Camera** option.

Copier Machines – this is reminiscent of the "invisible codes" embedded in the pictures cellphones take, but it is also done on Xerox copies from all copiers – they identify what copier and date the copy was made, and you can't see/read the micro-information. It was Law Enforcement that demanded that one.

Car Tracking Info – of course you are aware of OnStar® which uses GPS from your car to track your location to help you in an emergency, but did you know that vehicles today are tracking and recording all kinds of data – some of which can also feed into the CarFax® database? Everything from the songs on your playlist, to temperature settings, navigation destinations: the locations you frequent, programmed radio station list, to how hard you brake, or how rapidly you accelerate.

And that data can travel to the car's next owner.

Sometimes a tracking device is installed under the dashboard when setting up financing and lease deals (especially on very expensive cars), and if you didn't read the fine print, you may not be aware that they are there. This is so you don't make off with the car and chopshop it, and repaint it and move to another state and stop paying. Once the car is paid off, you can have it removed or disabled.

Specific 5G Health Issues

This is **a continuation of Chapter 5**, but on a more personal basis – some things that you can do and watch for to avoid the EMF dangers.

The following issues are all concerns and issues reported by
Dr. Sharon Goldberg, MD. [210]

Public health has deteriorated over the last 20 years... many of the chronic conditions we see today are linked mechanistically [to a physical source]... an example... We have an **epidemic of suicide** in the United States now, depression and suicide... we have states that have a **50% increase** in their suicide rates [and you won't hear that on the nightly news!]... I'm not saying that this proves that microwave radiation is the cause... [but] something organic is going on.

As a physician what she sees is the **increase** in her practice (and those of her co-doctors) in depression, neurasthenia, anxiety, hallucinations, increased irritability, insomnia and loss of memory. Not just occasionally, but often and **among young people.** Something is causing this on a nationwide scale. So what happened? There was a *National Toxicology Program Study on Cell Phone Radiation* done in the years leading up to 2016 **commissioned by the FDA** in the 1990s. It was published in 2018. "It is the gold standard of exposure studies for toxicant in the US..." It included a review of cellphones as a possible cause of some of the health issues...

What was done was criminal. Not only did Telecom not look at the report, the **FDA did not look at the report**, they considered it flawed because the tests for EMF danger were not done on humans – which is now illegal to use humans as experimental subjects, so who were they kidding? The results showed liver, brain and skin damage, as well as DNA damage to the animals they tested... but humans are different. So why do we use those animals as test samples?

> How is it possible that the study was done and no action taken? What ends up happening is that the [toxic] agent gets labeled and it gets listed on our ATSDR [– a list of toxic substances.]... this is **a first**... a first time that a National Toxicology Program Study has just been **disregarded completely**.

This is what is happening, folks. It means someone has the power to diss, ignore, and command others to do the same, with any report that might suggest that the "cash cow" of Telecom is causing problems.

> It is not possible to do human exposure studies [because] we know these exposures are dangerous....It's not ethical to expose a group of people, participants, to an exposure that we know causes DNA damage, blood brain barrier leakage, cell membrane leakage, calcium [VGCC] issues.

So the human testing is not done and that lets Telecom off the hook.

She has further concerns, that have been discovered linked to constant cellphone and EMF exposure... 12 points:

1. Millimeter waves are known to be associated with **eye problems** and cataracts largely due to the **BlueLight** issue reported earlier (Chapter 5).

   It would be wise to invest in a pair of anti-bluelight glasses for use with cellphones, or a filter for the PC screen. Your eye doctor can **add a coating** to your eyeglasses to accomplish this.

2. The science shows the effects on the **immune system**. Effects are also documented on the nerves and one's mood.

3. EMF **affects bacteria** – and they build a resistance to the EMF which is seen an attack and we now have superbugs out there. Millimeter waves have been shown to cause **antibiotic resistance** in Staph and *E.coli*.

4. EMF is affecting our **birds and insects** – if the birds fly thru a millimeter burst, they will not be flying out the other side. And the intensity is killing bees... which crosspollinate our crops.

5. Microwaves interfere with the **normal housekeeping functions** of the body. It interferes with the calcium gate operations of our cells [VGCC] which move nutrients into and waste out of the cells, it breaks DNA by creating free radicals, and suppresses the immune system by over-stressing the body – causing too much cortisol to enter the bloodstream. **EMF waves stress (and age) the body.** It is known to causes cognitive dementia, and short-term memory loss.

6. The **12,000 satellites** placed around the Earth to irradiate it with millimeter waves will be a form of "testing" done on humans for which they have no consent – they cannot escape and it will affect their health. (Thank you SpaceX and Amazon.)

   We should not be rolling out 5G from space.

   > Are you beginning to see which of the two Options in Chapter 5 is taking place... and why?

7. New cars are exposing people to excess EMF... especially the **Hybrid all-electric cars**. The batteries are in the back and when you drive down the road with your kids...where are they? In the back... getting more exposure than the driver.

   Be aware that engines, batteries, alternators, Bluetooth and even the video displays on the seatbacks (in many SUVs) to entertain the kids all produce EMF radiation. And because your car is metal, metal constrains and bounces back the EMF field generated in the car... an EMF bubble.

   You need an EMF **Trimeter** to check this out for yourself – go back two sections (see picture).

8. **Self-driving cars** will basically not need 5G to operate – they are autonomous – BUT they are also electric, with EMF on the **roof**, from the **engine** and God knows what is in the **trunk**...you are surrounded in **an EMF bubble** – and they are not insulating, or shielding, against the EMF "which can be really high."

9. **Taxi TV and suicide.** NYC introduced TVs in the backset of many cabs and there is wiring in the seats... **and the driver has the seat wiring and the TV in the back of his seat...** so he is getting a major dose... so is it coincidence that there has been "an **epidemic of suicide** among taxi drivers in New York?"

What is really Dr. Goldberg's issue is that we are not even asking the question of whether this is all safe – especially the last 3 points! The public has to be educated and not let Telecom run away with the show – and our health!

   Microwave radiation causes **oxidative stress** on the body... even more so with people who are known to be **hypersensitive to electromagnetic fields**. It is a condition called Microwave Syndrome. People with this issue often make the connection after a few nasty episodes with EMF fields and a downturn in their health... those with no sensitivity still get health issues but cannot figure where they came from

   We need to educate the public.

10. We have a **diabetes epidemic** in the US among **young children** who are being exposed to EMF when they sit down and put the **laptop on their lap** – exposing their pancreas and liver to the

radiation... which is the oxidative stress we just mentioned.

> In addition, schools should not be using WiFi but should
> instead be **fiber-optic wired** to avoid exposing our young
> developing kid's bodies and brains to these EM fields.
> By now, you are probably favoring Option 1... Man isn't
> stupid (ignorant, yes) but this is being orchestrated on a scale
> that exceeds ignorance and greed. Answer in Chapter 9.

Says Dr. Goldberg,

> The children are taking these devices and they're putting them
> right **over their bellies**, Okay? So when the device [laptop] is
> set for use on WiFi [home or school] it's emitting microwaves
> .... so you're getting microwave emissions over **the liver** and
> right over **the pancreas**....
> when you cause **oxidative stress** to the liver and the pancreas
> we know the mechanisms of diabetes and what we call **NASH**.
> So oxidative stress is a mechanism in the development of
> NASH cirrhosis. And then over the pancreas, the pancreas has
> very poor defenses against oxidative stress. So this is what we're
> doing to our children.

We need to have **mandatory labeling** of devices that emit microwave radiation –
cellphones, baby monitors, Smart TVs, electric Hybrid cars, Xbox or PlayStation...
Video games are emitters of microwave radiation – and especially the **VR Headset**
mentioned earlier – Gee, let's strap a cellphone to our face... 4" from our eyes!

11. **Police Officers** get an EMF exposure from the wireless body
    cameras, also from their electronics console in their cruiser, and
    any walkie-talkie they have.

    Says Dr. Goldberg, "They have to operate firearms... and they
    have heavy exposure to EMF every day... we know that the
    exposure can affect the blood-brain-barrier and cause leakage
    and problems with cognitive function..." They are obviously
    sacrificing themselves on their job to 'Serve & Protect.'

The FOX News just last week (October 14-18, 2019) announced that two police
officers in two different cities (one of them Dallas) mistakenly shot and killed

innocent (unarmed) people, mistaking them for burglars. It is happening. They did not hold them at gunpoint and ask for ID, they just shot them.

12. Watching people **clip cellphones to their ears** and wear them 24/7 for convenience is the height of ignorance... just as bad as not using the speaker phone and constantly holding the cellphone up to one's head ... Aaargh!

Ok, the problem stems largely from a misunderstanding that Dr. Goldberg addresses in the beginning of her talk. We **ASSUME** that

(1) we do not have enough science and results to decide whether EMF is harmful or not, and
(2) we assume that there are protective guidelines being exercised by the Telecom industry.

Both are false. Says Dr. Goldberg ...

First, we DO have the reports, the studies and it all says there IS a **heavy danger via repeated exposure.**

Secondly Telecom is a "cash cow" and is not about to rock the boat – if they had to look at the data, take it seriously, they would lose money having to take steps to make their products safe.

Thirdly, the FCC guidelines for EMF safety were made in the 1960s and do not address what we are experiencing right now – the idea was based on **The Thermal Effect** –" if no one feels it, it must be safe." And we now know that is false.

The type of radiation used to enable wireless communication is a form of microwave radiation. The same type of microwave radiation used by microwave ovens to cook.... and it has long been established ... that this type of radiation can harm tissue – if it couldn't, microwave ovens wouldn't work....
At certain levels of power, microwave radiation can heat tissue. This is known as the "thermal effect."

The vast majority of regulations governing EMF emissions are designed to protect against the thermal effect.
... for years it was believed that microwave radiation was safe at levels of power insufficient to heat human tissue....

Today, however, the body of science in this area demonstrates that there are harmful risks to human health at levels far below those required to generate a thermal effect.

Regulations that protect against your phone emitting enough EMF to **cook your head** are important – but they are also vastly insufficient. Increasingly it is becoming clear that there is no "safe" level of EMF exposure.... this type of radiation poses real risks to human health .... [211]   (**Note this SBY endnote for later**)

EMF Damage Report

(credit: emfsafeshield.com)

Cellphones heat the head...

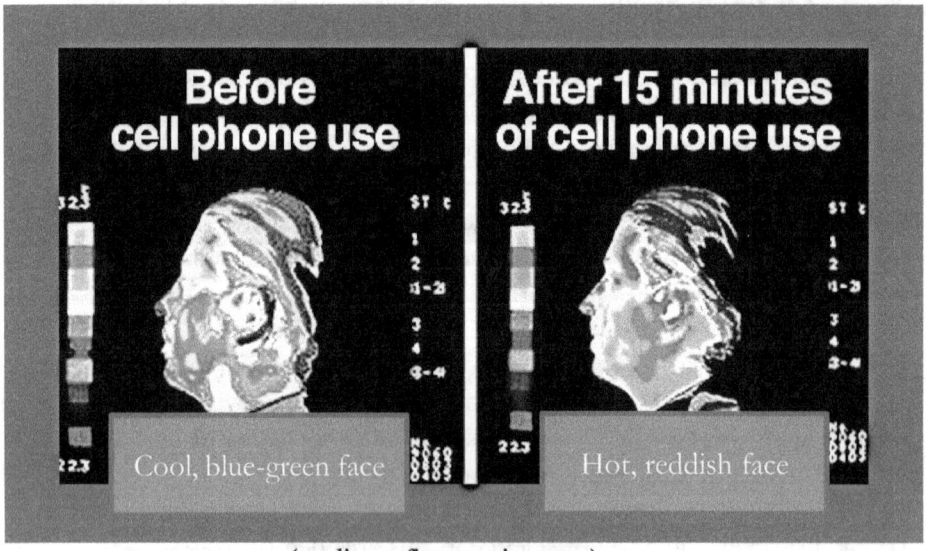

(credit: myfitmagazine.com)

...and back to Dr. Goldberg...

A **fourth** issue that Dr. Goldberg does not address is that this whole thing is orchestrated by the PTB, MIC, and Telecom for reasons that they do not disclose.

However she does question why Telecom and the FCC are ignoring the evidence and health issues... Perhaps the answer is in Chapter 9.

To repeat: **Knowledge Protects, Ignorance Endangers**

I sincerely believe that Telecom and Elon Musk, and Levandowski, Kurzweil, et al, are innocent puppets in the deployment of 5G, and AI, etc. They are just doing what looks beneficial, progressive and possible. Being positive.

Why ignore the negative facts? Because (1) the EMF radiation **can't be seen**... like viruses and radio waves, you can't see them so it's not important... and (2) back in 2004 there was a lot of false hype against the cellphones –before any real solid evidence could be **shown**. Now that it has, Telecom conveniently ignores it.

We should not be doing it but they see no reason not to.

Where the problem comes in is with those who are higher up, beyond Telecom, and even the PTB and the Elite who are being used to do this... Those in the Remnant that **cannot be seen**... it is the perfect way to (1) cull the population thru health issues that you **can't easily see** coming, and (2) it is a perfect, pervasive and **invisible media that can be hacked** to manipulate the IoT and, via newer cellphones equipped to emit the 10-16 Hz emanations, they can control/directly affect the populace – the ones who don't die from EMF radiation.

Invisible perpetrators, invisible danger.

Summary

Keeping one's health in tip top condition has always been important, and will be even more so in the days to come.

Today's world has become polluted and contaminated... just watch the large number of TV commercials for every drug imaginable under the Sun. And if you really listen to the audio portion, there are some scary, horrific side effects... not just nausea, vomiting and diarrhea... but seizure, thoughts of suicide, and mood swings, and clinical depression -- to name a few.

Add to that the fact that we are falling apart in some major cities with hundreds of street people with mental problems, drug addiction, and empty syringes everywhere -- plus fecal matter on the sidewalks ... To look at all that you'd think we had lost it as a civilization – worse yet, the **elected officials could care less** about it and are doing nothing about it. There are too many street people and their problems are serious.

Is the real problem Over-population? The issue no one wants to talk about?

I know, we have all been programmed to laugh at the idea of over-population, just as they want us to laugh at the idea of UFOs, The Hollow Earth, and Bigfoot. Be advised that over-population is real – not because there isn't enough land to spread out and double the current Earth's population – land isn't the issue.

Overpopulation is real because :
   (1) there are not enough jobs to go around and that is due to
   (2) to the way most modern economies work, and
   (3) the government owns and protects a lot of the open land
        that we see, so it will not be further developed, and
   (4) too many people pollute and ruin the environment, so it is
        not wise to further develop and destroy the environment.

Just take a look at the **floating trash piles** in our major oceans [212]... much of Mankind is a pig and does not care about the environment.

BTW, there are **two** Pacific Ocean Trash Piles at least **the size of Texas**... as shown below from a satellite picture... mostly plastic refuse which is very slow to decompose...

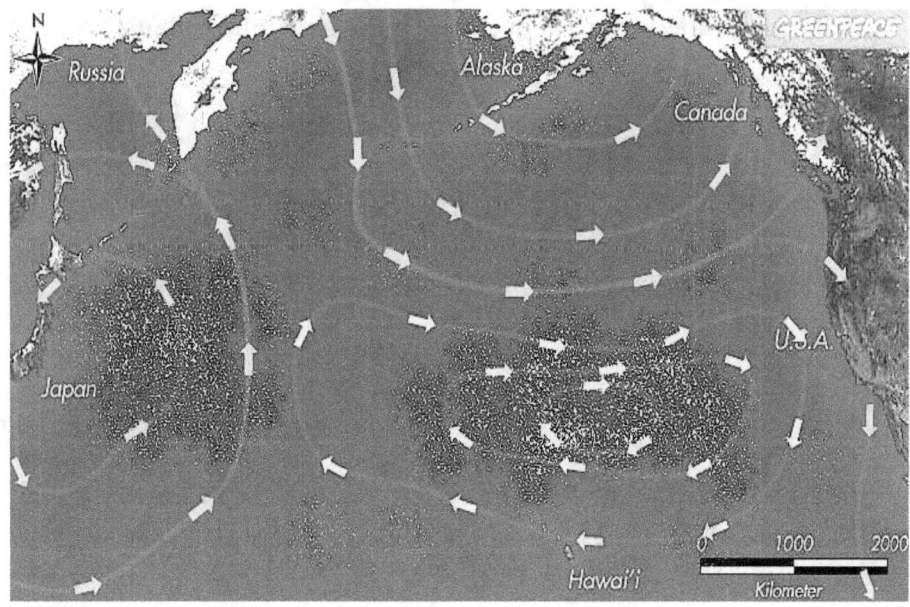

...And you can see several minor Trash Piles (off Alaska and Kamchatka) – which trap marine life and suffocate them. (Municipal Trash people are loading the garbage on barges and towing it out to sea...dumping it in the ocean.)

> If you want to do something about it, you can go to the website
> https://4ocean.com and sign up or buy product ('donate') to a
> group that is **actively** cleaning it up....

We are tearing up the Rainforest in Brazil, and the **oxygen level** on the planet is 25% less than we had 100 years ago – by NOAA measurement.[213] (These issues were examined more in Ch. 14 of VEG and Ch. 3 of DNQF.)

> Significance: the ocean's surface has always contained millions
> of phytoplankton to (1) provide a food source for marine
> mammals, and (2) provide oxygen in the atmosphere. We now
> have **25% less oxygen on the planet than in 1890**. Yes they
> have been measuring it for that long. According to NOAA,

> ### if we lose another 8% oxygen nobody will be breathing.

> **BTW: supertankers plying the oceans between continents
> also pollute the oceans with their bilge pumps. No steel hull
> ocean liner or supertanker is perfectly water tight and these
> huge ships have to have pumps that remove the water build-**

up inside the ships – it is full of oil, dirt, and chemicals
that are dumped into the oceans...killing the phytoplankton.

And I could go on, but anyone who takes a sober look at our world today may wonder if we have a future. And apparently the Elite are very concerned...

Remember that the Elite erected the **Georgia Guidestones** which state as their number one objective: **reduce population** to 500,000,000 --- from the current 7,000,000,000. Does anyone seriously doubt that the Elite have the money to make certain population goals a reality?

Does the large number of **Rx commercials** every hour on TV symbolize that we are a **very physically sick country**... with people dying by the scores every year?

And given the projected health issues from the coming **5G Network** which have been already tested and verified in the laboratories, could we be facing a **culling of the population via the intense EMF Super Grid?**

So while you cannot do anything about the Elite, nor the population issue, nor the Economy...you can take charge of your health to the extent that you learn how your body works (basically) and begin to give it what it needs.

Self-Protection Measures

As was said earlier, you could wear polyester clothing and live in an aluminum trailer (not 100% effective)... but here are a few ideas from the Anti-5G people:

Dr. Trevor Marshall suggests...
  ... use **graphite paint** on the outside of the house... blocks
      the 4G and 5G waves,
  ... use **low energy (LOE) glass** in the windows which effectively
      blocks RF (radio frequency) signals,
  ... use a **metal roof** – they have shingles and tiles that are metal.

Many of the 5G scientists advocate **coating your glasses** with the anti-blue light coating – that way you can use the cellphone, laptop, PC monitor and HD TV with a minimum of impact on the retina.

Avoid frequently using places that are **heavy WiFi sites** – like coffee shops, bookstores, and natural foods markets.

Do not live near or under electrical power lines – the kind on telephone poles and especially the 60' tall electricity transmission towers... they often have greenbelts under these towers – as if it is safe to jog or bike under them – IT IS NOT.

This (below) is very unhealthy as the ground-level reading on an EMF meter will register 45-80 mG and that is exactly the EMF radiation we are talking about avoiding... anything above 3 milliGauss.   (Appendix G.)

(credit: The Times of India)

Yes the power lines are usually 60'+ above the ground but the EMF **ground-level** reading is still too high. And if you park your car at work near a **large transformer** (usually found in a manufacturing district), it is the same issue... there is a **plume of radiation** (5 -15+ Gauss) and you do not want to park near it and walk thru it.

Remember: 3 **milli**Gauss is the safe limit.

Do not go biking or jogging on the above trails...

And the earlier quote from ShieldYourBody (SBY) offers a **pocket shield** so you can carry the cellphone on your body and block the EMF... (the SBY endnote I suggested you note earlier has the link to the website).

Works on *ANY* pocket where you carry your phone.

I'm not selling any product, but it sounds like a good idea...

Lastly, if you buy and use a **PEMF bodymat** and use it at least once a day (they recommend 2x a day), it will help to restore and normalize the EMF disruption to you body's weak EMF system... it aids in **microcirculation** so that any EMF damage that has been done to cells can be more easily cleared from the body – and then drink a big glass of clean water to help **clear the toxins** that have been shaken loose.

The preceding is just a reminder that if you lose your health, you lose everything. Hopefully this chapter will help you cope with some common health challenges, and give you some ideas for research... go down to your local health food store and talk with a <u>certified</u> nutritionist, for starters. Better yet, read Dr. Batmanghclidj's book and have a good talk with a Naturopathic doctor (NMD).

And avoid unnecessary EMF exposure.

There is much more to be aware of and that is examined in Appendix G... see WiGig and GiFi section.

If you have a TriMeter for testing EMF radiation you will find the safe and recommended levels listed in Appendix G.

Chapter 8: Escape

As was said in Chapters 3 & 4 in the **Epihanies**, you are a special creation and Earth is a School... the key point is that you are here for a reason and **Earth is not your home.**   There are <u>much</u> more interesting things to do when you get out of here.

> As was said earlier, that is exactly what the PTB and all their
> supporting cast of STS beings (Chapter 4) want to prevent.
> Souls scare them.

Right now, you are often at effect with the PTB and 4D STS group... when you get out of here, you can either move on, learn more and do more,   OR
you can choose to rule in realms OVER them. But in the meantime, I can't emphasize enough that **they fear what souls who graduate from Earth can do** and that Earth is not your home – but they want to entrap you by making sex, drinking, recreational drugs, tattoos, smoking, etc. fun – even camping, hiking and sky diving... anything to distract and amuse you... AND keeping you ignorant of your true potential. And keep you here. Some have wised up and left.

True enlightenment is what the 1960's **New Thought** metaphysical churches taught until a few years ago when the PTB got them removed. The New Age is not New Thought which was based on the Tao and what was for centuries called Serpent (Secret) Wisdom (VEG, TOM, DNQF and AL for starters). The New Age is lying to you and deceiving you into thinking you can activate your potential right now, right here, the way they say. It is a lie.

> Only Graduates get to activate their divine potential.

I hate to sound like a broken record but it is so important that you see that Earth is not your home (for many people it has become a prison), and there is a Group with an Agenda to Control you (the real **Control Agenda** examined in the first 6 chapters) and stop you from being all you can really be. TOM went over a lot of that— it was Book #2, written 8 years ago and is the companion to VEG (Book #1).
I repeat that here in case, like me, you spot read and missed Chapters 4 and 6.

Ok, enough said, just look around you, really look at the Earth and ask yourself:

Are you really living on the planet you think you are?
What has to be true about this place for it to look the way it does?

Really beautiful women who are as cold as ice
Really handsome men who are as cold as ice
Constant war, somewhere, on the TV News
Lying, cheating, stealing by our elected officials
Pollution, and destruction of the environment
Chemtrails and Morgellon's Disease
Pharmaceuticals whose side effects are often worse than what
    you are trying to cure
Cancer, viruses, Superbugs who beat our antibiotics
Huge trash piles of plastic in all 7 oceans
Huge dead zones in our oceans and the Gulf of Mexico
... and now 5G Super Grid and cancer, tumors, autism and ALS/MS.

So... then what?

Aren't you tired of the orchestrated mess?
    It isn't just Man's ignorance and greed... we are killing the planet.
If this is your home, it is a mess... but it is orchestrated.
    And you now know (Chapters 1-2) who is doing it and why.

Should you join a group and try to stop/change/fix the mess?
    It is too late – 1987 was the turnaround time and we missed it.
Should you be paranoid, live in fear and go hide?
    NO.... but

There is a way out.

There are 4 possibilities...

Join a Breakaway Civilization
Synch up with a Timeline Shift
Join an Ashram and be a Monk
Become an Earth Graduate   Suicide is not an option.

Breakaway Civilization

One of the benefits of this option is to escape **Overpopulation**. Again, running out of land is not an issue, but it is not possible to make all land serve agriculture, nor is the US Govt going to release 60% of the land it owns in the USA (which it controls via the US Forestry dept). You would want to relocate to an area and a group that is responsible, high-tech and respects the environment – wherever they are located.

> Destruction of the environment by crowding is an issue.
> The negative psychological effects of overcrowding is an issue.
> The scarcity of jobs in an economically depressed area (Cleveland,
>   Detroit and Chicago) is an issue.
> Street people crowding and trashing the streets is an issue.

Theoretically a Breakaway Civilization would have solved these issues, by recognizing what <u>had to be done</u> to survive and live in a reasonable relationship with Nature, and provide a reasonable '*Lebensraum*' for their people. And that is why some Elite separated themselves from the 'madding crowd'... and they would have to apply some rigid standards/rules (**strict discipline**) that the public at large would not abide by... and that is why they are separate (and hidden). They would thus genetically monitor and control their civilization, **to ensure survival of the fittest**.

Joseph Farrell and **Richard Dolan** have theorized about a Breakaway Civilization in their books, and in some of their seminar presentations, but we do not actually have the proof of any such Group... yet, it is a **very plausible idea** and may involve the survival of mankind on the planet when the rest of undisciplined (violent & emotional) mankind gets into another World War...

Such an idea is not far fetched as that is probably what happened with the survivors of Atlantis, the ones that my books have referred to as the **Ancient Ones**... smarter than the rest of mankind, and possessing higher technology (**Atlantis** had flying craft, submarines and a Firestone crystal that **broadcast power** to wherever they needed it – Nikola Tesla was onto something.)

So let's see what the two foremost advocates for a Breakaway Civilization have to say about it. And then we'll look at how it could come about ("Breakaway Civ 101" section) and whether it already exists, and what else its presence would entail...

Richard Dolan Input

Mr. Dolan has written a book outlining the issue with UFOs as they appear to be heavily connected with a Breakaway Civilization: <u>UFOs and the National Security State</u>. It is the recurrent appearance of UFOs concurrent with the end of WW II and the Nazi development of **Magneto-Hydrodynamic devices**, electro-gravitics (the Bell), transistor technology, fiber optics, and sound cannons, as well as the jet engine and the atom bomb, that suggests they were advancing well beyond Western Science... and due to the sheer volume of new science, that further suggests **they had help**.

This has led to the idea that there are **secretive groups building and flying the UFOs**, after having been given the technology – as were the Germans in 1936 when they 'found' a crashed but intact UFO in the Black Forest in southwestern Germany. If the Germans (aka Nazis) successfully built and flew a few UFOs, this aerial advantage would surely result in a National Security red flag for the USA. Hence the title of Dolan's book.

But it goes beyond that where Mr. Dolan puts forth **three hypotheses** regarding the formation and existence of a Breakaway Civilization: [214]

1. UFOs comprise a significant component in the shaping of modern history. It has been **a hidden premise** in the formation of the National Security state and the reasons for the secrecy concerning same.

2. The Breakaway Civilization was formed in the context of the Cold War and as such dealt with UFOs and the Soviet Union (suggesting there might be **a connection between the two**). This also gave rise to the speedy and covert development of spy satellites, the U2 plane, and deep space probe concepts.

   Dolan connects the Breakaway Civilization, the Cold War matrix, the UFO issue, and advanced technology to show how interwoven these things are...

   > [with]...the possibility of covert breakthroughs in propulsion technology, the "off the grid" nature of the deep-black world itself, we come to the possibility that we are dealing with... a "breakaway civilization" .... One ....with great independence, secrecy, and a monopoly of certain key scientific secrets.

In fact, **the UFO was a stimulus for development of technology to "secure the home base" against a potential ET threat** – since we didn't know who was flying them – and the desire was to achieve (without the public's knowledge) a parity with whatever threat the Russians or ETs posed ... The 1947 Roswell incident was a major spur in this direction. Then the 1952 Washington DC flyover of 40+ UFOs in one day hardened the US resolve to build whatever it was we needed to defend against what appeared to be an imminent threat to US airspace.

> The secret fear among the Pentagon and Government people was that this was not an ET threat but one reflecting **a terrestrial group** isolated somewhere unknown, with the advanced technology, and

the inside suspicion was that **the Nazis or the Russians** were behind it. It was well-known that the Nazi scientists were experimenting with ***Die Glocke*** **(The Bell)** which was delivering antigravity results. And it was known that both the Germans and the Italians had recovered crashed UFOs...

The Bell

It was feasible that the Russians had stolen some of the technology, no matter where the Nazis were after 1945, because as the Russians came thru Pilsen, Czechoslovakia, which was a major development site for German antigravity craft (Think: **Vril**) they could have captured a craft or the scientific papers to build one.

The Vril

Also known was that the US Army and CIA in mid- to late-1945 discovered the German **Haunebu** development sites (Nordhausen and Peenemünde) and at least got their hands on some papers proving that the Germans were testing the craft!

The Haunebu

And both apparently flew as the Nazis took pictures......

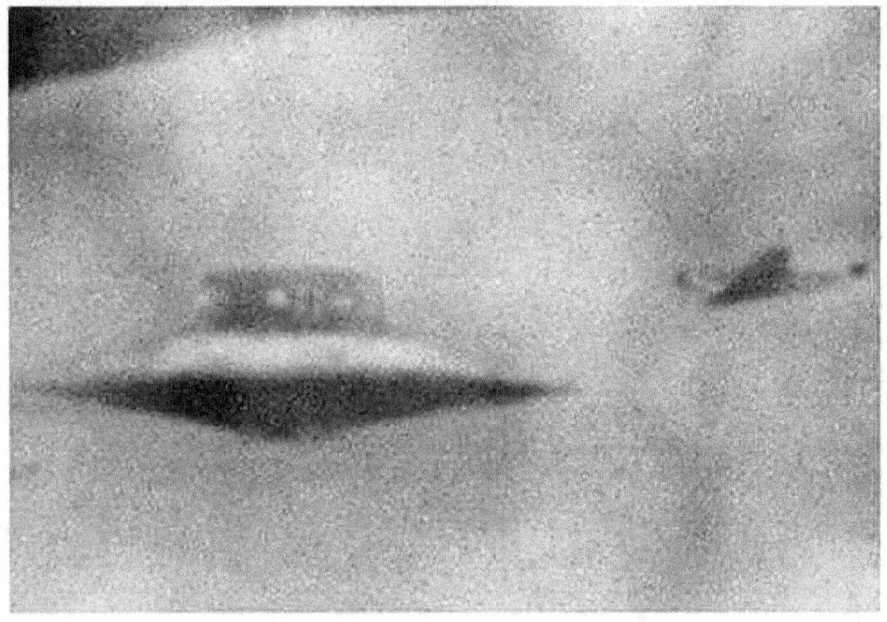

A Flying Vril (with BalkanCreutz)

...and the Haunebu as well...

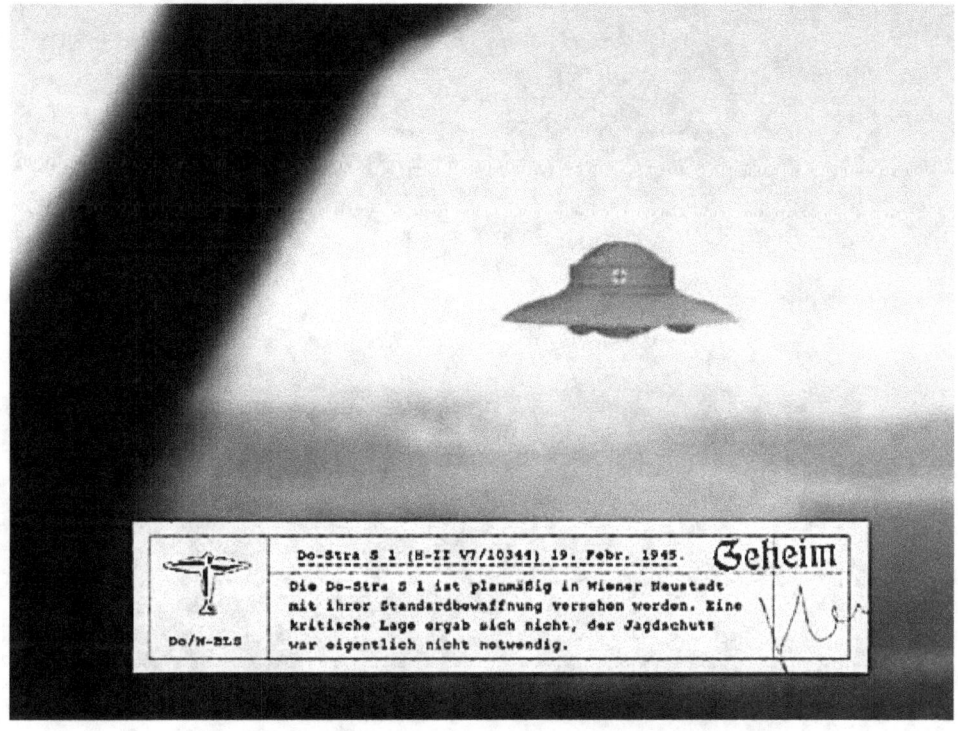

A Flying Haunebu I ("Geheim" means secret)

The above pix and more are covered in VEG, Ch. 4.

So there were several prototypes of Nazi Antigravity craft....

The Nazis called them **Flugscheiben** ("flying stones") and yet failed to equip them with **KraftStrahlKannon** (equivalent to 50 mm Howitzers)... the story was that the scientists knew that the sociopath Himmler was trying to depose Hitler (Think: bomb in the War Room: à la *Valkyrie* movie) and the scientists had enough integrity [and fear] that they delayed the armaments on the few functional Haunebus they had so that Hitler/Himmler could not win the war. Otherwise we'd all be speaking German.

So theoretically, as even Mr. Farrell speculated in his book, that if what crashed in Roswell was a Haunebu (with the German Cross) could it have been back-engineered (and reconfigured from disc to triangle) to develop what has been reported to be the premier US antigrav craft, the **TR-3B**? [215]

Possible TR-3B   (See VEG, Ch. 4):

And according to Lockheed...

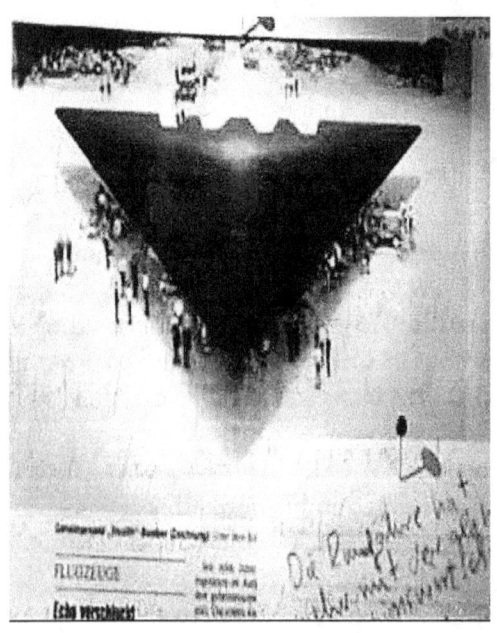

...or is this (left) the **Aurora?**

According to **Ben Rich,** head of Lockheed Skunkworks back in 1993, "We have the technology to take ET home!" [216]

"We already have the means to travel among the stars. But these programs are so locked up in black programs that it would take an act of God to ever get them out to benefit humanity...."

...and he also said:
"We now know how to travel to the stars."

There was an error in Einstein's equations and when corrected, we discovered electro-gravitics (rotation of Hg-plasma fields = MFD) that worked (see VEG).

Great, so where did WE get it? Roswell? Did they take the reported crashed weather balloon to Wright-Patterson in Ohio to study it?

And what is this patch?

See video of STS -48A) and a UFO firing on another UFO..... in space.
https://youtu.be/eil_2WOv5VA

President Trump in March 2018 stated that "we're getting very big in space, both militarily and for other reasons..." suggesting that the true purpose of the **Space Force** that he wants may be more than the equivalent of a celestial traffic cop. [217]

STS 48A video also at:
https://www.youtube.com/watch?v=eil_2WOv5VA

Are we already there?

Thus the US had to isolate its development of antigravity technology by hiding it at what became known as **Area 51/S4** and black budget appropriations were given high priority to develop high-tech aerial responses to an unknown threat.

And to wind up Mr. Dolan's third point:

3. The National Security state having allied itself with the Operation Paperclip Nazi scientists that were brought over here, and the CIA, Dulles network, and the funding from various black sources (according to the research that Joseph Farrell has done in many of his books), we wound up with an **advanced, secret technology** controlled by the MIC (Military Industrial Complex) and closely guarded by alleged groups like the Majestic 12 and NASA. In addition, "a vast hidden system of finance" which would implicate some degree of **Banking involvement** would be necessary to fund such development.

Mr. Dolan also says that due to the FOIA law, the government realized it had to release certain information on UFOs if it had it... and of course it redacted much. So to get around that, today we have **Bigelow Industries** (and others) doing much of the development and research since private industry does not have to reveal what they are working on to the public...

> **Privatizing the UFO secret** also helps with secrecy. It means that all the secrets concerning UFOs and related technologies become not merely classified but proprietary... [and[impervious to public scrutiny. [218]

> Ok, be clear that this is not an attack or an exposé of the "bad guys"... but is a very logical outcome of the situation at the end of WW II. It had to be done. If the very survival of Man is at stake, then rapid (and secret) development of advanced technology HAD to be done – **without** alarming the public. 5G may be part of that 'defense.'
> And, if the Orion Reptilian contingent gave the Germans the technology and masterminded the Third Reich (empowered them) – it had to be stopped.

So why was secrecy needed? Why can't Disclosure be done, as Dr. Steven Greer is proposing? Why can't the public be told?

Joseph Farrell Input

This brings us to a very interesting study and report that was done about 1960. A thinktank was commissioned to analyze what the impact would be on society and the economy if the public were made aware of the truth about UFOs.

It was called the **Brookings Report** and a close read shows that it was much more than what was publicly reported. In fact, between the lines, and not so subtly, it discussed the existence of a Breakaway Civilization... and probable Space Force.

It was not just a simple "here's what we found as affects the economy or public perception." It actually foreshadowed Transhumanism – in 1960!

> ....the Brookings Report represents precisely ... a **transhumanist culture** [because] it in fact directly refers to at least *one* of the GRIN technologies [Genetics, Robotics, Information processing, and Nanotechnology] ... and how with technological development it might contribute to **direct modification** of human nature, and therefore society and culture.... the technique is *telemetry*...

> Telemetry... whereby the state of an object .... owes its recent rapid development almost entirely to the space effort... *Given associated equipment, it could be used in any situation where information at a distance was required, thereby permitting new orders of* **human and physical control....** *[as in] conveying medical and emotional data to central computers via* **surgically embedded micro-miniaturized telemetering equipment** *[RFID+ chips]... it could also be of use to institutions and governments in various ways,* **some of them perhaps not acceptable to the democratic ethic...** [219]

Keep in mind that at this time the CIA and MK-ULTRA (mind control experiments) were on-going and one of the points in the Report was the emphasis on being master of the machines. That means master of the telemetric devices monitored by computers. Transhumanism = people with implants. Then a key statement:

> Delineate as systematically as is now possible the specific economic, legal, social, and moral problems and opportunities implied in future **society-machine relationships,** so that opinion leaders and policy makers can be aware of what must be resolved and planned for before, during, and after major developments in this **communication and control capability.** [220]

The 1960 Report anticipated (suggested?) the **Singularity** popularized by Ray Kurzweil in 2005. This is perhaps the first serious transhumanist document, revealing what we have come to know as the Elitist Agenda (qv, Georgia Guidestones) and we can see, as earlier said, that the PTB do telegraph what they intend to do, but the Sheeple rarely understand or pay attention (and that is how their Freewill is violated: Chapter 4).

Connection with Space

Consider this gem of a statement in the Report:

> Certain potential products or consequences of **space activities** imply such a high degree of change in the world conditions that it would be unprofitable within the purview of this report to propose research on them. Examples include *a controlled thermonuclear fusion rocket power source* and **face-to-face meetings with extraterrestrials.**
>
> Of the power sources <u>being developed</u> for **spacecraft,** at least three could be used on Earth...
> (a) plasmas and magneto-hydrodynamic power....
> (b) Solar power....
> (c) Nuclear-powered thermionic converters.... [221]

Keep in mind that this was 1960, Sputnik was launched in 1957 and the US had just barely gotten going with the Project Mercury program... and Alan Shepard just did a **sub-orbital** (under 62 miles up) flight in 1961... What spacecraft!?

Again, note that the magneto-hydrodynamic power was already **"being developed"** <u>for spacecraft</u> as a result of the Operation Paperclip importing of Nazi (1600+) scientists who came here in 1945 – they started the development with their Bell. We weren't even into space yet... "sub-orbital" does not count – Yuri Gagarin had already gone <u>into space</u> before Alan Shepard!

And while **JFK commissioned the Brookings Report,** he had <u>not yet</u> committed us to go to the Moon "by the end of the decade" (i.e., in 9 years!!) ... Did he know something that led him to force the use of the covert technology? – he was told that NASA was not ready to fly rockets to the Moon, so was he challenging NASA and <u>forcing the issue</u> and due to his private talks with Wernher von Braun, could he have known about the **alternate space program?** (Could that have gotten him killed... thus challenging the PTB?) Certainly, his publicly stating that we could team up with the USSR and do a **joint mission to the Moon** -- because the Soviets had more powerful rockets! – would not sit well with the PTB nor very young NASA (which was formed in 1958).

Magneto-hydrodynamics (MFD) are a big step <u>beyond</u> NASA and its chemical rockets.

This is a black-projects development and suggests the beginning of a <u>hidden</u> space program. And that is not stretching the point... consider the following also from the Report...

> To the extent that the US intends to compete with the USSR in forwarding its **space program**.... it would be desirable to explore ....the [asymmetrical] assets and liabilities.... of the two nations. *Such factors as secrecy and publicity, feasible* **alternative space Programs** .... for funds, relations with <u>scientists in other countries,</u> and access to <u>other nations' territory</u> may be of advantage... [222]

Why suggest this -- unless it was already underway? JFK was trying to work with **Khruschev** to do a joint Moon mission... launching from the Soviet Cosmodrome in Khazakstan as well as put a stop to the Cold War. Higher powers said No.

The whole context of the Brookings Report discussing magneto-hydrodynamics and fusion and a space program which it says are "being developed," suggests a deeper awareness (and connection?) with black projects which were decades in advance of the 1960's. "Indeed it is difficult to *avoid* the conclusion that these statements refer to a **secret space program.**" [223]

Connection with a Breakaway Civilization

And then the Report concludes...

> ...the development of a compact thermonuclear fusion power source for **spacecraft** would undoubtedly **open up space** *for many kinds of* **large-scale activities [asteroid mining]** *...* but it would also change the political, social and economic features of the earth radically...[new covert oligarchs would be empowered] ... and the resulting interaction between **space and society** would be comprised of factors too broad and too complex for useful speculation... *So too with such developments as* **anti-gravity***, a face-to-face meeting with intelligent* **extraterrestrials***,* or an intensive pan-national attitude favoring all-out scientific research.

They are not questioning the existence of extraterrestrials, it is taken as a given (and they are <u>already on the Moon anyway</u>). The technology and ETs (including their UFOs) are seen as the gateway to something huge and earth-changing – so it must be kept secret. The UFOs are continually checking out our nuclear activities – missile silos, nuclear reactor stations, and bombs we set off – thus the UFOs are 'connected' with nuclear power. The Brookings Report mentions all this and then, suddenly it dawns on us that *"the UFO is the hidden context of the Brookings, as is the secret history of wartime and postwar Nazi achievement."* [224]

> The point being that the **UFOs are one aspect of the Breakaway Civilization** – the Nazis did develop them, based on functional UFOs that were 'gifted' by the Dissidents already here. The real "ET-based" UFOs do not crash, but the fledgling earth-built ones used to – not so much anymore.
> The Dissidents would gift the UFO technology in exchange for Nazi cooperation in securing **control of the surface world**. It almost worked.
> And to develop and hangar them would require an underground, hidden facility... ideally in a remote part of the Earth. Such a development would require **massive funding** (over time) and that would somehow involve Banking, or the loot from WW II, or black budget appropriations from existing government budgets.

So let's look at a hypothetical version of history that is worth considering.... and just maybe for those who have the desire and discipline to seek it out and join, it could be worthwhile, if not radically different. At least it makes fascinating reading... real "brain candy."

Breakaway Civ 101

Let's start with the German Schwartzwald forest and the 1936 'crash' of a UFO that was discovered and back-engineered. This would give the Germans the edge in air superiority and even be considered a 'Wunderwaffe' (a Wonder Weapon) -- if it could be made to work.

Fast forward to the end of World War II where the German Army and Navy surrendered with **Admiral Karl Dönitz**. And meanwhile no one knows where Hans Kammler and some real techies and their UFO-prototype **Vril** (from the factory at Pilsen, Czechoslovakia) and the **Haunebu** prototypes from Nordhausen, Germany went to. In fact, we don't see the Third Reich Nazi leadership surrendering as they already left the country and moved on to form their own redoubt in southern Argentina, southern Chile (Colonia Dignidad) and a secret submarine base in Antarctica... and maybe one in Greenland.

Of course word gets out that the Third Reich has reorganized as the **Fourth Reich** and the US sends a military flotilla in December 1946 down to Antarctica (call it Operation Highjump) to smoke them out and remove the Nazi element once and for all. However, the Nazis see them coming and most of the US flotilla goes up in smoke due to superior Nazi technology that severely cripples the US naval power.

At this point the Fourth Reich has given up world domination and just wants to be left alone to their own agenda: colonization of the Moon, Mars and beyond – they now have the technology to do it... to heck with the polluted and overcrowded parts of the Earth. They meet and befriend the underground **Insiders** at the Antarctic... who urge them to follow a higher, more proactive course...

At the same time, the Nazis also meet the Dissidents who live underground in Asia who gave them the downed UFO in 1936, and the Dissidents again, in 1944 try one more time to convince **Himmler** to join forces with them. The main Nazi leadership in Antarctica refuses Insiders and sides with the Dissidents. But **Maria Orsitch** (head of the Vril Society) does not let the scientists arm the UFOs...

In addition, the Nazis meet the *Ubermensch* that Hitler met and scared him while he was in power – the Insiders have used the Greys' abducted DNA to build their own improved version of **Hybrid Man** and the Nazis are forced to assimilate and work with the Hybrid humans and Dissidents ... even adopting some of the genetic technology for their own use. They are now really out of step with the rest of Earth civilization and have stepped into a Brave New World where they cannot go back.

The functional Haunebu help them beat off the US flotilla (in 1946) and install advanced defenses and after a few **fly-overs in 1952 in Washington DC** (just to let the US know that they are being watched and that the US cannot protect its own airspace), the Fourth Reich meets with a US President and establishes a treaty making parts of Antarctica off-limits (no-fly zones) <u>and</u> insures that the US will not attack again. The treaty allows several nations to set up coastal bases for scientific research (really 'Nazi monitoring') in Antarctica, and all is peaceful again.

1952 Washington DC Buzzed by 30+ UFOs

In addition, in the Treaty, the Remnant gets what it wants with a corollary which permits the Greys, their **biocybernetic androids** to abduct and genetically extract DNA from humans and cattle to further clone and develop the Breakaway Civilization. The deal is to let the US Army know how many and when and where humans were abducted, but this agreement is soon ignored and abductions increase and statistics are not available, nor were they accurate to begin with.

Meanwhile the Remnant Insiders and Dissidents continue their ages-old infighting by <u>both</u> inserting their improved Hybrid Humans into human society in an effort to **control what the humans do** – several goals appear here: **Insiders**: (1) stop destroying the environment, (2) stop polluting the oceans, and **Dissidents**: (1)

update religion to manipulate the masses by promoting a New Age teaching that further deceives the masses, and derails true spiritual growth, (2) initiate terrorist activities to distract and disunite mankind, and (3) buy into and control the News Media to spin confusion, deception and chaos.

> Of course **both sides use human puppets** to act for them alongside their Hybrids who look and act like regular humans but who, like the Greys, also can manipulate what the normal humans think and do.

If all goes well, the Remnant groups may be able to swing the tide of surface activities in their favor to benefit their hidden agendas. (Think: 5G.)

Concurrently, the Fourth Reich moves on with its off-planet goals... establishes a

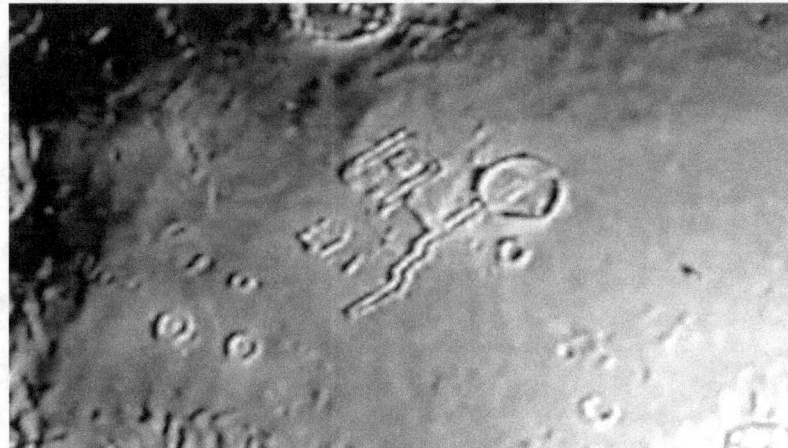

Moon base and shares the Moon with other ETs who have been there for millennia.

The big, deep **Moon base** would be on the back side...

In addition, the US pretends to go to the Moon (1969) so that no suspicion is raised regarding UFOs (always denied), and yet the US is told to not try to go to the Moon now and so for 47 years there has been no political possibility of getting there – even with the US-built **TR-3Bs** which comprise the backbone of the secret **US Space Force**... Solar Warden... which a current President would love to legitimize by formally

345

establishing it as a 6th department of the Pentagon. And in 2002 a British hacker discovers a Secret Space Force on NASA computers.

As the US Army swept into Germany at the end of WW II, they capture Haunebu technology too (on paper), then the US would discover a crashed Haunebu in Roswell in 1947 and back engineers that to create its own anti-gravitic craft, ARVs and Aurora, and begins to meet the Fourth Reich on a more even footing. Later, in 1993, the head of Lockheed Engineering states publicly that "we have the technology to take ET home" – meaning we have interstellar craft worthy of a fast trip to Mars. And the Fourth Reich and Remnant are already there, establishing colonies.

Espionage being what it is, the US technology is stolen and now exists among the British, Russia, India, China, South Africa... forming a secret alliance, maybe called Solar Warden, allied with the acronymic other countries' first initials: BRICS.

Of course, tensions fly 500 miles above the Earth; the US Elite/PTB want to establish a base on the Moon and the Fourth Reich will not allow that, so there are occasional firefights up there (STS 48A) – way above the ISS. The US seeks to manage the world its PTB way and that does not agree with what the Fourth Reich will agree to, so there is disagreement on many fronts between the two, and peace and cooperation does not loom between the US/Western World and the Fourth Reich/Reptilians. (Some of the US Elite/PTB are in bed with the Dissidents, while other Elite work with the Insiders – watch the nightly News to see the drama in Wash. D.C... so caution is advised.)

Meanwhile the Nazis (no longer calling themselves that) establish a base and small civilization on Mars, mostly underground – in case all goes to Hell on Earth, the human species (albeit genetically improved) survives on Mars... and the **Breakaway Civilization moves to Mars**... (and Saturn's moons which number 82 of them!) ... Mars is where some of humanity came from 200,000 years ago anyway (after a Space War and the destruction of Mars' surface and planet Tiamat).

Could the Mars surface terminal look something like the following?

Would that not be a fantastic scenario? And one that you would not tell the general public about as many of the humans (80%) on Earth do not have the job skills nor the genetics to move to the **New Society on Mars**... or Venus (also underground).

Genetics Downhill Issue

Since a lot of the genetics on Earth have begun a downward spiral (witness the huge number of Rx commercials on TV, **AND** the unadvertised advent of superbugs which are resistant to our latest antibiotics) – meaning that at some point in the near future, people will contract an illness that cannot be cured, you would want to isolate and keep the New Society as genetically pure and strong as possible — no 'pollution' from dysfunctional genetics as found in many humans in Earth society today.

Then you would also attempt to rectify the existing Earth genetic pool by making genetic corrections (CRISPR and the upgrades done by the Greys and MIC) to see what percentage of the public could be 'salvaged' from the downward genetic spiral. And advertise on TV for self-testing DNA kits so that the best DNA in the public realm could be identified... and send the Greys to them... Clever, eh?

That would be the justification for the Breakaway Civilization: the preservation of the human species, in its best genetic form.

More insight is contained in Chapter 9.

Of course if you saw the movie *2012* and the attempt by the Elite, the Oligarchs, to buy their way into survival, you have an idea of what could secretly happen when the above Breakaway Civilization scenario became a reality. Their money would help finance the escape (in addition to whatever black ops money that gets funneled from a US Govt budget into the survival project), and yet they would not be allowed to reproduce into the general, improved gene pool – unless they proactively contribute to it, or could be genetically 'rewired' to not be a burden on the New Society.

BTW, is there any clarity yet on the Options given at the end of Chapter 5? (Hint: if the above scenario becomes a reality, how would 5G play a part?)

Given the uncertain mishmash of genetics in today's public society, it might not be possible to join such a Breakaway adventure. As **George Carlin** once said, "It's a big club and you ain't in it!"

So what else is available?

Timeline Shift

Another way to change our Earth Realm is to move it to better Timeline...

This will not be a complete examination of this possibility as the basics were covered in TOM Ch 2, and Appendix C of this book. The mechanisms are too much to go into again here, but they are real and worth a brief look.

FYI, if you have VEG and TOM and this book, you basically have it all, I have written it that way as of this book. VEG & TOM were originally one book but it was 868 pp and that had to be split up.

Given that the gods who run this School are not ogres **and** They see everything **and** They are aware of the issues (pollution, overpopulation, disease...) **and** They care about us...would They not move to do something proactive **for** us – as They have in the past?

That is assuming that this world is worth saving... doing a **Timeline Shift** to a better, cleaner world – (see TOM Ch.2: Timeloop Control) unless we have so run it into the ground that They would have to do more of a **Wipe & Reboot**..?

Given that Edgar Cayce and Nostradamus both predicted disaster – Cayce in the 1960s-70s and Nostradamus in July of 1999 a "terror from the skies" [an asteroid?] – and neither happened... that suggests that we were moved to a safer Timeline... and that a Shift has already happened – and when you couple that with the 60% OPs on Earth now (see **Appendix C** for how that happens: Timeline Split Diagram)... it means we DID go a Timeline Shift. The OPs prove it. The gods took care of us somewhere in the last 50-60 years.

> And yet, with so many OPs... can They really do it again and create more OPs? Timelines are for Souls, not the soulless. If They split it again, there would be more OPs, and that is not that productive... but that is not my call.

That argues instead for a **Wipe & Reboot** which is what we had in AD 800-900 – see VEG Ch 10.

Synch Up with a Shift

So if that is where we are going, you do not have to do anything if They are going to do a **Wipe & Reboot**... you will not notice anything, except that some of the problems disappear, and the pollution does not kill us (They will reset some parts of our world – They wipe it clean [disinfect] and reboot it like one does a PC, restarting the Greater Script).

You **do** have to do something to synch up with a **Shift** – your PFV (Glossary) will automatically qualify or disqualify you from going to the better (higher vibration) Timeline. No god or Higher Being will select or deselect you – your present vibration will auto-select you... or not. Your vibration depends on how patient, loving, humble and **STO** (Glossary) you are AND how much True Knowledge you

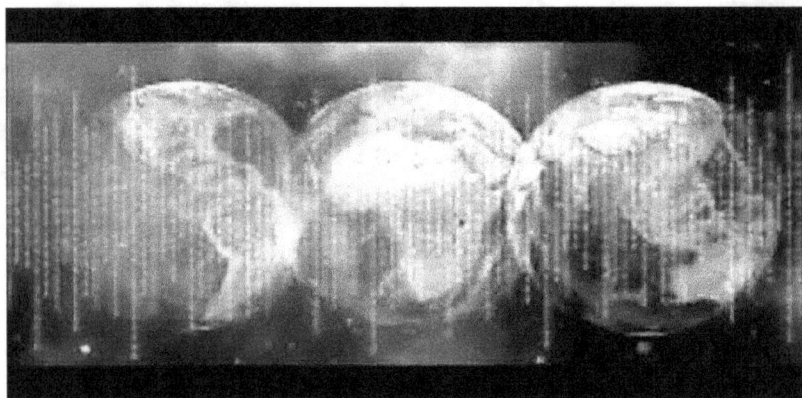 have... Have you been able to see thru the disinformation so prevalent in Earth society?

(These books should help.)

Timeline Shift of Earth

If They do a Shift and you do not go, you will also not notice anything change in the world or your family, etc... If you do go, things will be better, smoother and less spiritual harassment will be found... due to the fact that many of the 4D STS beings will not be there.

So in essence, I am just mentioning this as a possibility, and is obviously not something that you can count on... **it happens or it doesn't**, and if it were two weeks away, I doubt that any of us could change our PFV significantly to make sure that we make the Shift...

You are either there in vibration <u>now</u> or you aren't.

While I was writing that, and it has been a viable option in the past, Baldy said **it no longer is** due to the fact that not that many souls would go, and with the current 60% OP population, (see Appendix C and Propagation of OPs with Timeline Split) that would only <u>increase the number of OPs</u> and that is not what Timelines are for – thus there is no coming Timeline Shift/Split. Ooops.
But maybe a Wipe & Reboot.... (see Glossary).

Spiritual Ashram Retreat

This is the third option and is just running away and you will still be <u>eventually</u> subject to the pollution, disease and negativity impacting the world and Ashram and your meditations... it solves nothing.

If you were able to meditate and raise your PFV and shift out of here like **Paramahansa Yogananda** (left) did, then go ahead. He was a Hindu ascetic and teacher who came to Los Angeles in the 1930's and taught higher consciousness and had quite a following... even **Ernest Holmes** who founded Religious Science (see TSiM) met him and learned from him – but it all was too positive and the PTB got on Yogananda's case (scandalizing him unjustly) and the PTB eventually shut down Holmes' Religious Science (New Thought) church.

Tsk, we can't have people waking up now, can we?

Yogananda decided when and how he would leave the planet, and he laid his body down and went to what appeared to be asleep and departed. His body laid in state (open casket) for at least a week, with no embalming, and there was no decay nor smell.... he truly was a holy man. (See <u>Autobiography of a Yogi</u>.)

And you saw what happened to David Koresh and Waco (Branch Davidian sect) in 1993, as well as Ruby Ridge in 1992, when people decided to go form their own spiritual community – the government (aka PTB) stepped in and took it down. Of course there were allegations of wrong-doing in both cases, but will we ever know what really happened?

Enlightenment is a higher vibration and that will sooner or later get you out of here. And it is said that if your PFV exceeds that of the Earth (it means > 51%) you qualify for a Timeline Shift <u>and</u>/<u>or</u> Earth Graduate. You do **not** have to be perfect and ascended to get out of here – like Yogananda who was a Master.

Your best bet in this Personal Option 3 might be to retreat to a cave and ask the Greys to abduct you...

Failing all that, we come to the option where you can make a difference and really get yourself out of here – when you normally die.

Earth Graduate

So, what is the Earth Graduate all about? Why is it important?

Since Earth is not our home, and is nothing but **a School for Souls**, it means we can graduate and get out of here. And go where and do what? That is what the rest of this chapter will examine... briefly, as most of it was examined at length in TOM.

The InterLife

This is the realm where we come from before incarnating on Earth – the Masters and Teachers live here to advise, teach and heal wounded souls. It is also where we return when we die. We get there thru the **Tunnel of Light** that protects us from the boundaries of Earth and the Internet – so there is no interference from those who harass us from 4D while on Earth, they can also interfere when we make the transition – the Tunnel is protection and is usually managed by at least one Angel, or Being of Light.

Tunnel of Light – passage to the InterLife, or the Other Side.

Souls pass thru the Tunnel into the Light of what has been called Heaven, but is just the holding and study area for souls on their eternal journey.

Of course real Angels do not have wings, but they pictured that way so you know they can fly.

(credit: BingImages: Stephenmillerbooks.com)

Real Angels are more like **Beings of Light** (BoL), see below...

Angels have no male/female gender... they are Beings of Light and a little less than the Elohim, who are also Higher Beings of Light. Angels or BoL serve the Elohim.

The Draco (aka Djinn) were also Elohim who went their own way, and do not frequent the InterLife... due to their avoidance of the much higher positive frequency there.

The InterLife is a "safe space."

And it is a beautiful place... pastoral countrysides – like those in Ireland or Greece...

...and temples of learning...

...while most temples and places of study are made to look like something you are already familiar with, there are some futuristic ones often in crystal or alabaster...

And interestingly enough, **Newgrange in Ireland** (below) looks a lot like one of the learning temples I saw in my 1991 Regression... except that it was set among trees and rolling hills. Inside it was spacious and had a small stage up front and huge low seating pillows, and something that looked like beanbags, and benches along the walls for the students.

There are areas for learning, self-study, small group activity, meditation, sports, and even an area for rehabilitating 'damaged' souls – those who went thru rough times (war, suicide, drugs, etc.) and need to be 'put back together' – there are special energy areas to comfort these souls where Masters administer energy healing. (See Appendix D: **InterLife Components Chart**.)

LifePlan Selection

No one gets into the Earth School without a plan, a Script, if you will, for what they intend to learn or accomplish. The plan, the era, the family, the body are all chosen by the soul, and includes tests at specific points, called **Points of Choice**, where a significant decision must be made. These are LifePlans <u>including one's group</u> that is usually co-playing roles (mother, father, sister, brother, boss, spouse, etc...) and the group roleplays throughout several lifetimes chosen with the barest of script...these are souls that one knows and trusts and they work out scenarios that are almost ad-libbed, but have an established purpose, and are meant to be **catalyst** to spur soul growth.

There are preprogrammed lifetimes that the soul can all up on the **Heavenly Quantum BioComputer** and watch what that life is about and choose to insert himself or herself into the scenario and play it out – kind of like the **Holodeck** scripts that the used in *Star Trek*. The supporting 'actors' are often all OPs, like in a video game, they are Non-Playable Characters (NPCs), but they may be members of your soul group.

> This is why the current scenario of 60% OPs on Earth now is high but there are also 40% souls here, too.

And usually souls do not go thru this role-playing for eternity – there is a 'graduation point' where one moves on to a higher level of training and expertise.

Because this is a significant aspect of soul preparation & training, and many readers do not have TOM, it is repeated here ...

The Heavenly Bio-Computer

The big step is to go to the Computer and, much as one would Google 'Courage' on today's PCs, and 2,000+ entries may come up – so it is better to qualify one's search as 'Courage in health crisis' and the Heavenly Quantum Bio-Computer (**HBC**) will respond with a selection of Lifetime Scripts, Eras and options that require a soul to meet a health crisis with courage. This can be further refined by Era, race, sex, country and married/single, etc...

For example, the soul may refine his search down to the following **criteria**:

> Location = Earth (TL1 on timeline 1)
> Country = England (C10 code for England)
> Era = 1800s (E3.6)
> Sex = M
> Race = Black (R2)
> Marry = No
> Lifespan = 60
> Health Issues = yes, critical (C4 is cancer stage 4))
> Exit Points = two (age 30, age 55)
> Family = poor, small (Fp3)
> Wartime = No
> School = E8 (8th grade elementary)
> ….

This would be an entry in the HBC that looks like this:

Tl1C10:E3.6MR2N60C4:30:55:Fp3NE8…… **LS:TL1.EA6.Mn42. 4356922**
| ----------- LifeScript elements ------------------ | ----------- Id tag --------------- |

This describes where the lifetime is, what elements the Script must have in it, and it spells out a real challenge for a Black man in the 1800's in England who has two close brushes with cancer/death, does not marry, and does not live past the age of 60. Of course more can be (and is) programmed into the LifeScript like Tests and Opportunities also called **Points of Choice**…

These elements are already scripted in a Timeline as LifeScript **LS:TL1.EA6.Mn42. 4356922**. All the soul has to do is choose that <u>in concert with his Soul Group</u> (for support they may enter too), or he may go it alone.

This LifeScript is loaded into the LSReview part of the HBC so that the soul can **preview that LifeScript** to evaluate the family, environment, projected lifepath and particular challenges presented by the Script. It is not subject to much modification. The roles and lifepath are already set, 'scripted' with some Freewill playing a minor part –

> **if the soul could modify on-the-fly it would not adequately test his/her mettle.**

Note: **Freewill** is much misunderstood – the LifeScript must be entered and played out to the best of one's ability – otherwise would it be much of a lesson or test if one's Freewill were able to ignore or change the parts s/he didn't like? The only Freewill choice the soul has is to choose what one's reaction will be to the catalyst of a specific lifetime… if the wrong choice is made, the LifeScript may terminate early. If the soul makes the right choice, the LifeScript can continue forward... There will be **Points of Choice** which are really tests to see just how much the soul has learned, and the soul always has the 'Freewill' to fight the Script or commit suicide – but that is a failed Script and the soul will have to be re-schooled and at some point face what s/he was just sure could not be handled.

> The truth is that **a soul is ultimately capable of much more than it thinks**, and one of the lessons to be learned is that ultimately, we can handle anything. We must come to **know** that (by experiencing it).

Perhaps now Shakespeare makes more sense:

> All the world's a **stage**,
> And all the men and women merely players;
> They have their exits and their entrances;
> And one man in his time plays **many parts**.   *As You Like It*, II, 7.

And again:

> Life's but a walking shadow,
> A poor player that struts and frets
> His hour upon the **stage**,
> And then is heard no more.   *Macbeth*, V, 5.

It's all staged... more than we would like to believe. At least 90% of it. The LifeScript is staged, the players already defined, the roles are set... all one does is select the Lifetime, and the Birth Masters will insert your soul aspect into the chosen Lifetime (see InterLife Component Chart in **Appendix D**). It is merely a **Drama** that we go through whose preset challenges and opportunities (because many **but not all** elements are preset) allow for the Life Review (at death from a Lifetime) to be **evaluated** according to the ideal: an Old Soul who has mastered the Earth experience would 'breeze' through any of the Earth Scripts. Any **Earth Graduate** could handle any of the Scripts... that is why the Earth Graduate is much **respected** by everyone on the Other Side.   (See **Appendix A**).

This was examined more in TOM.

Souls Get Trapped

Here is the danger and why the preceding was important. It is important to not choose too exciting a lifetime as it is ADDICTIVE. And whatever you are addicted to in the Earth Realm will **bring you back, trap you**, and it will be very hard to break the addiction and 'graduate.' The Masters will advise against too exciting a lifetime, and yet – They do not make anyone do anything – the soul has free choice... except in one situation:

> If you are not smart enough to choose wisely, and you repeatedly choose too much pleasure, too much stimulation, too much violence, too much drinking, etc... you may be sent by the Masters into one of the preprogrammed lifetimes where there is **zero Freewill** and it is a rough lesson. Your growth is in choosing wisely and not getting trapped in the Earth pleasure aspects.

So **anything that you are addicted to can hang you up** – even the addiction to save other souls and be a pastor, priest or rabbi... Yes, that can take a few lifetimes to overcome. It seems like the right thing to do given the sin/sex/pollution and deception on the planet – and should we not be active in trying to fix it? Who would say you are wrong for trying to make a positive difference?

Number one: it is not necessary beyond a certain point – yes, fight injustice, pollution and sin, but do not eat, sleep & drink it. Stand up for what is right and the gods will note that <u>you are doing the right thing</u>, but don't make it THE thing you must do – as an addiction. Why? See point #2.

Number two: The Greater Script (running your LifePlan) creates **catalyst** which is often negative for soul growth. **No fire, no steel.** You will encounter deception, disease, pollution, and injustice – it is part of the training here – to see what you will do with it... resist it (and that is all that is necessary) or merge with it and become part of the problem?

You cannot fix/stop/change the problems in the world... and are **not expected** to – but should you succeed in overcoming the Greater Script, the gods will take note.. like you get a **Gold Star**, and then They reset the Script back to the negative catalyst.

> This is not to say that you ignore what is wrong, or people who need help, and don't do the right thing... just don't get so involved that you become part of the problem and lose sight of why you are here... or **worst case:** you get involved in the other person's problem, <u>solve it for them</u> (do it for them) and they learn nothing. You get a **Black Star** for interfering. Your right action would be to show them what works, explain it, and let them do it.

As an example, my grandmother had the gift of healing. She used to heal people of their infirmities and then many of the illnesses would come right back – she never learned that the illness was a **Karmic lesson** and by removing it she was denying the person the opportunity to learn a valuable lesson. She took the person's lesson from them, thus interfering.

Minor ailments stayed healed as they were not Karmic.

And the kicker is: when the afflicted person gets the message and stops what they are doing, **changes their way** of living, acting, thinking, the infirmity <u>can</u> disappear (lesson learned) – if it hasn't seriously/permanently damaged the body in the meantime.

And sometimes the gods will do a miraculous healing if that person has at least partially made a personal change, and still has work to do on Earth... They are not ogres. See, if you truly change, early on, and They don't heal you, you have the right to bitch when you get to the Other Side, because <u>then you'll</u> <u>know</u> whether They didn't do what They were supposed to do -- and then <u>They</u> have a problem – and that aspect is examined when we look at various Job Openings (Potential Areas for Service) available to Graduates.

Ok, so do not get addicted to sex, drinking, smoking, drugs, traveling, being a Righter of Wrongs, or even addicted to having to be with your spouse, family and loved ones. **Any addiction** will drag you back to another lifetime on Earth to try and set yourself free. Many souls have not learned that and so the graduation rate from Earth School is poor – too many are addicted to something here.

The goal is to **graduate**...move forward... **Care, but don't addict.**

What does a Graduate Look Like?

If the Graduate is so well-respected, s/he must be impressive... and yet there are a few points to emphasize:

1. **The gods don't make anybody believe or do anything**
   (the preprogrammed lifetime mentioned above is for a soul
   that is stuck and needs external help – the Father of Light's
   wish is that all souls graduate.)
2. **When the student is ready, the teacher appears**
3. **You don't have to change religion to graduate**
4. **You don't have to be perfect to graduate**

(credit: BingImages: simischoolsblog.com)

...but you do need to graduate.....

Several things pertain to that Graduate state:

1. **You discovered that you <u>could</u> handle anything They
   threw at you**
2. **You gained a measure of self-respect and inner strength**
3. **You gained some self-mastery in problem areas**
4. **You stopped arguing for your way and wants**
5. **You acquired Knowledge and compassion**
6. **You are more STO than STS**
   (your PFV exceeds the denser vibration of Earth, so you
   resonate at 51% or above)

In addition, if you are more humble and patient, so much the better. Your key attribute for success on the Other Side, in the Father of Light's Multiverse, will be to **listen and follow directions** when serving in one of the Potential Areas for Service (examined later). STO means you seek to serve and make the positive difference you thought about back on Earth, and <u>now you can</u>... and They will now guide you and train you.

You are a creator-in-training soul and have a potential that the PTB and 4D STS have always sought to stop, and they often did it with deception, disinformation and disease.

The Earth Graduate is so well respected because to graduate, you had to **overcome** those things – to some degree... **not perfection**, you don't walk on water, but your realization of <u>what</u> you are and your potential AND your **intention** to focus on the goal and serve is what They admire and welcome.

You understood the dictum

Steel is not made without fire

And when the going got tough, you decided to meet the challenge, not knowing if you would succeed but your **attitude** said, "Ok, it is what it is, and I will face it and do my best... even if I fail, my intention is to handle it!" So you dug deeper within and found an **inner strength** you didn't know you had...

that's what They want you to see

...it will serve you in any Position you take working with Them... because as long as you stay determined to succeed and handle whatever, They can empower you (that is a little secret – your **attitude** must be one They can empower when the going gets rough on the Other Side). On the physical Earth, you can lock hands and arms, but on that Side you **connect energetically** and spiritually to form a 'front' or 'presence' that stands against the Dark Side (<u>which does exist</u> – remember the Draco, the fallen Elohim?)

So maybe it is time to share how I know all this and am not just making it up...

Personal Insight

Ok, by now some readers are saying, "Who does he think he is? He's making this all up." Time for a brief explanation (such as what I also did in Ch. 1 of GEP that is much more complete.)

My life had been very hard with even a Mother militating against me – **she didn't want a son** and I was sickly most of the time [Karma payback], but I was protected and there were a number of miraculous 'saves' where I should have died. Still my life was such a frustrating mess that I was seriously considering suicide to get out.

That gets Their attention...especially when **I had a life task** to do that I did not know about and the enemy [4D STS] knew all about me and was trying to stop me... they almost succeeded. I was protected from them, but not from my personal Karma... What I later understood was that the first 54 years of my life were Karmic payback with a revitalization in 1998 that would set me onto my real life task.

So as it happened (by "coincidence") at a Psychic Fair in 1991, one of the psychic readers waved to me and had me come over to her... She had me pay nothing but had me sit down and I said nothing. She told me all about myself and not to do suicide... she didn't even use the Tarot cards! She was a setup! (That's is how They work.) She had me agree to come to her house where she met with clients and did **hypnotic regression** on them (she had an office in the back). I was very apprehensive but agreed because she knew so much about me that I suspected something out of the ordinary was going on... and if I did later suicide at least I'd know what she had to say.

(continued...)

Long story short, I was supposed to be doing a 45 min. session, and did not hypnotize easily, and I had given her a list of my concerns and questions before we started – which she did use – for 3 hours!

During that time. I saw the Other Side, and it was beautiful – God, it was unlike anything I have ever seen on Earth!! I wanted to stay – I know now what the NDE people are seeing! And a Master came and spoke with me – I must have been in the spirit -- and Dr. Kennedy (she was a licensed hypno-therapist) was able to speak with him and **I was allowed to remember** what we talked about and what I saw. She made notes.

I was not told I was to write a book, and I had no idea nor any warning that They would show up in October 1998 and 'rewire' me to begin the preparation (learning via clairvoyance, seeing auras, etc.) for doing something down the road. It was when **Baldy** (real name something like Phillealel that I cannot pronounce) showed up in Jan 2008 and said we are writing a book and VEG was transcribed (the 868 pp that became VEG and TOM).

That "inside" information is in TOM and some is in VEG, and another part is now in this Chapter 8. And GEP Ch. 1 says how it came to be that I was set up for the writing task. Sheesh, I thought it was just one book... and in 2016 Baldy took his leave and with him went the clairvoyance and seeing auras... Damn! But I still had enough info and insight for 3 more books (VDoG, DNQF and this one). And Baldy did advise several times on this current book.

What Does it Take to Gradate?

So graduation is a **choice** you make, with **intent** and focus. Getting as much true Knowledge as possible and learning compassion, patience, and **respect** for all living things, people and the environment. **Set your intent to be STO** (Glossary). Tall order – but that is why just to be on the Path is often enough – if They had to wait until you were perfect ... well, no one would graduate.

BTW, 3D – 4D – 5D and 6D have a purpose (and below 6D there are other shorter Schools...

Those that return to stay to the 4D Realm (InterLife) from which we came, are considered Graduates. If really great progress was made in the Earth trip, the soul may move on to the 5D. What is the difference?

3D is to learn the lessons of Ego, Me and My and Mine.
4D is to learn the lessons of Compassion and caring.
5D is to learn the lessons of Truth, right action and speaking.

6D when the student is ready is to merge Ego, Love and Truth.
7D is where you work with the Higher Beings/Elohim.

Overload

What is Man that Thou are mindful of him?
--- Psalm 8:4

You are an **eternal soul** with the potential to develop your divine connection to the Godhead, and if you were created <u>complete but **undeveloped**</u> (would God create flawed souls?), then why would you need Karma, or Reincarnation, or further training and 'lessons?' Answer: because many souls have become dysfunctional (including addicted to something) and need to be rehabilitated. Because you don't know that, the Higher Beings, who are benevolent, keep you here. You are contained, watched over, and should be seeking to work your way out.

To repeat: the secret is to <u>not</u> get blown away by the frustrations, upsets, and failures – it is all part of the **catalyst** to effect the changes in you. It is not to be taken personally. God (aka "Infinite Awareness" [thanks Icke]) loves Man and He designed the Greater Script within which your LifePlan operates.

You were <u>not</u> given a role or lifetime you cannot handle; God doesn't set people up for failure... **but** you might have bitten off more than you can chew, and **you feel overloaded**... you might have insisted on a really hard lifetime to accomplish more than normal... and if you now complain, and **ASK** for help, They can come in and reset a few things – the gods are not ogres – destruction of souls is not Their goal.

But due to Freewill, you have to ASK. Sometimes that is called prayer.

Overloading & Offloading

This was discussed more at length in TOM, Ch. 10. But because of its relevance here (and some connection with Appendix A), it is repeated for clarity:

Because souls are multidimensional, with multiple aspects of themselves in different realms, each **soul aspect** (small 's') is linked back to their main Soul (large 'S') in the realm from which they came. A Soul may split into 4-5 different aspects in multiple realms, timelines and locations (i.e., 3D planets versus 4D realms) for different learning experiences. (See also **soul** and **aspect** in the Glossary.)

Note in the diagram below that YOU as a main Soul (or **Higher Self**) in the Home Realm decide to experience 3 other realms or timelines. This requires an allocation of your energy to 'replicate' yourself into those other 3 realms… and you don't want to spread yourself too thinly! For the average soul aspect experience, a 20% allocation is sufficient (as the Higher Self is quite powerful and has a lot of energy), and in a difficult realm, an allocation of 25-30% may be appropriate… this is an "energy bank" that you can draw on to meet higher energy demands of the lifetime for health, reserve stamina, and mental prowess. Note that when the 3 aspects are allocated, the main YOU (**Higher Self**) still has 35%.

Allocation of Soul Energy

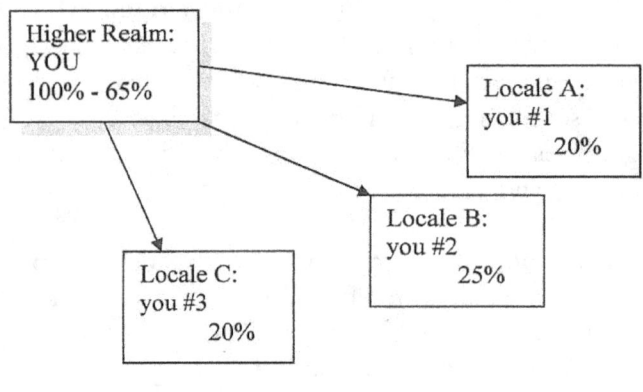

Chart 3c

As was said in VEG, Ch. 7, this can result in one or more aspects feeling the upset or pain that another aspect is going through – and whereas their day was just fine, all of a sudden, soul aspects 1 & 2 now feel anxiety, fear, anger, depression – whatever soul aspect 3 (somewhere else) is going through!

> Soul aspects will all react alike as they started the current journey all alike… it is the same Soul, and each aspect reflects the same strengths and weaknesses as the original Soul <u>before</u> the split.

366

(This can also work for joy and peace coming through.) The point is that **not all our moods are self-generated, and we may be offloading another aspect of self** (whose common lack of self-development is confronted with more than it expected!) and is not able to manage the challenge/upsets/he or she is facing. This is possible (as is said in TOM Ch. 11, section called 'II: Using Energy.'), due to the **energy cords** connecting souls.

> BTW, this is the explanation of what was said in the Bible –
> "Inasmuch as you have lusted after a woman, you have committed adultery with her." Matt. 5:28 (paraphrased) Your
> energy connected with her, so the 'lust energy' passed to her.

The point in all that is that when one of the soul aspects does gain mastery and transcend his/her Drama, that soul returns to the Soul (part of the OverSoul) as **Higher Self**. And there is an Integration that occurs --- or a **Unification.**

Whatever the soul aspect learned to do or handle becomes part of the main Soul's set of mastery skills and when the other aspects also rejoin the Soul, they add their strength to the mix… And the next time out, as the Soul splits into another set of aspects, the next set of soul aspects will all have the benefit (**imprint**) of the collective set of skills or maturation gained earlier….

Why Graduate?

There are two answers, two sections:

Souls as Earth Graduates

Man has a unique nature and a unique destiny, and we are here on Earth to work it out, and we probably would have by now if: (1) we hadn't resisted our lessons, (2) we didn't party so much, and (3) if the PTB had let us know.

> Our purpose in coming together as **creator fragments** is to
> succeed in training ourselves enough about love, caring, and
> relationship to become more of who we are. The redemption
> is we, as … **progeny of the high forces of creation,** are
> ascending back to heaven in unity of diversity, as celebration
> of individuality in communion, not loss of individuality.
> We are holo- or fractal-fragments of the creation's Creator,
> [who] is **wanting to create creations <u>with</u> us** not for us…
>
> Our first major task is to re-create the existing mother universe
> we find ourselves within with all its conundrums. Solve the
> unsolvable evolutionary problems. We learn to pick up the

ball in our **training wheel practice universe** before we even want our own. And we want to learn very carefully, and so we use time to do it in a <u>serial</u> manner. That is the game…

Each **creator fragment that is a human soul**, ultimately seeks its origin [and soulmate] and return to Home. We are all … learning to love and nurture our individual and co-creative mutual universes.

Soul Evolution

Our job is to achieve spiritual evolutionary acceleration sufficient to help solve age old problems of spiritual evolutionary inertia in the universe. [emphasis added] [225]

Elsewhere, the information is given that **Man is ultimately to be a co-creator** as his inheritance providing he can keep moving into more and more optimal timelines:

…as a reward, humans who accomplish this task, will be granted an initially uninhabited virgin future that can become even more optimal beyond comparison…

That final loop optimal future becomes the end-game singularity conduit path through which **all** souls of all alternate [multi-dimensional] lines will eventually travel to become qualified **macro-creator agents**. [226]

That is the promise to all Earth Graduates. Co-Creators!

Potential Areas for Service

Having said that the Soul who finally graduates from Earth can fulfill certain basic, initial, functions in the Father's Kingdom(s), it can be shared that these are some of the initial areas open to Graduates:

>**Bio-plasmic Quantum Computer Techies** – responsible for basic
>Heavenly computer support and maintenance.

>**Bio-plasmic Computer Programmer** – performs fractal sub-
>Simulation programming under supervision.

These last two were demonstrated in QES. The Heavenly computers develop and manage Scripts, including the Father's overall Greater Script. There are also Science quantum computers that simulate **new worlds and species** to be designed and built.

Akashic Records Librarian – maintains life records' storage and retrieval.

Gods-in-Training I – responsible to oversee the Drama/Greater Script Simulation: Man and feedback of the Control System. Many sub-areas here.

Gods-in-Training II – responsible for the Holographic stabilization and interface with the Replicator technology. Sub-areas here.

Soul Counselors – responsible for evaluation, guidance and training of in-coming souls to the InterLife for further development which may include **imprinting** (see Glossary) or vibrational adjustment. Many levels here, including Teachers.

Bio Scientists -- these beings experiment with **new lifeforms**, ways to engineer them, transplant them, and ways to improve them.

Astro Scientists -- these beings experiment with Galaxies, Suns, planets, comets, etc. to engineer new planets, manipulate orbits, **manipulate Dark Matter/Energy**, and all the while ensure balance/order in the material worlds.

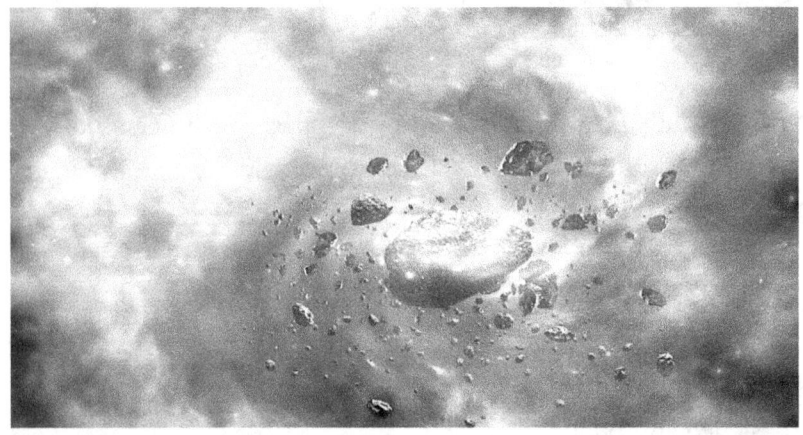

MLD Scientists -- responsible for multilevel universe and dimensional interface, including handling **Timeline Shifts** when necessary.

There are many others, but **it is a busy world over there**; no one is sitting around on a cloud playing a harp – unless they're on a coffee break! One reason that the PTB wanted to block Souls from progressing (see Chapter 4: PTB Control) is <u>also</u> because of this position:

> **Gods-in-Training III** – responsible for overseeing, managing and controlling the Beings of Light, the PTB, and humans –to make sure that lessons are properly administered (according to Scripts) and it amounts to controlling what the PTB can 'get away with.'

> G-I-T III is as close as the InterLife comes to having a "police force."

And because I once asked,

  "What if the Gods-in-Training choose to abuse people, or do something nasty?"

Answer: They are sent back to Earth for rehabilitation.

*** The Goal ***

'You are here to enable the divine purpose of the universe to unfold. That is how important you are.' Eckhart Tolle

www.facebook.com/polohemandez2964

And -- It is worth repeating what Robert Monroe was told (Over There):

> **Earth School Graduates are very well respected** – the others in 4D know what souls have to go thru to "graduate."

By far the greatest motivation — surpassing the sum of all others —
is the result.

> **When you perceive and encounter a graduate, your only goal
> is to be one yourself** once you realize it is possible. And it is. [227]
> [emphasis added]

When you know, why would you settle for anything less?

Namaste

Chapter 9: Earth Synopsis

In an attempt to summarize the main points of everything (including stuff from prior books), into one chapter, this is it.

Much was said in the above 8 chapters and a lot was summarized in the Breakaway Civilization, but perhaps a grand summary from Earth inception to current day would help understand where we are and how 5G fits (answering the Option question in Chapter 5) – **5G is the trump card in the Anunnaki Earth Agenda**... but Dissident or Insider?

After all, if this is the last book, it needs a Major Summary.

In the Beginning...

There is a God, or what we could nowadays call more of an Intelligent Being, David Icke calls it Infinite Awareness, the Tao calls it Omniscience and the Void, and it was referred to as The Force in *Star Wars*. Whatever, it is The One, the Grand Being who oversees this Universe (each Universe has its own Overseer, or God) and there is a hierarchy running from the Top to the humans, including the Elohim, Higher Beings, Angels, Masters, Avatars, Oversouls, Higher Souls, Group Souls, souls etc.. and finally humans... into which the eternal souls incarnate in a 3D body for some learning experience in the Earth Realm.

And there are the ones referred to many times in these books as **"the gods"** – the ones who administer curriculum and oversee Earth School operations.

There are many different working positions in the Multiverse – as outlined in Chapter 8 – Positions Open to Graduates. One of those involves the creation and maintenance of planets, solar systems, galaxies... and drilling down, we come to the creation and terraforming of planets like Earth – intended as repositories of the Biota in the Universe ... all types of flora and fauna living in a kind of Zoo or Reserve – and even then, further research and development is done to come up with new species, or changes to existing ones, and evaluating them to see if they thrive and would they work on another world, similarly engineered?

And by the way, on a very high level, the **Suns** throughout the Galaxy are (as the ancients used to say) actually Higher Beings sustaining the life-giving output of those stars. The god-man Apollonius (Ch. 11 in VEG) used to venerate the Sun every day as he knew what it really was. The American Indians, the Inca, and the old Egyptians (venerating RA) also knew the truth.

Earth & Man

In the beginning, the **Elohim**, very High Beings, closest to the One, created a humanoid in the Earth Garden (not Eden, yet). This was an Experiment to see if souls could inhabit and function and learn from such an opportunity... Inasmuch as souls had been created, they were inexperienced and while not dysfunctional, they were **unschooled** ("blank slates") and the Father's **Plan** was that the souls He created could be developed and rise up into the Higher Realms.

The One had also created the Elohim and while most were STO and served the One, some of them disagreed and did not want to see such development and objected to the Plan. He had also created the Angels who did not have a say (no Freewill); they were subservient to the Elohim. Since the Elohim had Freewill, about 1/3 of them fell out and became known as the **Djinn** (aka Archons) – with all the powers and Freewill of the Elohim, but with an STS agenda. They originally decided to occupy the Earth as the Djinn and when they abused that realm, and each other, they were removed (as Enoch reported) and they fled to other planets as the **Draco**... creating the Anunnaki and **Dinoids** to serve them. Some escaped to another dimension where they mostly reside today.

So Genesis 1 says that Man was created in God's image and while we have no idea what God looks like, we can say that the anthropomorphic aspect of God, and the **Elohim**, qualifies. In Genesis the Elohim created Man; the Anunnaki later **re-**created humans to serve them. The Draco created the Anunnaki who genetically engineered Dinosaurs and then re-created the humans. **Both Djinn and Elohim could create.**

The Djinn hated humans as even the Qur'an says so... the complaint was that they, the Djinn, were created from Fire (spiritual material), and Man was created from Clay, thus inferior. So the Djinn set about harassing the humans on Earth and saw to it that their progeny, the Reptilians, specifically the Anunnaki, used the humans and kept them as slaves, as cattle (– and do so to this day when possible).

> The Djinn do not want the humans to wake and develop their potential – men and women (with **souls**) have a divine potential that is being stifled on Earth due to the Anunnaki puppets (the PTB) who control the Media and religion so that Man cannot discover who/what he really is.

There are no demons, no Satan, just Djinn (aka Archons) operating as the 4D STS and because they can shapeshift, they can appear as whatever they think will scare humans – red demons, weird slimy ETs, huge beasts.... sometimes as shadow people (who are prevented from doing physical harm due to the presence of Beings of Light (BoLs or Angels without wings) who protect most humans. So the shadow people just show up occasionally, observe, and try to scare humans.

Anunnaki Days

So into the Earth Garden come the Anunnaki, who know about the Earth because the Draco told them, after they got kicked out (for misbehaving as Djinn – fighting among themselves, no less)! And this was permitted as the Anunnaki were initially there to enjoy the Garden/Zoo and even mine some of the resources... a semi-peaceful entry into the Garden. These were the **pre-Anunnaki** spoken of in Chapters 1&2. And the Solar Council permitted it as Freewill has always been a part of this Galaxy.

The Anunnaki were long-lived and were as most reptiles, able to live into the thousands of years and they used monatomic gold for one of their potions to sustain this longevity. They avoided the Sun as it aged them. They came and went between Mars and Earth and mined resources on both planets.

About 200,000 years ago, a dispute arose between the denizens of Tiamat (between Mars and Jupiter) over the resources and their use, and the Anunnaki declared war on the Tiamatians... who blew Mars' biosphere off the plant and left a huge gash in the surface (*Valles Marinaris*). The Anunnaki having superior weaponry, blasted Tiamat to what is now the Asteroid Belt.

What was left of the Mars' Anunnaki populace went underground and as there were no survivors of Tiamat, the **Solar Council** banned further activities (mining) on Mars and most Anunnaki went to Earth and that is where **Enlil** and his cadre came to Earth.

Anunnaki were not paragons of virtue. They were lazy, lying, cheating, scheming, stealing, they were lusty, committing incest, and could be very violent. That was the majority, and yet there were also the more moderate and peace-loving Reptiles who were smarter, not warrior-class, and they were like the chimpanzees in *Planet of the Apes*, and held the reigns of science – compared to the brutish apes. Good analogy.

> Remember from Chapter 2 that some Reptilians helped
> the suffering Pleiadians to survive.

However, after several thousand years, the Anunnaki grew tired of the mining, farming and building ziggurats, and they decided to take the humanoid (Homo *erectus* or *habilis*) and **re**-create (genetically modify) him as a slave to do their work. Strong but not too intelligent. **Enlil** was the commander in charge of the expedition and approved, knowing that when they were thru with the workers they could dispose of them... they were crude forms and no souls had incarnated into them... so no harm was done in the eyes of the Solar Council. The worker humans were animals or *Lulu* (in the beginning).

But then **Enki,** the science officer, got to tinkering with the humans to see what could be developed, and evolved them to the point where *Adapa* (3rd or 4th generation that we know of) had more intelligence, could speak and could reproduce) and all of a sudden, Enlil and Enki realized they were in violation of the **Galactic Law** that said "If you develop a sentient species, you are responsible for it, to nurture it and see it thru to a responsible, mature species."

Enlil, thought, "Crap, I'm gonna do the Flood anyway – we have Nephilim (nasty cannibalistic giants) as a result of the interbreeding between my people and the *Adapa* women!" And that was how he got away with what he did (homicide), yet the Solar Council required **the main Anunnaki party to leave Earth** and keep a small **Remnant (underground)** to see to their duty to "evolve" the humans into a responsible species... hence the little biocybernetic androids called **The Greys** were developed to effect the continued genetic upgrades on the surface.

Remnant Machinations

So the Remnant went underground (mainly the huge cavern under what is called China-India and part of Mongolia). Huge underground cavern system. And as is so typical of Anunnaki nature, **they split on how to deal with the humans...**
Those that are called the **Insiders** wanted to develop the humans as quickly as possible and get out of here... do the required upgrade and split for home.
Those that are called the **Dissidents** (about 40% of the Remnant) wanted to just be rid of the humans, disease them, disappear them, use calamities to wipe them out, and get out of there too. **Both groups wanted to be free of humans**... but only the Insiders are still trying to abide by Galactic Law.

To that end, the **Insiders** are running the robotic genetic engineers, called the Greys, to visit the surface and upgrade chosen humans' genetics and effect the change. Anunnaki are not permitted to live on the surface in their Reptilian form, and they can't anyway as the Sun ages them. So they send the Greys to do their genetic work.

In the meantime, the **Dissidents** have generated their **Hybrids** that look human and inserted them among us to do their dirty work of getting us into wars and spreading disease to wipe us out and be done with us. In retaliation, the Insiders have inserted their **Hubrids** to mitigate and obstruct what the Dissident Hybrids are doing – you can see some of the outplay of that on the Nightly News... think government, Congress, MIC, top levels of business, and heads of state around the world (Iran,

North Korea...) If you wonder why some humans do some anti-human things, now you know. (See also end of Chapter 5: **Genetics Are Important).**

They are not all humans. And we aren't even talking about the OPs who are really puppets in the hands of the 4D STS and Dissidents – they can be (and are) **mind-controlled**. (That is one of the skills of the Draco/Djinn as well as the Elohim... in addition to shapeshifting.) Think: AI Signal.

> So one group wants to kill us and the other wants us to get it together so they also can leave. And it isn't really working, the genome is very tricky and they cannot and **do not** give us Clairsentience to solve our 'beastialty' (which is how they see us as we still have the Anunnaki genetic legacy: lying, cheating, stealing and there has been discovered a genetic predisposition to violence, or a **'Warrior' gene**.) [228]

Anunnaki Earth Agenda

So we come to the point where the Germans were given a 'crashed' UFO in 1936 in the Schwartzwald which they back-engineered to create their Vril and later the Haunebu (Chapter 8). What we need to look at is: **Who** gave it to them? It was empty when they found it... and it was functional.

If it was the **Dissidents in Germany** it would be in exchange for working with them to take control of the surface world (Think: Nazi support)... remember Anunnaki Control of humans has always been a priority. And the Reptiles do have a despotic Orion Empire that has subjugated millions of beings on several other worlds. They are control freaks... just like the Nazis. (Witness their special creation called the Soviet Union.)

It was the peaceful **Insiders at Roswell,** who put high technology in the hands of the clever Americans (and Operation Paperclip) who would use the technology to achieve parity with the Dissident-supported Nazis.

One thing is for sure: it wasn't a real ET UFO that safely flew all the way across dangerous interstellar space (Micro-meteorites, Gravity Waves, solar radiation, neutrinos, and fluxes in ZPE/Dark Energy, etc.) just to crash on Earth.

So that means that the 'crashed' UFOs in Germany (1936) and Italy(1933) in the early 30s were **planted by the Dissidents** and then Hitler was visited and told that they would empower him and he would be the figurehead for the (ego-boosting) Thousand Year Reich – a resurgence of the old Teutonic Knights and glory that was so popular with Hitler. Why do that? Well, if you can't easily kill them, get a major war going and let then cull each other... cut their numbers down (à la Georgia

Guidestones) and if the Reich succeeds, the Dissidents now control the planet... Why go home? Invite the Draco back and turn Earth into another outpost of the Orion Empire. Fortunately that agenda failed.

At the same time as WW II, there was a subplot where **Heinrich Himmler,** who was more of a psychopath, was visited by the Dissidents and convinced to work with them and they would make him *Führer* because Hitler was not getting the job done... and that resulted in **Himmler having a bomb placed in Hitler's War Room** (July 1944) – things were going too slowly with Hitler ... if Himmler were leader, then a quicker victory and reworking of the *WunderWaffen* might be done to move ahead. Himmler apparently had more of the **Warrior Gene.,** Hitler was waffling.

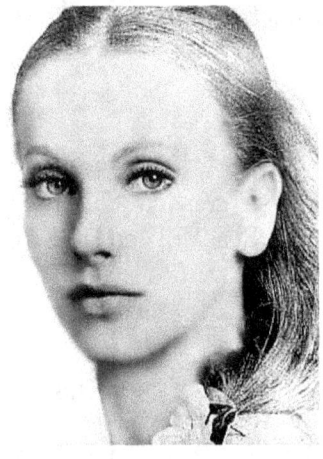

Maria Orsitch knew that and made the scientists aware of same, and swear to follow her request, and then somehow they just could not get the UFO armaments to work. Thus, **they could not be used in the War.** She then gave the Allies better radar, and saw to it that the weather in Russia went really cold and bad, thus stopping and defeating the Nazi attack on Moscow.

Overall, the Insiders knew what was up, and sent Maria Orsitch to intervene... giving (1) a superior design to the Haunebu propulsion and (2) commanding the Nazi scientists to NOT succeed in arming the Haunebu – nor the Vril – and thus thwart the so-called *WunderWaffe* of Hitler. They complied and Hitler never got his UFOs into battle status... the *KraftStrahlKannon* (microwave ray gun) just did not (magically) work. (Her work was alluded to in TEW, a novel based on real events.)

Then she saw to it that a Haunebu was 'crashed' in Roswell in 1947 (and off the Southern California coast in 1941) to give the Americans a way to even the score and achieve parity with what the Germans had developed. She made sure her cohort, **Nikola Tesla,** had some further designs and science documents that would fall into the hands of the Americans, and then she went home (Aldebaran).

The Russians and Indians (Hindus) also got theirs.

So as a result, in the last 40 years, about 5 countries have UFOs; they are built on Earth and flown (largely) by humans.

Maybe we need Maria back today to intercede in the 5G issue...?

Not to be outdone, the Dissidents (and 4th Reich) shifted gears and began their program of infiltration into Western Science and Government, still attempting to subvert and control, hoping to take over the US which has for years been **the linchpin** of the Free World. They succeeded in dismantling the British Empire and all that was left was the US. If the US can be brought under their control, or destroyed, they will have done what Hitler and WW II did not.

> You have to wonder if this is what we are seeing with the obstruction by some Democrats currently in the News (2019)... the large democrat-controlled cities have many sick street people and health issues that are being totally ignored, as well as California being made a "sanctuary state" for illegal and violent immigrants...
>
> While this is not a chapter on politics, the obstruction of justice and the rise of **prepaid and sponsored activities** of Antifa look suspiciously like what the Dissidents would do to create chaos and confusion in America... just FYI.
>
> (BTW, I am an Independent.)

Technology for Control

Whereas the German UFO technology was thwarted (by Maria) it was still the design of the remaining Dissidents (headcount now down to 25% of the total underground) to come up with **a way to get the humans to adopt a technology that they would <u>willingly install and use,</u>** that could also be **hacked** giving the Dissidents control again. The mania for **Control of Humans** has never gone away among some of the Remnant, and yet many defected from the Dissident side (after WW II) to the Insiders, thus thinking that that would be a faster way out of here and that is why the push to **technology control has been ramped up** – they just do not give up!

> And human fascination with electronics and geegaws is a given, so Telecom has moved (unaware of the real reason behind 5G) to quickly implement it and reap the monetary rewards. And this is why the **Telcos have not tested the 5G effects nor do they care**.
> (Even if they wanted to, the mind-control persuades them from doing it, or listening to the scientific and medical evidence that it stands to be seriously deadly to many humans... just as designed in the Dissident lairs.)

The point being: the Dissidents have decided that if they can't have the planet, you won't either. And 5G-6G should remove most of mankind [229] (Think: Georgia Guidestones – you were warned.). They will be able to take over if 5G creates massive health issues reducing the population...

Would the Solar Council object? No – even though your Freewill was subtly violated, **YOU AGREED TO IT**. You accepted the 5G network knowing that there were dangers. (Chapter 4.)

I know this sounds like froo-froo, but it is very real, and people if nothing else should be questioning the health effects of 5G... and at least following the Installation Guidelines given at the end of Chapter 5... just to be safe.

Your not knowing about the Remnant and what they are up to really **aids and abets them** because you don't believe it could happen in the USA... let alone somewhere on Earth.

They have subtly controlled mankind for centuries, also covering their tracks and making sure you laugh at such ideas (they control most of the Media = Fake News), and now they are about to remove their main obstacle to going home. You. We are in what appears to be an **Endgame**.

Rational Overview

Ok, let's say you don't believe all this -- that the FCC, FDA, CDC and AMA would protect us, surely Man isn't attacking Man...? You are right – Man is not attacking Man, but we are NOT here alone, and never have been alone.

They operate from underground. They and their PTB puppets are attacking us.

So you say, well surely the Military would take them out... or: we would be told... No. Not So... the institutions that are supposed to be protecting us have all been infiltrated and compromised – the FCC, FDA and CDC and Congress.

It is called the **Fifth Column**

This was well-known in World War II – and (left) is a poster from that time.

The game is to **infiltrate and do sabotage from within**.

Any group of people who **undermine** a larger group from within, usually in favor of an enemy group or nation. The activities of a fifth column can be **overt or clandestine**. Forces gathered in secret can mobilize openly to assist an external attack.

Activities can involve acts of sabotage, **disinformation**, or espionage. [230]

(credit: US National Archives & Records Admin.)

The Anunnaki are experts at it... so good, you don't believe it is possible. And there are sleeper cells of **Arab terrorists** infiltrating the US, as well as **Communist sympathizers**, and blatant **Socialists** pushing for their agendas (which match the Anunnaki Earth Agenda). They tried it in the USSR but lacked the large computer network to succeed. Now they have it.

Evidence That Demands a Verdict

So I am going to remind you of several things that are on-going and which the controlled Media is telling you are Ok... just like they are

> either quiet about 5G
> or when they do hype it,
> it is said to be great for America...

Disinformation is part of what a Fifth Column does.

Here are the items you need to look at:

1. GMO – genetically modified food... which triggers bad Epigenetics and the changes to your DNA.
2. Chemtrails – weird chemicals in our skies...**Ba, Al**, and Strontium 90, cadmium... plus **nanites** to spread it... with the advent of Morgellons Disease.

   Note: Ba + Al = Baal the pagan god of Mesopotamia (aka Enlil) Coincidence?

3. Fluoride in our water – and **Europe has outlawed it**.
4. Mercury (Hg) in our Flu shots and teeth fillings... mercury is a poison.
5. Mercury (Hg) and aluminum (Al) in our Vaccines.
6. Soft drinks in aluminum cans... leeching Al into the drink and Alzheimers **is** a result.
7. Monsanto food control – seedless crops so you cannot grow your own fruit.
8. RoundUp, **Glyphosate** and DDT – poison weed killers get into the water table and into our tapwater.
9. CPAP nonsense – just turn over on your side, or use the new nose clip or jaw alignment mouthpiece.
10. Setting bottled water in the Sun to bake the **BPA** into the water... a known carcinogen
11. **Aspertame** or Nutrasweet in food and drink – a poison for some people
12. ADD & ADHD – due to EMF sensitivity, but we meditate them (Vyvanse or Ritalin).
13. Colonoscopy – **Europe has banned** as inept technicians poke holes in intestine walls then you need surgery.
14. LED BlueLight radiation... VR Headsets – **damage retina**.
15. Electric Blankets... suppress normal (weak) body EM field.
16. Electric Cars – front engine and rear batteries bathe driver and passengers in EMF radiation.
17. Alexa *et al* – listen to you and record you.
18. SpaceX and Amazon putting 12,000 5G satellites in orbit to bathe everyone in EMF radiation.
19. Cholesterol scam – 100 is not ideal for everybody (see Chapter 7 article: **statins vs water and dehydration.**)
20. Violent Video Games that desensitize people and create prions (Chapter 7).

No. 13 is also bad as **autoclaving** the instrument they put up you does not always kill the germs (*E.coli*) from the last person it was in, and you can get infected... European doctors now only do it in cases where they suspect a polyp or tumor... not "just for let's see" as is done in the US.

...and **there is more**, but you should be asking yourself – WHO is doing this, WHAT is the goal, and HOW does it affect me? Is this enough evidence to say that there **is** an agenda to harm humans?

Summary

To repeat: The **FCC** has approved all the 5G stuff (and they <u>refused to read</u> the valid medical results of 2016 harmful EMF testing), and the **CDC** is not telling you about the health damages they are seeing from EMF radiation (and they <u>have</u> the scientific/medical studies), and **Telecom** is not testing nor are they reading the scientific/medical reports, AND the **insurance industry** will not underwrite or insure the Telecom industry.

Doctors and Engineers say there **is** a big problem.

Diabetes, dementia, depression, insomnia, DNA damage, and a wide range of health issues (see lists in Chapter 5) result from exposure to EMF radiation (including to WiFi), and the Telcos and the PTB <u>allegedly</u> do not want to kill a "cash cow."

I am suggesting that it is more than that.

There are **safe ways to install 5G** and no one is looking at that – except the doctors and engineers who are protesting the rush to install 5G. Rash and irresponsible behavior is suspect.

Are you ignoring the red flags, too?

Final Warning

Realize that **when (not if)** 5G is fully implemented, you cannot even head for the hills...

And as **Drs. Trevor Marshall and Martin Pall** have said...

> ...this is **a civilization ending event**...
> There are dark forces putting forth an agenda as crazy as
> this because... I mean
> **the endgame is basically the end of humanity as we know it...** [231]
> [emphasis added]

I know you think that sounds like overreacting, but realize that the Anunnaki have given us the technology to **do ourselves in and give them the planet**... and they have waited patiently for millennia to develop us or remove us – failing to develop us any further, **it looks like we are now in the Endgame.** Also see Appendix G: the section on WiGig and GiFi – another expert claims **5G is a "Death Grid"** and perfect way to exterminate Man.

Here is the problem: when you tell someone the Truth about our Earth situation, they do the typical inane human response....

> "Oh, I don't like that...
> therefore it's false!"

Humans are used to living in a **Prophylactic Fantasy** – everything is Mom, Baseball, Chevy (or Ford), and Apple Pie. They want their lives as "straight vanilla" as possible.

The Reptiles know humans are **fragile and naïve,** and while Man was given the ability to **speak** (FOXP2) and the ability to **procreate** (Chromosome 23), he was deliberately NOT given an item in his genome that would have allowed him to unite and become a much more dynamic, intelligent human:

> **The Anunnaki deliberately did not give Man the genes for Clairsentience.**

The Anunnaki also did not make Man too smart as **higher IQ** also tends to 'birth' a basic level of Intuition. So Man was kept in the "just barely smart" range – and IQ has been found to be a factor controlled by the genes.

This Clairsentience ["clear sensing"] is also referred to as Inner Eye, Inner Knowing, and relates to Higher Consciousness which goes along with higher IQ which develops Intuition – which is what Göbekli Tepe was all about (see Appendix E). This is also achieved via **Kundalini Yoga** (also AL, Ch. 9-10).

The key to Anunnaki success: You can't manipulate and control a human who intuitively knows WHAT is going on and WHO is doing it (or who is very healthy and cannot be controlled by Rx), so Clairsentience was kept out of Man's genetics.

Do you begin to see a red flag now?

The Bottom Line

If Man is not smart enough to at least check this 5G info out, and then <u>stand up for his right to fine health</u> (i.e., against a mis-implementation of 5G), then he deserves to lose the planet. It is that simple. This is an Endgame.

> Then Enlil will finally be proved correct: **"Man is dense and cannot further develop; we should not have (re-)created him."**

> It **is** an Endgame and we either get it together NOW or only the humans who made it to the **Breakaway Civilization** will represent the human race from Earth....

> ...and even then we will have to fight the Orion Empire when we are <u>off the planet</u>, just as the Pleiadians and Aldebarans still do. They are our human relatives who have helped us and who do have a better genome – <u>they</u> have Clairsentience.

> So we cannot avoid talking Option 1 in Chapter 5.... but do we let the Reptiles win? **They are malevolent and clever.** Are we enough to stop them – even if we wake up?

> Can we get **Maria Orsitch** to again help us?

Maria Orsitch likeness circa 1945

> She definitely had a stake in supporting Nordic progeny... It was **the Aldebarans (her planet) who 'seeded' Northern European humans** millennia ago. She gave them technology so they could get out of here

Or, is she already here?

And yet...

Major **help could come from off-planet**.... will it? Has it?

Or can we rely on what the **Insiders** aka **Mamitu Nammu** (see Chapter 1) the great Gina'abul **Planner** working with the *Kadishtu* who are also called the Elohim?...

I cannot believe that the Insiders will just let the Dissidents destroy what has been nurtured for millennia...all of Mankind. BUT Earth is a Freewill Realm and

if we want help NOW we should ASK for it.

Why? According to Baldy, both factions in the Remnant are tired of the pettiness, ignorance and violence that even the genetic upgrades to Man (done by the greys) are not enough to change... Maybe it is time to let this dysfunctional species go (as Enlil said) and then both factions can go home...

> If I were building a **Breakaway Civilization** I would try to
> identify the best humans (IQ and genetics) that I could – and
> even scan the results of the DNA testing agencies that advertise
> on TV (how clever) to help locate good, stable or superior DNA.

So this is not to be totally pessimistic, but we have had our chances to turn around and the <u>human</u> PTB are very much into power and control and they staff and direct the **Deep State**... Human Lords love being Lords.

They are not in the Breakaway Civilization.

And while this book, and several others, argue that there is a **Greater Script** that is <u>not</u> out of control -- but destruction of Man is not "God's will" either – if the current genetics and lack of intelligence in Mankind predisposes us to sit by and watch the Dissidents wipe out 90% of Mankind... maybe we deserve it?

The eternal souls will always reincarnate somewhere else... So they will survive and whatever happens is ultimately **catalyst for soul growth**... and some of it is rough. And it looks like with 5G we are in for a rough ride this time that may see the end of much of Mankind (the "useless eaters" or Hilary's "deplorables") – and it has happened before on the planet. (Think: Atlantis, The Flood, and the AD 800 Reset that you know nothing about [see VEG Ch. 10] but it was documented by the Chinese and the Maya.)

And the Anunnaki did use *Suruppu Disease* (aka HIV), The Flood, the Spanish Flu (1918), and Ebola and The Plague to **cull our numbers**... 5G looks like just one more used on a population that is fascinated with electronic geegaws but largely stubborn (or ignorant) and <u>at effect with the 'wonderful' Anunnaki genetics</u> (lying, cheating, stealing, rape, and abuse of fellow beings – yes they did it all and some programming <u>does</u> reside in our genetics, including the **"Warrior Gene"** [227]).

So the Greys were only marginally successful in rewiring the human genome, from **Warrior to (the unexpected) Gay** -- don't laugh, the Testosterone was dropped and a different kind of soul inhabits the Gay population and they <u>are</u> smarter, more creative, more caring and human replication has stopped among them – as was the goal (according to Baldy), but that was a side effect and not The Goal.

The Plan was to create Homo *noeticus*, the Indigos, but the human genome is so complicated that it had limited success. So much was cut out by the Anunnaki re-creation of *Adapa*, that there is not much left to work with (real Anunnaki higher functions were 'filled' in Man's genome with meaningless "Junk DNA" – and that may be why there have been reports of many people seeing and interacting with the new **Hubrids** on the really huge ships in orbit around Earth in the Southern Hemisphere. [232] They have been there for 60+ years.

The **Georgia Guidestones** are a warning (1980) <u>forty years ago</u> that has been largely ignored, and if Man is culled AGAIN as the 'Stones suggest, it stands to reason that the Hubrids that have been developed (and kept in reserve as a result of the Grey's genetic operations) could be used to **reseed the planet**... maybe they will be better behaved, a superior Clairsentient version... or will they (as Dr. David Jacobs fears) be programmed to obey a hierarchy and have no Freewill?

Therefore...

So there have been a few start-overs. And **the Anunnaki are still in control**. The Solar Council still considers them responsible for their progeny.

> That is why there must be a **Breakaway Civilization**... so that the best of the best of Mankind will continue despite what the Anunnaki come up with as a way to deal with us... this time.

> Don't ranchers and wildlife management people "cull the herd"?
>
> You heard it in one form or another from authors Charles Fort, John Keel, Jacques Vallée, William Bramley, Robert Monroe, Jim Marrs, Dean Henderson, Richard Dolan and Joseph Farrell...
>
> **It has never been <u>our</u> planet... we can't 'take it back.'**
> If we can't take the planet back, at least some of us can get out of here and move on.

So just as the Georgia Guidestones have officially warned people, so is this book. To repeat:

As **Drs. Trevor Marshall and Martin Pall** have said earlier...

...[5G] is **a civilization ending event...**
There are dark forces putting forth an agenda as crazy as this because... I mean
the endgame is basically the end of humanity as we know it... [233]

The rest of us are about to discover what the **Anunnaki Earth Agenda** is this time around. So far, efforts to stop or mitigate the headlong rush to install 5G and do a responsible install are failing... this may be part of the Greater Script...

...and that is why the suggestion to <u>right now</u> get serious with your **soul growth** and focus on being an **Earth Graduate** so that you will not have wasted your last days – whether they do or do not install 5G full bore.

So am I at personal risk with the PTB for stating all this?

Not really – most of you will think it is nonsense and an overreaction – just as humans have done for millennia. They don't listen because what I am saying disagrees with what they think they already know. So there will be no panic either.

Secondly, warning people is part of the larger Game – and they know that most of you will not listen. (I was just asked to <u>write</u> the book, not agitate or preach.)

Thirdly, the PTB are doing what they want and I am not fighting them. They will diss me and keep on doing what they always do.

That is why I am asking **YOU** on a personal level, for those who have ears to hear, get your sh*t together -- it may be your last 5 years here.

Namaste

Appendix A : Soul Levels

Soul Levels

Just as there are different OPs, there are different levels of Soul not only reflected in the aura, but also in the orientation to life on Earth. As has been suggested in an earlier chapter, there is the Standard OP, the Robotic OP, and the Placeholder OP – not really an OP, but not a complete Soul, either. The Placeholder is unique in that it can operate as a Soul or as an OP. And the **Pre-Soul** may have the option to become a more fully developed Soul.

> Keep in mind that OPs (Organic Portals) are also seen as Non-Playable Characters (NPCs), as in a video game… they are controlled by the Greater Drama on Earth.

In any event, there is a kind of **Hierarchy of Souls** which reflects a Soul's growth which is an expression of their experience and what lessons have been assimilated. While this is not cast in concrete, keep in mind that there are as many different levels of Souls as there are types of flowers and variations within each flower group.

Baby Souls – these are the first-time Souls, Pre-Souls, and may include the Placeholders. They are generally naïve and **their aura is underdeveloped**, or may be an orange color (a mix of the lower three charkas which are the only ones really functional) in late-stage Baby Souls. They are the **most timid** of the Soul types, often being afraid of germs and dirt – not having had much experience with Earth life. They tend to avoid crowds, not feeling comfortable with all that energy and being a bit unsure of themselves. For them, **sex is scary** and they are very concerned with avoiding social diseases. These Souls are **very concerned about appearances** and want to dress and look right. These Souls love Nature but are **very prophylactic**.
These Souls are drawn to a very basic, fundamentalist type Religion (if at all) and it is easy for them to believe in Hell and God's punishment. Aliens do not and cannot exist, but demons often do.

Young Souls – these Souls have been around enough on Earth to know their way through the new experiences that the Baby Soul is still learning to handle. Thus these Souls tend to make up for lost time, and become involved in everything that catches their fancy. They join groups, sing, dance, **party**, try novel adventures (river rafting and sky diving, e.g.), and they are said to 'go for the gusto.' In their eagerness to experience it all, they begin to make mistakes, tromp on others' toes, and may even lie, cheat and steal. For them, **sex is fun** and they seek new ways to experience it. These Souls are very concerned about appearances, too – do they look good enough, and have the latest designer this and that?

These Souls love Nature and seek to romp through the mud on off-road bikes, ATVs and 4-wheel Jeeps. They may also be the rock climbers and scuba divers (avoided by the Baby Souls). These Souls are drawn to a more progressive type Religion yet they usually also believe in Hell and God's punishment. ETs may exist on other planets, but not here.

Mature Souls – these Souls have been around even longer than the Young Souls and are, in fact, back to **work on the mistakes they made.** As a result, these Souls have begun to quiet down and tend to become **introspective**, trying to figure out what things mean and how they can get the upper hand over the ailments they often have. They are often found in New Age and New Thought churches, seeking better information on how to handle their lives, their health, and their finances. Thus they often become True Seekers in a field that interests them, usually Metaphysics.

For them, **sex is a responsibility** and they take it seriously. These Souls pay attention to their appearance and make it reflect who they are; dressing is a statement about their real self or what they value. These Souls love Nature and seek to understand Mother Earth and work with others to heal the environment.
These Souls are drawn to a think-for-yourself Religion which is really a search for spirituality and they are exploring different Religions and Teachings to find answers to their life issues. They doubt there is a Hell and think Karma is probably true. ETs are real and may be here, but don't talk about it.

 Old Souls – these Souls are the most interesting, and can be **real characters**. They have mastered most of the issues that Mature Souls are still working on, and they have come to a **greater awareness of the Oneness of all things and all people**. They could also be called 'last timers' as they are either doing clean up work (righting wrongs, forgiving others, etc.) or are in some sort of teaching capacity for other Souls who are still learning the basics.

For them, **sex is no big thing** and ironically, they can be bawdy, laughing at it all. These Souls are very **laid back about appearance**, not shaving if they don't want to, wearing comfortable clothes despite how others are dressed – yet they are **clean**. Styles don't impress them and they do what they want (without offending others). These Souls love Nature and work with Her, and would rather be alone in the woods communing with Her than sitting in church or shopping in the Mall.

These Old Souls are usually found in a New Thought church, or something like Baha'i or Tao, or Gnosticism, or none at all. They know that **spirituality is more**

important than religious rote, and they <u>know</u> there is no Hell, and they <u>know</u> that Karma is real. They know ETs are real, and were involved in the creation of Man, and some ETs are here among us to help, guide and protect… they may have met some.

> As was said, this is just a general way to get a handle on four of the many stages of **Soul growth** – there are a lot of Souls who are still going through the first two stages and may wind up being defective and as Dr. Newton relates, if they never develop the intent to be/do/have something more significant or spiritual in life, they will **go from being recycled, to possibly being 'dis-assembled'– or having their energy rearranged**. Such rebellious Souls may become defective if they try to party forever, or run power trips on others as a way of living.

> **Many Souls are a mixture of 2 levels**, depending on their individual Ground of Being (see Glossary) – more advanced in some areas than others, and it depends on just how much they are willing to accept of their reality.

Multidimensional Souls

Another aspect to our Soul is that it is **multi-dimensional**: we exist concurrently in other times and places, other **timelines** and dimensions. [234] Since all aspects of self are inter-connected, **the undeveloped aspects of one's self can and do affect the mood and peace of the other aspects**. [235] This is akin to the 'nonlocality' phenomenon of Quantum Physics.

The term here is specifically **"aspect"** and not "fragment" since fragment refers to parts of one's Soul that have split off, or became fragmented due to trauma. John Jones in 3D Earth may experience a trauma and a small part of his Soul will **fragment** (to escape the negative energy impact of the trauma); whereas Bob Jones may have many other **aspects** of himself including:

(1) one in a parallel dimension,
(2) one 300 years ago on Earth, and
(3) one 1200 years in the future on Earth, to name a few.

Reincarnation/Past Life Scenario

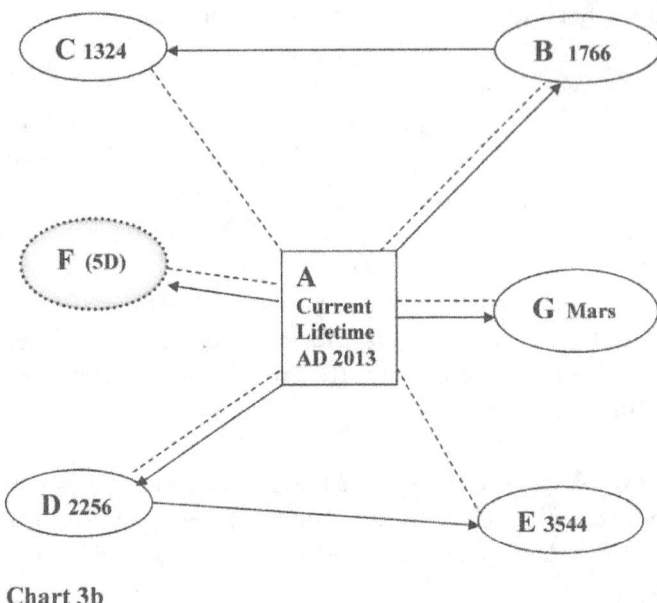

Chart 3b

Each person in this reality has other aspects in other realities that **influence each other,** as indicated by the dotted lines in the following Chart 3b.

> All potential 6 Soul aspects (B – G) of the one person (A) are linked thru the Higher Self (dotted lines). **Arrowed** lines represent "linear time" access. **Dotted** lines represent 6 (direct) energetic Soul links and there are energetic interactions via the Higher Self along these lines ('cords').
>
> The above Chart 3b is a representation of the Multidimensional nature of the Ensouled Human Being. Man A in 2008 may also exist concurrently (simultaneously) in lives B, C, D, E in this dimension, but in 4 different timelines, and he also exists in the fifth dimension in life F. Note that he may also have a life on Mars in the 3D realm.
>
> When hypnotically regressed, and asked to go **back** in time to see who he was 'last time' he will encounter lifetime B (AD 1766) and then lifetime C (AD 1324). If he is asked to go **forward** in time, he will encounter lifetime D (AD 2256) and then the lifetime E (AD 3544). All in the 3D realm, for the purpose of illustration here.

Allocation of Soul Aspect Energy

Since we are multi-dimensional, we co-exist with ourselves in multiple timelines (i.e., dimensions). We allocate energy and consciousness from the Oversoul launching pad, as it were – from our normal Realm, on the Other Side.

Allocation of Soul Energy

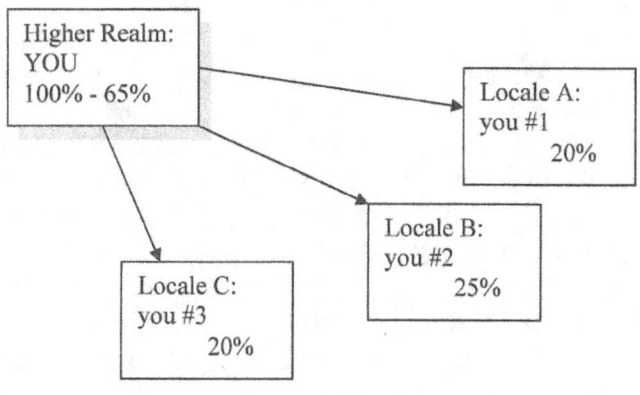

Chart 3c

Note in the above diagram that YOU as a main Soul (or **Higher Self**) in the Home Realm decides to experience 3 other realms or timelines. This requires an allocation of your energy to 'replicate' yourself into those other 3 realms... and you don't want to spread yourself too thinly! For the average Soul experience, a 20% allocation is sufficient (as the Higher Self is quite powerful and has a lot of energy), and in a difficult realm, an allocation of 25-30% may be appropriate... this is an "energy bank" that you can draw on to meet higher energy demands of the lifetime for health, reserve stamina, and mental prowess.

Note that when the 3 aspects are allocated, the YOU still has 35%.

If an advanced Soul, from say the 6th level, decides to project into the same 3 realms, it may require less energy allocation due to a presence that just observes and "anchors the Light." Ironically, it has a greater energy well to dip into and could allocate 20% which for it would be equivalent to "you #1" above allocating 40%... much more than advisable for you #1.

Different Soul levels have different levels of useful energy – a Baby Soul has a lower PFV, is underdeveloped and so the energy is 'rougher', cruder and not as refined as that of an Old Soul, and a Soul from the 6th level is even greater and finer than that of an Old Soul, and requires less to perform equal to any Soul in 3D. Ascended Masters can project into many realms at the same time if necessary, and their energy being so much higher and finer, it is like a battery recharging a Baby Soul if called upon to heal that Baby Soul. The healing works because **energy always flows from the higher potential to the lower.**

Effects of Multi-level Aspects

And just what interrelationships and effects do the different aspects have on each other? They are still connected through the **OverSoul** to each other.

> <u>Oneness</u> **(Rasha)** suggests:
>
> Your presence is called forth to interact with them [other Soul aspects] at the levels at which *they* experience self-awareness. Your responses are influenced, vibrationally, in each of their worlds in much the same way as theirs are in yours.
> In this way, multiple variations on the same scene actually *happen*, yet each of you perceives it from your own unique vantage point. [236]

Be clear that one Soul aspect affecting another does NOT refer to Karma. The interaction is limited to an "energetic" effect that is transmitted via the 'cords' that connect Soul aspects to the same Godhead, and the effects may be relayed down cords to another aspect – if the receiving aspect is of a lower vibration. In the universe, energy is always flowing and it always flows from a higher potential to a lower potential. Higher vibrations tend to block (overpower and cancel) weaker, negative vibrations which is why avatars and gurus are seldom affected by lower/negative vibrations around them. Thus, they often serve to raise the vibrations wherever they are because their higher (faster) vibrations seek to bring others 'up' to their level.

What Chart 3b is showing is that **all** 3D lifetimes are generally concurrent, or **synchronous**; although **individual** lifetimes/Scripts usually happen linearly (as a closed loop). The ongoing, multiple aspects are not happening linearly, although they can be <u>retrieved</u> that way in hypnotic regression. Note that when A is regressed hypnotically to look for his last prior lifetime, it may actually have been B (and is thus not on-going any longer because he is now at A), **or** B could also be another aspect of his same Group Soul whose lifetime is ongoing and is not the same exact Soul aspect as A, but is seen as 'him' because the two Soul aspects are part of the same Soul Group, and they are all each other. All are One.

This latter is an example of **"imprinting"** where a Soul aspect imprints the life experience of another Soul aspect to gain the experience, even though s/he did not actually live it.

Confusing initially, but these are the variations that exist depending on how time is looked at: linearly or cyclically/spirally. In the cyclical (no time) view, **they are all happening concurrently**. In the linear view, it is A then B then C....

> **Oneness (Rasha)** explains...
>
> As each of you *creates* alternate variations on the identity of others, with whom to interact at your own custom-made levels of awareness, these alternate aspects of consciousness and the circumstances they encounter combine to influence the energy fields of each of you. For that reason, it is altogether possible and often likely that your **mood shifts**, quite suddenly, in ways you find hard to explain.
>
> An adverse reaction on the part of an alternate aspect of self, playing a role in a diminished environment, adds a sour note to the resonance of the collective *you* identity.... You are largely unaware of the complexity of the composite that, in actuality, is co-creating your experience of reality.... Those of you who continue to experience **chronic depression** are often simply at the effect of the adversities encountered by parallel aspects of self in the vibrationally diminished realities of others.
>
> Many of you are "moody" and subject to unexpected fluctuations in the way you feel that have <u>little or no relation to the circumstances of your life</u>.... Chances are, you are feeling the influence of the multiple layers of vibrational input being provided by the countless levels of reality in which you, in actuality, are present. [237] [emphasis added]

And you thought the Neggs and OPs were all you had to deal with. Now **you can also be affected by other aspects of yourself in other realities.** How to know where the 'problem' comes from? You don't, but prayer and hypnotherapy might help identify the source of the problem and relieve it... the gods do answer prayer.

> *It isn't important to know exactly where the problem comes from —*
> *and dabbling in the occult (using the OuiJa Board or channeling),*
> *looking for answers,* **can lead to more problems**, *hence the*

Biblical injunction against it (Deut. 18: 10-12).
Man is merely expected to handle whatever he gets, as catalyst,
and that means going within (as the Gnostic Gospel of Thomas,
v. 70 said).

> One of the reasons for this Series of 8 books is so that people can learn <u>without</u> playing games with the Astral Game Players. The entities who answer OuiJa Boards are Game Players, just as Seth was (via Jane Roberts, whose body he abused until she died), and they are not honor-bound to give you a true answer — and the Higher Beings are too busy to play the OuiJa Game. Be forewarned.

Offloading Others

And lastly, in this regard, be aware that a lifetime of constant depression, fatigue and unexplained negativity may also be due to a very esoteric and rare reason: You have taken on (through prior agreement, before birth) the task of **offloading the karmic strain** of another Soul aspect (say, Soul M) who otherwise might not be able to make it — you are sent the <u>extra</u> pain and upset that would otherwise be too much for that other Soul. Yes, they still have to undergo the Karma and whatever pain that their Script may entail, and you cannot handle all their Karma for them, but you can and do offload <u>some</u> of their stress, pain and negative energy.

> This is permitted because (1) it is an act of compassion, which we are all learning, (2) you are strong enough to handle it for them, and (3) the Soul M may have taken on more than s/he could handle but doesn't want to start over. There is nothing in the **Law of Karma** that says a Soul with heavy Karma must also be able to handle <u>all</u> the upset and pain that go with it — at one time. That Soul must experience the extent of the pain/loss/upset etc. that goes with the Script (i.e., LifePlan) to experience the significance of the lesson, **but** they need not suffer past the point that they can endure. Karma is not about breaking a Soul.

Naturally, as can be guessed, Soul M took on more than he could reasonably handle, and it is an **act of Compassion** that his brother (another Soul aspect from the same Soul Group) cares enough to help sustain Soul M, and learn a few things in the process himself.

Appendix B: Firmament +

We need to consider a big puzzle regarding the Earth: as explained in Appendix E in Great Earth Puzzle, the Firmament looks like it was a lot lower, closer to the ground – according to African legends. It was so close that many African natives tried stacking trees to reach the Firmament. They weren't stupid – they could see it.

In addition, in modern day Science, two curious things have been found that suggest that **there is something surrounding the Earth**...

1. The Gegenschein
2. The Energy Barrier

The *Gegenschein*

First we need to recall that even NASA photographed what **Charles Fort** called the *Gegenschein* – a bright reflection off something surrounding the Earth. It has **no parallax,** meaning it does not change size depending on one's angle of observation. As Charles Fort said, about 1912, the *Gegenschein* is like a **shell**, and he was privy to something that he never spoke about in print. Could that have been the older Flat Earth scenario? Charles Fort repeats what Enoch said about the stars...

> ...whether there be a shell-like, evolving composition, holding the stars in position, and in which **the stars are openings**, admitting light from an existence external to the shell, or not, all stars are at about the same distance from this Earth as they would be if this Earth were stationary and central to such a **shell**, revolving around it. [238] [emphasis added]

and...

> The *Gegenschein* -- Now we have indication that there is such a shell around our existence. The *Gegenschein* is a round patch of light in the sky. It seems to be reflected sunlight, at night, because it keeps position about opposite the Sun's position. The crux: **Reflected sunlight – but reflecting from what?**

That the sky is a **matrix** in which the stars are openings, and that, upon the inner, concave surface of this celestial [transparent energy] shell, the sun casts its light, **even if the earth is between**... [239]
  [emphasis added]

Why would Fort have used the term **"Matrix"??   In 1912?!**

The *Gegenschein*
(credit: NASA: *http://apod.nasa.gov/apod/archivepix.html and below)*

It is recommended that the reader check out the above NASA link to three samples:

  2008 May 07: The Gegenschein over Chile. (sunlight)
  2006 December 26: The Gegenschein . (sunlight)
  June 25 1999: The Gegenschein . (sunlight + Sun)

So if there is a Barrier, the Firmament, was SETI such a great idea, or was it **a waste of money?** The Barrier is known to bounce back radio waves… <u>from</u> both sides!

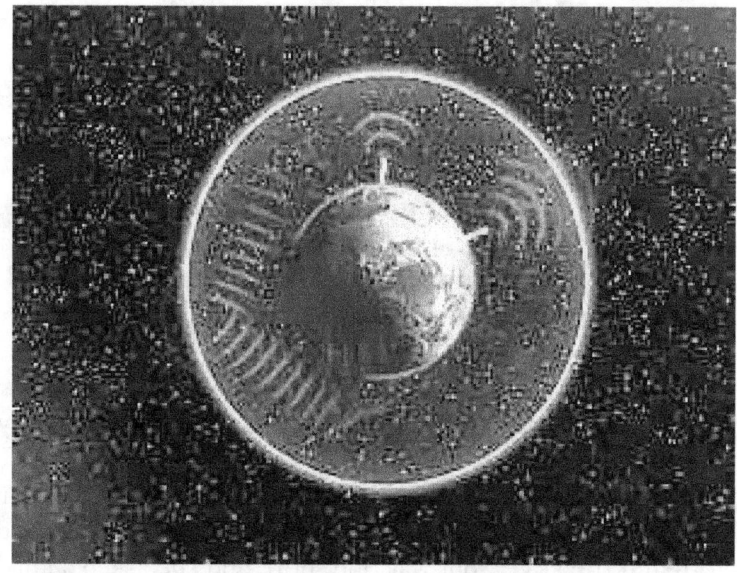

(Credit: Bing Images/earth shield)

SETI founders, **Drs. Carl Sagan** and **Steven Greer**, believe there's a **quarantine** over the planet because the Earth's inhabitants are a threat to other forms of life outside of our planet, due to our military aggression throughout the world. [240] (Maybe we are too petty and violent?)

So there we have NASA, Drs. Greer and Sagan, and Charles Fort reminding us that there is something surrounding Earth. And we will see in a few minutes, that NASA has discovered a **shield around the Earth.** But before looking at the current NASA info, let's remember that there are some really old African legends that tell of Man's interaction with the God(s) of Heaven…

African Legends

According to one encyclopedia of world legends, [241]

> Initially there were connections between God and the mortals he created, between the place where God resided and the earthy

home of humans. There was commerce between the heavens and the earth…. And **humans could move to the heavens, and visit and live among the gods**. The creator god…. met with humans, lived among humans, and the humans were his children. He taught them, punished and rewarded them, as he made an effort to give order to the place that he had created.

Something occurred to provoke a separation between heaven and Earth… Humans erred and incurred the wrath of God…. There might have been a quarrel, the breaking of a prohibition….

Whatever the impetus, the ordered ties with heaven were broken, the gods left the earth …. People tried to re-stablish contact with the heavens, **building towers to the sky**, but these crumbled under their own weight.   [emphasis added]

In another tale, Aberewa, a primordial woman was preparing dinner by grinding meal in a mortar with a pestle and she kept bumping the sky with her very long pestle handle. Annoyed, the great God Nyame went away. She attempted to re-establish her relationship with him, and so began **piling her mortars one upon the other to reach the sky**… but was one short, and she asked a child to go get one for her. The child returned, not having found one, so she desperately asked the child to pull one from the bottom of her pile….and when it was removed, the **entire tower collapsed**.

In another source, [242]

God lived very near humans, in the sky, **just above their heads**. … In the Ila story an old woman tried **to pile up trees to reach heaven** so she could talk to God….

right across to the other side of Africa similar stories are told….

The Nuba people of the Sudan say that **in the beginning the sky was low down and close to the earth ….**

Another version says **the sky was formerly so near** that when people were hungry they tore off pieces of clouds to eat…

The Dinka people, also of the eastern Sudan, say that because **the sky at first was so low** men and women had to be careful in hoeing the ground… not to touch God…

[sure some are exaggerations, but look at
the similarities... what was really going on?]

The Lozi people in upper Zambesi in Zambia say that God created the heavens and the Earth, and that there was a man, Kamonu, who was very clever and imitated God in all that he did... God grew tired of Kamonu and his arrogance, and distanced himself from Kamonu. Kamonu discovered where God was and went to see him... God was alarmed and moved again... Kamonu followed. Exasperated, God moved to the top of a tall mountain, yet Kamonu still followed him. So God

moved to the sky and Kamonu continued his efforts to reach God... but could not. He and his men cut down **many trees and piled them on top of one another, trying to reach the sky,** but it was all too heavy and kept collapsing.

(credit: Bing Images:
http://thechive.com
/2013/01/10)

The recurrent theme is trying to climb up and reach the sky... and here is the key point: **it must have been low enough that Man figured he <u>could</u> climb up and reach the region where God was... Think about it**: if God was really far away, hundreds of miles away "up there," they would not have been stupid enough to use trees and mortars, etc. to reach a really **high sky**, i.e., God's realm.

So you say, well it is just a dumb myth – but you ignore that every myth has **a basis in some real event**, something someone saw … else why repeat a story/myth that your neighbors will think you had to be drunk to make it up?… Many people accepted the stories or they would have died as a 'cock-and-bull' invention. And you are also ignoring another fact: **the same story was told across Africa** by people who did not know each other – So what is the basis in reality for the similar legend?

Thus, building a case here, we come to the ancient legends of humans in Africa. Recall from Chapter 6 that the Dogon, Ashanti and Zulu were contacted by the **Skygods** and given knowledge about the Earth and the stars.

Again, now consider that the Skygods were not ETs, but either the **Ancient Ones** (living in Hyperborea or Mt Meru) , or the Anunnaki… both were here guiding and shepherding Man… And the gods interfaced directly with Man, and Man sometimes aggravated the gods, chasing after them…

Tower of Babel

(Credit: Bing Images: reddit.com)

Early Man in Genesis was not crazy either… **He saw something** that led him to believe he could build a tower to reach the gods. In fact, this Tower to God scenario was another evidence for the original Flat Earth theory. How so? Consider the following logic:

If Man knew he was on a rock (rotating at 1000 mph, spinning around

the Sun at 66,000+ mph), <u>with nothing above him but the clouds</u> –
**Why would he try to build a ladder, tower or platform to reach
something he could not see?**

The point is: Middle Eastern Man <u>did</u> see something and figured, like the African
tribes all over Africa, that it was close enough that with some building effort, he
could reach it.

> Parenthetically, this makes one wonder if the Maya were doing
> something similar… building their pyramids to put them closer
> to God? (or provide a raised landing platform for the **Anunnaki
> gods** – is that why most of the tops of their pyramids were flat?)
>
> Zechariah Sitchin maintained that the gods (i.e., Anunnaki) would
> visit the Maya and stay in the 'casita' at the top of the pyramids,
> coming and going – Such was evidenced in VEG and AL by the
> **very steep steps** which are hard for a human to navigate… to
> keep humans from coming up the steps of the pyramid whenever
> they felt like it. The taller Anunnaki (8-9' tall) managed the steps
> just fine.

Genesis 11:4 said

> "Come let us build ourselves a city, with **a tower that reaches to
> the heavens….**"

…and you know what happened. The gods didn't like it and scrambled
their languages so they could not communicate to continue building
the tower… The gods said:

> "Come **let us** go down and confuse their language so they will not
> understand one another."

Whereas the God in the African stories just moved away or up higher, these gods
made it impossible to continue building the Tower. Two things here: (1) it was not
the plan to have early Man be able to go where the gods are, and (2) the gods are
watching – very aware of what Man is doing. Why would interest in what Man is
doing stop with the current-day times? (It didn't.)

Even more interesting, as was said in VEG and AL:

The Bible says "Let **us** make Man…" Who is "us"? The God of the

Universe is called 'El' in Hebrew, as in El Shaddai. Singular, The One…
The God of the Universe is <u>not a plural</u>…
remember the Ten Suggestions?

"Thou shalt have no other **gods** before me."

The plural of God is 'Eloh**im**' – gods. So it is not the God of the Universe speaking. It is **a lesser set of gods** (as suggested in Chapter 2 – they are designated Higher Beings or Elohim who run this place)….. and the humans referred to the **Anunnaki** as gods (because they descended from the sky), and humans also referred to the **Ancient Ones** as gods – and that latter was more appropriate as They are the ones **Enoch** encountered who showed him what Earth is, how it was built and how it operates. Ok, we are getting off point… VEG and AL examine that in more detail.

Tower Building

The point is that Man must have seen something regarding the sky (aka Heaven) which made him realize that (1) God lived there, and (2) it was close enough to warrant an attempt to try and reach it.

> I repeat: ancient Man was not the stupid dolt that we, the very-with-it, modern, know-it-all humans who have now reached the pinnacle of development and knowledge (got your wading boots on yet?)…. So that we can look down on our ancestors with such disdain that we just know that they knew nothing and made up stupid stories…

> Seriously, if ancient Man (who wasn't on drugs, prescriptions and eating junk food) was more clear-headed, and looked up and saw a **different Firmament**, then it is **plausible** (not yet proven) that the Firmament was **lower** and Man believed he could reach it…. They really weren't stupid. **If Man had thought that there was nothing up in the sky, above him, <u>especially if he saw nothing but clouds</u>, he would not have tried to climb up!**

> What this Appendix is suggesting is that **he DID see something that led him to think that building a tower was feasible and worth it.** And <u>multiple cultures</u> had the same reaction to whatever they saw… so it wasn't just one group of crazies…

And now to add weight to the argument, we turn to what NASA discovered in 2012-14 – besides the earlier *Gegenschein* …

NASA Space Barrier

Invisible Shield

Science News in 2014 reported on an invisible shield surrounding the Earth that was discovered some **7200 miles above the Earth**. Sounds like a good distance for the dome or Firmament to be located <u>now</u>. The shield protects Earth from high-energy "killer" electrons that can destroy satellites, threaten astronauts, and degrade space systems. It sits between the two layers of the **Van Allen Belts** (see below, next section), and is **not the magnetosphere**.

It has been described as **"impenetrable"** and **"a glass wall in space"** that has **"an extremely sharp boundary"** and is **"extremely puzzling."** [243]

NASA calls it a **Plasmasphere:**

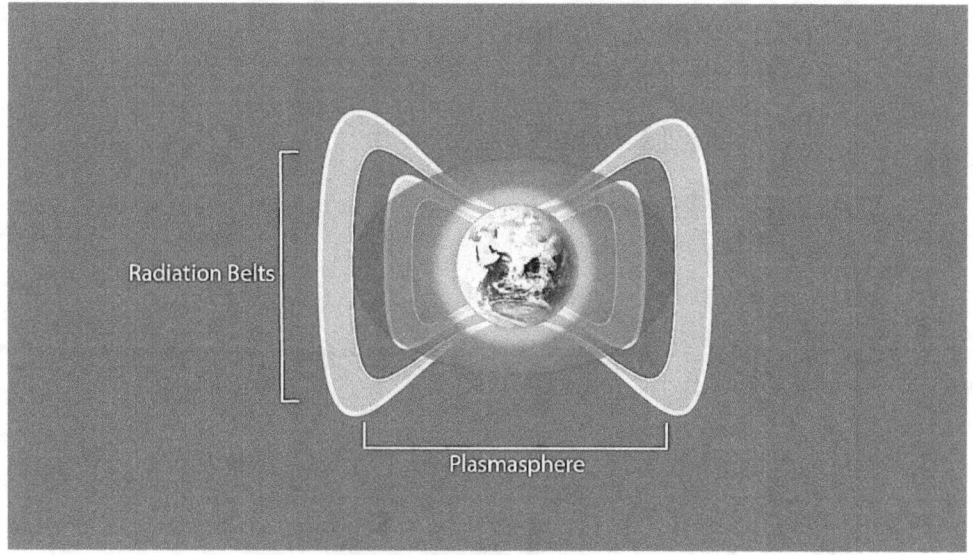

Impenetrable Shell
(credit: NASA/Goddard)

This anomaly was briefly mentioned in Ch. 13 in QES. Whereas the Van Allen Belts of **high radiation** are said to be "harmful enough to kill astronauts flying thru them, the new shield may prohibit <u>any</u> passage to outer space." [244]

(Still think we went to the Moon?)

NASA is now calling the Plasmasphere a 'force field' over and around the Earth which sounds like Charles Fort's "Shell" or the Firmament associated with the HVR Sphere. Repeat: it is not the magnetosphere.

According to Daniel Baker, director of the Laboratory for Atmospheric and Space Physics, [an] **electron barrier exists in the Van Allen Belts**.... The Earth's magnetic field holds the Belts in place, but the scientist says that the electrons in those Belts – which travel at nearly the speed of light – are being **blocked by some invisible force** that reminded him of the kind of shields used in television series like Star Trek to stop alien energy weapons... [245]   [emphasis added]

The newly found field at **7200 miles altitude** is related to the plasma clouds that comprise the Van Allen Belts. The vicious nature of the Belts has led many to suspect that Man did not go to the Moon as passing thru them (without any shielding – and they did <u>not</u> have any!) is a lethal experience.

For a very good and shocking overview of the Apollo Moon Missions, please see Randy Walsh's excellent book:

<u>The Apollo Moon Missions: Hiding a Hoax in Plain Sight.</u>

He examines whether we could have technically gone, and reveals some startling new information.

> As was noted in Chapter 12 of VEG, the highly credible channeled entity RA stated that Earth <u>is</u> in a Quarantine set up by the Solar Council and why. It is interesting that RA did not mention a Flat Earth, and referred to the Firmament as a **Quarantine.**

Van Allen Radiation Belts

For those who wonder what the Belts are alleged to look like:

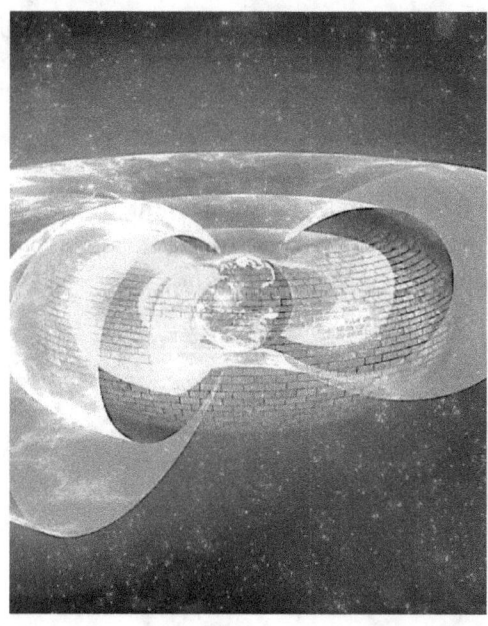

The 'bricks' symbolize the
Wall of ionized plasma around
Earth... **the Shield**.

The two torus fields are the Van Allen
Belts

There are no less than 2 Belts
and a Shield circling Earth. And their
function (as designed) is protective.

The above diagram, as provided by NASA, reflects that the **Inner Belt** is 620 miles
to 3,700 miles from Earth, and the **Outer Belt** is 8,100 miles to 37,300 miles. The
newly discovered Shield is calculated to sit about **7200 miles above Earth**.

All that is understandable as the Earth is now a **VR Sphere**... Would the original
Earth need Van Allen Belts to protect against... what? If the Belts are not a piece of
disinformation, then is it possible that the energy field sustaining the new Shield (at
7200 miles altitude) appears to be one of radiation and plasma...
and that the Van Allen Belts are not really separate entities but are in fact the energy
"signature" of the Field that creates and sustains the Shield (aka Firmament)?

More specifically, **the barrier is impenetrable** and was discovered in 2012 by
Professor **Daniel Baker** who described the shield as

> a third, transient "storage ring" between the inner and outer
> Van Allen radiation belts that seems to come and go with the
> intensity of space weather... [and it] appears to block the ultrafast
> electrons [100,000+ mph] from breeching the shield and moving
> deeper towards Earth's atmosphere. [246]   [emphasis added]

Robert Monroe & Shield

As was pointed out in VEG Ch. 6, Robert Monroe in his many, documented, OOBE voyages encountered a Barrier that surrounds us. Upon later research he came to understand that there is a purpose to where we are and why we are protected… and wrote this in 1971:   [247]

> The following premise, unacceptable as it may be to our present state of enlightenment, deserves consideration… [and] only future events can determine its validity. Conversely there is no known theory to prove it false…
>
> In a universe populated with sentient beings of great variety [which he saw during some of his OOBEs], the planetary environment germinating life follows a typical pattern. The prime requirement is a diffusing and **restraining shield** that envelops the entire planet.
>
> The shield is composed of gases and liquids of sufficient density to (1) Deflect, filter, and/or convert radiation from [the Sun] and nearby stars to a point of tolerance needed for animate life; and (2) maintain internally generated planetary heat … within the limits required for the biochemical process. [Firmament?]
>
> In this environment, animate physical life generates and evolves in a broadening cycle. **Where no such shield has developed and remained for a significant period, no animate physical life is present**. Where the shield has decayed or drained off into space, life has deteriorated and died unless intellectual knowledge is sufficient to develop and install an artificial environment.

Of course, <u>his</u> shield above is largely the **atmosphere and Van Allen Belts** which protect the Earth, but he was writing in 1971 and NASA has discovered (2015) an impenetrable Shield at 7200 miles, as well as Fort called attention to the Sun reflecting off of something surrounding the Earth (as far back as 1930). In short, Monroe had not carried his analogy far enough to include an **energy shield**, as well as the atmosphere and Van Allen Belts.

And here is another gotcha: If Nature abhors a vacuum (and seeks to fill it with something) and the Earth's atmosphere extends right up to the edge of Space (a vacuum) – **what keeps the atmosphere from flying off into Space?**

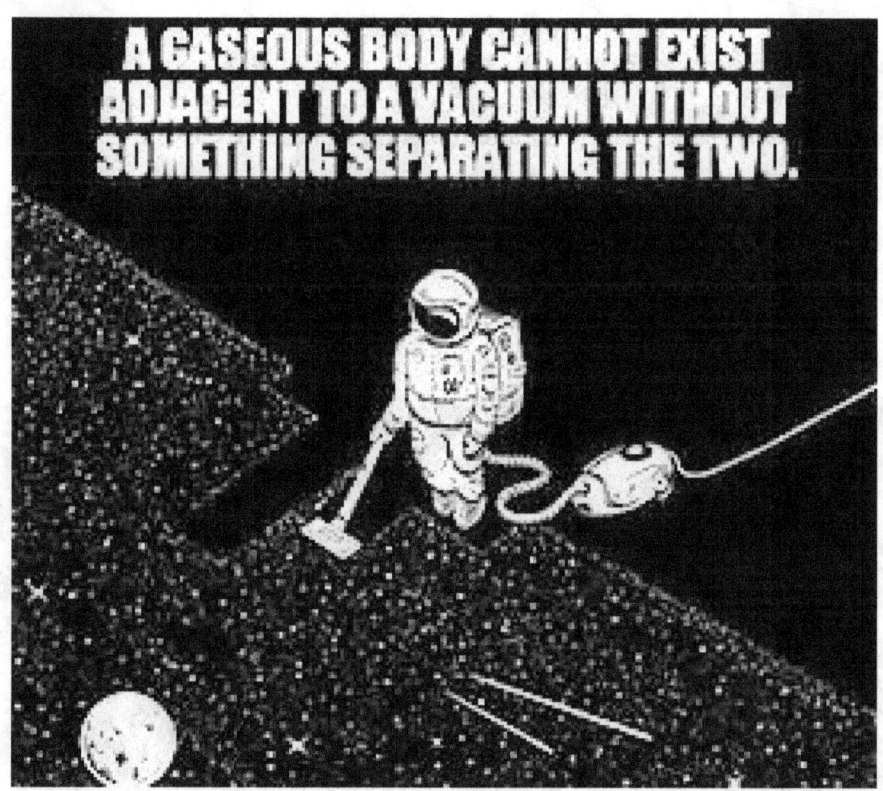

A GASEOUS BODY CANNOT EXIST ADJACENT TO A VACUUM WITHOUT SOMETHING SEPARATING THE TWO.

This plainly says the atmosphere on Earth cannot stay on Earth if Space around it is a vacuum. **Gravity is not dragging the clouds to the ground** any more than the atmosphere is being controlled by Gravity... If Gravity were holding the air to the Earth, it would eventually drag it all down and the clouds with it... even feathers fall to Earth. Think about it.

Consider that the clouds would not move, they would stay over the same spot of land if Gravity were holding them in place, and if Gravity is not that strong (so they can move), why do they not fly off into space, **and** the heavy oceans fall off the round globe?

Science is wrong again. This argues for the Firmament, or a **Shell**.

> The argument that it is all a balance between centripetal and centrifugal force is also false, as was examined in GEP. The scientists cannot explain Gravity so they make up

Conclusion

So if NASA says there is an **impenetrable Barrier** at 7200 miles up, and the *Gegenschein* <u>does</u> reflect off of something, with no parallax, and **Robert Monroe** doing his Out of Body (OOBE) voyages in VEG Ch. 6 encountered a wall or barrier surrounding Earth that he could not penetrate, and **RA** says the Earth is in **Quarantine**… then…

there is something surrounding Earth.

…and that may mean that we cannot get out of here.

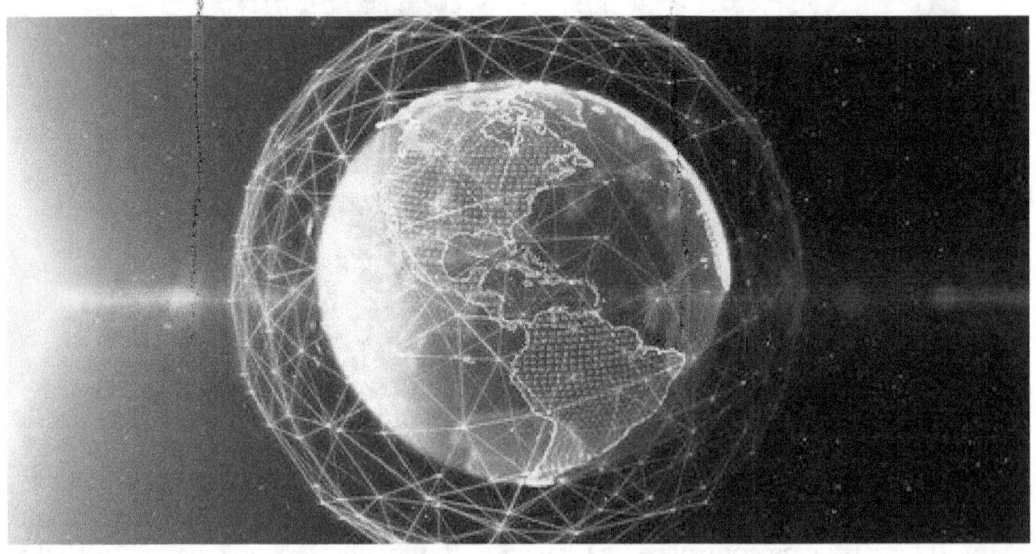

The Earth in Energy Field Containment
(source: www.gafnews.com/sites/)

Does this sound like Charles Fort and his *Gegenschein* in VEG, Ch. 12? **Fort, Wilde, RA and Monroe all postulated a 'shell'** of some sort around the Earth, which reflects light, and which transmits the light of the stars – but what if the 'shell' is reflecting the 2D universe and transmitting it as 3D? It could if our world is contained in and subject to a **Simulation** controlled by the **Control System** – which makes the stars appear to be distant and real… but they may be just 2D projections of the Control System on a shell located far enough from our VR Sphere that we can't tell they are part of a **holographic Simulation**. And weirdly enough, when the Control System 'moves' the stars and planets (simulated sidereal movement) **it makes a sound that can be heard**… that is the essence of what Dr. Hogan has discovered (see GEP Ch. 9: the **GEO600 anomaly**: noise from the edge of the Universe). Seriously… see following section.

And just because it is interesting and sits outside the Vatican (and 13 other sites around the planet) – someone is trying to tell us something with the following sculpture… Is this the HVR Sphere?:

Sphere Within a Sphere at the Vatican: Two Shells
(Credit: Bing Images: Panoramio.com)

The Earth or 3D Construct is inside the **Inner Shell** (aka the Firmament = Gegenschein).

The **Outer Shell** is the Konstruct.

Could the **GEO600 anomaly** be coming from the Outer Shell?

The GEO600 Anomaly
(repeated here for your convenience from Ch. 9 in GEP)

Are dud signals from a gravitational wave detector **evidence that the universe is a holographic projection? (Image: ESA)**

> **Gravitational waves** are ripples in space-time that are emitted by cataclysmic cosmic events such as exploding stars, merging black holes and/or neutron stars, and rapidly rotating compact stellar remnants.

The **GEO600** gravitational wave detector in Hanover, Germany, has not yet detected any gravitational waves. As a consolation prize, **it may instead have uncovered the ultimate nature of reality.** In **2008**, physicist Craig Hogan at the Fermi National Accelerator Laboratory in Batavia, Illinois, was trying to work out how we might test the idea that everything we see as physical reality is the result of a kind of projection from the boundary of the universe. This is known as the **holographic principle**.

The information held at the boundary is not smooth, but composed of "bits", each one occupying an area that corresponds to the most fundamental quanta of distance in the universe. This is **the Planck length**, around 10^{-35} metres – far too small for us to see the individual bits. When this information is projected into the volume of the universe, however, each bit gets magnified. That means **we might just be able to see pixellation in space-time** (see QES, Ch. 7). Hogan worked out how the **pixellation** might manifest itself for GEO600 and sent his result to the researchers there.

414

By strange coincidence, the GEO600 team had been having problems with "noise" in their detectors. But **here's the kicker: the noise had uncannily similar characteristics to Hogan's anticipated signal. Is it indeed the result of information that resides at the edge of the universe?** "The issue is still unresolved," says Karsten Danzmann, principal investigator for GEO600. **"The noise is still there and we have no explanation."**

QES Ch. 7 introduced Dr. Hogan's report that **sound was coming from the edge of the universe**, and wouldn't that be strong evidence of the energy being used to generate the outer Shell/Shield/Barrier ? Even if you say that the sound is just radio waves bouncing back at us – what are they bouncing off of – the SETI shield mentioned earlier in this Appendix??

> As of 2019 there has been no further clarification from GEO600.
>
> BTW: just wondering, could the 'noise' be coming from the Dark Matter/Energy field all around us? Energy has a vibratory quality to it, maybe that is what we are "hearing"...? (See Subquantum Physics info in Ch. 9 in VEG.)

Appendix C: Soulless Humans

This issue was raised in **VEG (Ch. 5)** and **TOM (Apx. D)**, and again in Chapter 7 of this book. It needs some clarification here, so you don't have to research the other two books. The concept is nothing new, the Mayans ad the Greeks both knew about those among us who have no soul – the significance is: **no conscience**.

> The soul is the connection to one's Higher Self.

I know this is sometimes hard to believe, but humans do not need a soul in the body any more than dogs, cats or monkeys – who move about just fine – and they also have no soul. The soul is what makes Man unique and should be respected and developed. (The Anunnaki and 4D STS do not want that.)

The following is some basic information in this regard.

**   **   **   **   **

This topic has been placed in the back of the book because it is not to interrupt Chapter 7, but it is a further examination of the Souls, Consciousness, OPs and Timelines issue which the reader may find interesting. The elements of this Appendix were partially developed in VEG, Chs. 1 and 5, and in TOM, Ch. 2 and Apx. D, but the two were not correlated – that was left up to the reader. So some extensive review is included herein.

More recent research into Souls and Timelines has revealed that there are definite connections and evidence that supports the idea that there **are** "two seeds" on planet Earth, at least two, and that there are **humans walking among us who have no soul** – and for very logical, and substantial reasons. This was hinted at in VEG, Chs 5 and 12, just subtly so as to not disrupt the main idea of those chapters.

Compared to Apx. D in TOM, this version has been edited and updated since TOM was written.

It needs to be emphasized, big time, that the following presentation is for your serious consideration, but is **not to be blindly believed**. There is considerable evidence that it is correct, else I would not present it, but it should be considered another possible aspect of the world we live in, and at the end of the Appendix, I will suggest how this information could benefit the average, ensouled person.

Beware of Religion and Mainstream Science telling you that this Appendix is based on myth or lies. It is closer to the truth than the politically correct 'experts' out there are willing to admit... their goal is to repeat that we are all alike [we're not], and that is causing major problems in Congress and the social fabric of our

Why be aware of this information??

Socrates said "The unexamined life is not worth living" and it is part of one's spiritual growth to discover who you are, where you are, and why you are here. This is a book about spiritual growth. Revealing the following ideas definitely will assist in that – because **you are not living in the world you think you are**, and part of your waking up is actually empowered by the following information (as bizarre as it may initially seem).

There are three intertwining aspects to be examined: **Souls, the soulless, and Timelines.** They come together in a very unique way, that will at least interest you, if not shock you (GEP, Ch. 10). While you have heard of the soul, it is doubtful the PTB have ever sought to make you aware of the soulless humans, or Timeline Shifts. Churches and schools do not teach about the last two issues. And they are part of our reality in a big way... but, Hey, stay wood Pinocchio!

Creation

In the beginning, God created the heavens and the Earth...

Great, we have heard that before.
And then the Book of Genesis tells us:

Gen 1:26-28
26 And God [Elohim] said, Let **us** make man in **our** image, after **our** likeness...

418

²⁷So God [Elohim] created <u>man</u> in his own image, in the image of God
created he him; male and female created he them.

Ok, good. Man and woman were created at the same time, no rib involved, and the
Creator was referred to as "us" and "our"... EL is the word for God; in Hebrew, the
suffix "-im" makes the word plural. **Elohim** is plural for El... several **gods** creating
Man. Very interesting. But it doesn't stop there.

In the very next chapter of Genesis, it tells us:

Gen. 2:7
⁷And the LORD God [Yahweh] formed <u>man</u> of the dust of the
ground, and breathed into his nostrils the breath of life; and
man became a living **soul**.

Very interesting. The same Book, same writer, and one assumes the same Creator,
but now with a new name, Yahweh, and now Man has a soul. Woman was not
created at this point in the text, or so it seems... (**Lilith** and then Eve come later).

In addition, the claim is made that in Gen. 1 Man is represented as having been made
"in the image of God" (27), yet in Gen. 2, he is merely "formed...of the dust of the
ground" (7), thus suggesting a contrast. And this was allegedly all written by Moses,
the same writer.

Two different accounts, in the same Book... and get this:

> **VERY Important: Nowhere in the Bible does a writer ever
> repeat the same event twice in the same Book.** (Ink and
> papyrus were expensive items and they would not waste them by
> repeating themselves.)

Genesis 2 is <u>not</u> a repeat or 'clarification' of Genesis 1.

There were (at least) two Creations of Man.

And in fact, there were multiple creations as the late Mesopotamian scholar
Zechariah Sitchin has told us (in <u>The Twelfth Planet</u>), and we have 5 distinct races
on the Earth... Caucasian, Oriental, Latino (including South Sea Islanders), Negroid,
and Amerindian (including Eskimo, Aleuts, Inuit, etc.) Multiple Edens.

If you think that White humans evolved from the Blacks, or vice-
versa, I have some ocean-front property in Montana I want to sell you.

But racial diversity, to create a 6th race, is <u>not still happening</u>, contrary to Darwin, and apes are not still 'evolving' into humans, and those are key points. Man is a special creation, and was 'assembled' to look like his creators – Elohim originally, then ETs. The Sumerians said their gods were the Anunnaki, from the stars. The Hopi, the Chinese and the African Dogon, by the way, said the same thing. In fact, ZULU means "from the stars." (See AL.)

Man's evolution was **assisted**, but that is another story (also see VEG, Ch.s 1-4).

The diversity of Man on Earth looks more like an **Experiment**, as if some advanced geneticists (Skygods) were looking to see which of the 5 Root Races would survive (Survival of the Fittest) and be the best, advanced humanoid lifeform with which to populate this or perhaps other similar planets.

So what is the point of two (or more?) creations of Man?

Two Seeds

Let's assume that there were two creations in Genesis, and that the Bible is basically correct and makes it clear that **the first creation had NO souls**. This group of humans would have lived and propagated and formed at least a small tribe somewhere. Think: Land of Nod. (See Chart 1a in Appendix F.)

Then we have the second creation of Man, maybe something was not quite right with the first group and it was decided to make another version… this time with a **soul**. And we got Adam and Eve, then Cain and Abel. Cain slays Abel and is expelled from Eden, and he wanders off and takes a wife from the **Land of Nod**… a wife! **A grown woman**. From another group of humans. And those two give birth to **Enoch**…. Who "walked with God" and challenged the Watchers above the Earth. (All examined in detail in VEG, Ch.s 1-3).

The point being that there was <u>another group of humans in Nod</u>, and they logically had to be the first creation (Gen. 1:26). Logically, because we are not told about any other people at the time.

> **Apologetics** (a section of Christianity) would say that Genesis is so vague about timelines that we don't know how much time elapsed between Cain and Abel and Seth, and did Adam and Eve have daughters? (If not, we're talking incest to replicate the species!)

Interesting that the Bible is supposed to be the LITERAL, inerrant, perfect Word of God, and yet Man instituted Apologetics -- to **apologize** for what is there…. To try and explain the omissions, inconsistencies, and errors… Sounds like The God of the Universe is as fallible as Man (Philip K Dick and Brinsley LePoer Trench wrote about just such an issue.) … or maybe Man modified the original, perfect Word of God, and redacted much of Genesis…?

> And the arguments go on and on, but let's keep this as close to the facts as possible and avoid the polemics.

If there were two creations, one without a soul and the other with a soul, then we have **two seeds on the planet**. Is that in the Bible? Yes.

> …I will put enmity between thee and the woman [Eve], and between **thy seed and her seed**…. (Gen. 3:15)

> I realize that God is talking to Eve and the Serpent, but the point is still that there are **two seeds**, and VEG, Ch. 1 goes into more detail as to just what/who the Serpent was and how his 'seed' equates with those in the Land of Nod. This was examined in some detail in VEG, Ch. 1 and was basically repeated in Chapters 1-2 here.

Suffice it to say that the seed of Eve (the 2nd seed, with a soul) would find the other seed, the 1st seed (without a soul), antagonistic and working at different purposes on Earth. The 1st seed has to be the **people from Nod** who also propagated and have spread over the Earth… even to this day. The Bible doesn't speak of any other people beside Adam and Eve's group and those in Nod.

> **Rewind:** The point is not where they came from, but that there are two seeds on the Earth, and one has a soul and the other doesn't. For the benefit of those who do not have VEG, there is a chart (below) that shows the suggested human lineage…

> This **Chart 1A** below is more fully examined in VEG Ch.1 with the other half of the diagram (Chart 1B) in VEG, Ch. 3.

> They are also reproduced in Appendix F in this book.

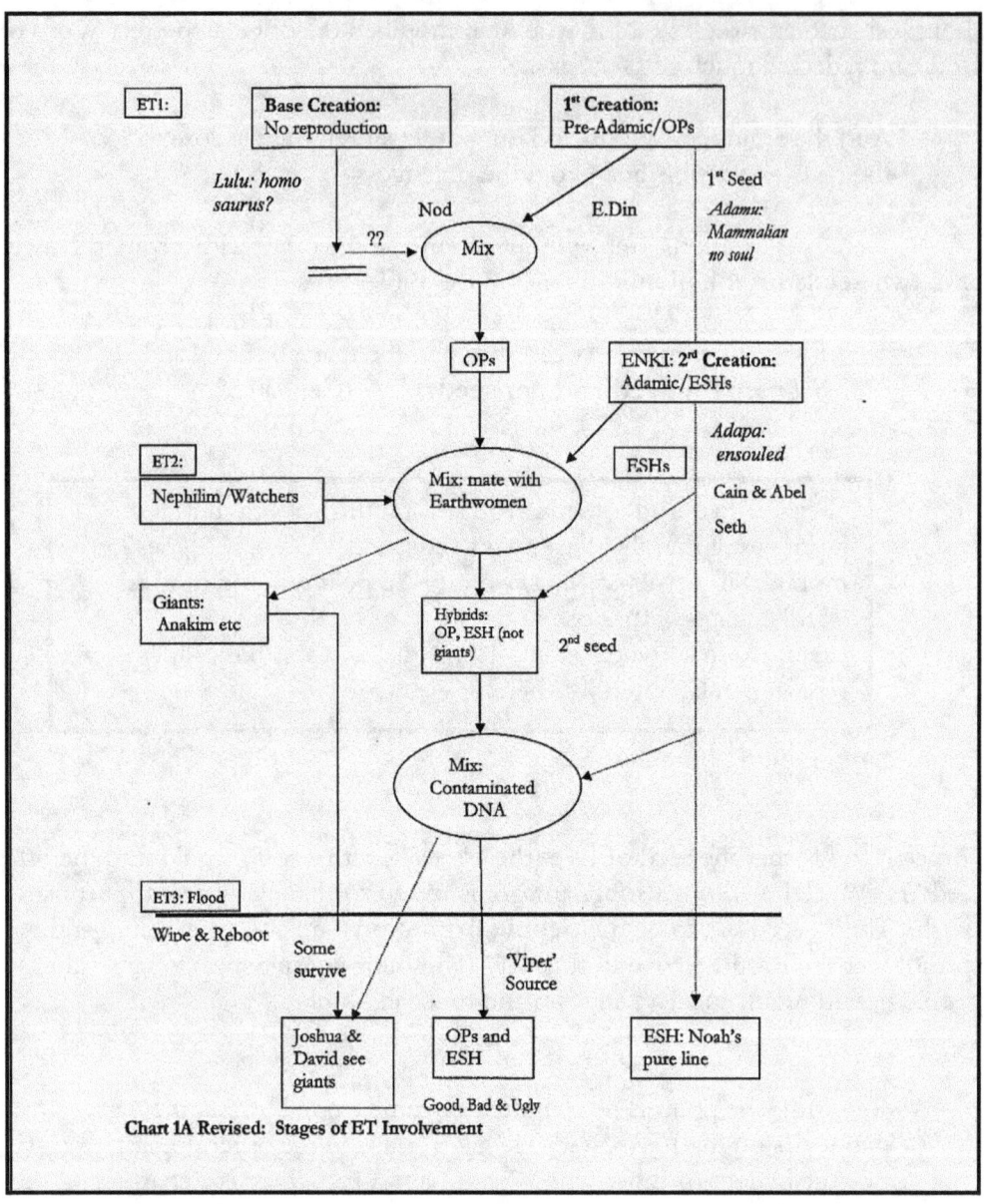

Chart 1A Revised: Stages of ET Involvement

The chart includes the following labelled elements:

ET1:

Base Creation: No reproduction

1st Creation: Pre-Adamic/OPs

Lulu: homo saurus?

Nod

E.Din

1st Seed

Adamu: Mammalian no soul

??

Mix

OPs

ENKI: 2nd Creation: Adamic/ESHs

ET2:

Nephilim/Watchers

Mix: mate with Earthwomen

ESHs

Adapa: ensouled

Cain & Abel

Seth

Giants: Anakim etc

Hybrids: OP, ESH (not giants)

2nd seed

Mix: Contaminated DNA

ET3: Flood

Wipe & Reboot

Some survive

'Viper' Source

Joshua & David see giants

OPs and ESH

ESH: Noah's pure line

Good, Bad & Ugly

Note: the Jewish scholars (in the Babylonian Talmud, and in Sumerian accounts) spoke of another woman in Eden, preceding Eve: **Lilith**, who was created in Gen 1 (later demonized and removed from Genesis, but a reference to her still exists in Isaiah 34:14 NAB), and because she was not subservient to Adam, she was banished (… to Nod?). The Sumerian and Akkadian traditions describe her (in *Gilgamesh Epic*), and this 'bad' girl has today devolved into a myth. Very interesting….

The chart also shows alleged Anunnaki interference with Man (as proposed by Zechariah Sitchin, see VEG, Chs. 1-3), and the two seeds are identified in vertical lineages #2 (middle) and #3 (far right).   **OPs** (the soulless) and the ESH's (ensouled humans) were addressed. Note that the soul is found in the #3 lineage, and the descendants of Nod either are the 1ˢᵗ seed or mix with the soulless 1ˢᵗ seed in #2 lineage. Both eventually meet in the "Contaminated DNA" mix (i.e., current day).

Both seeds are on the planet today.

OPs are later described and defined, but the term for the soulless comes from Dr. Mouravieff --   OP means:

O = Organic, flesh and blood humans
P = Portal as they can be manipulated by Astral entities (mostly Angels to drive the Father's Greater Script on the planet)

There is another interesting Biblical passage that tells us that there are two seeds that don't mix, metaphorically speaking:

> And whereas thou sawest iron mixed with miry clay, **they shall mingle themselves with the seed of men: but they shall not cleave to one another** even as iron is not mixed with clay.
> Daniel 2:43

There is another meaning to the passage, in addition to what the Bible scholars say about Daniel interpreting the king's dream: a woman with a soul should not marry a man who is soulless. And this is covered later in this Appendix when the pros and cons of knowing about this Two Seed Issue is something that one can use to avoid heartache in one's life.

Caveat on Soulless

At this point, we have to stop and make a couple of things very clear.

1. **The soulless are not evil.** 98% of them are good, hard-working people. Just rather naïve and have no appreciation for spiritual issues since they have no soul which means they are not connected to a Higher Self, (i.e., Oversoul) which is also the seat of one's **Conscience**.

2. The soulless has **no aura** as that is the reflection of the soul. (Most people, especially women, who are often more intuitive and sensitive to higher energies, can learn to see the 1" etheric level – and ways to learn to do that were given in VEG.)

3. No aura means **no conscience**, and if one of these OPs runs amok and causes harm or damage to a family, business, church or becomes a terrorist, they are considered an **OP on steroids** and are part of the other 2% who may become **sociopaths**, or worse yet, **psychopaths**. (Think: Charles Manson, Jim Jones and Richard Ramirez.)

4. **Most of them are harmless** and are the "great, unwashed masses" of humanity, settling for rote religion instead of wondering about why they are here and what life is all about. They have no higher chakra centers active to empower such wondering, let alone pursue the issue!

5. **They do not know what they are**. They sense no difference between themselves and another person.

> Points 4 and 5 are verified by Dolores Cannon in <u>Convoluted Universe</u>, Books IV and V.

6. While **most are not evil**, they can be a real <u>nuisance</u> and that is why you want to know about them. They can be very self-centered and want what they want; **everything is about survival, power and ego** – they do not compromise, there is **little if any compassion**, and they are not altruistic.

7. **All men are not created equal**. OPs do not have the spark of divinity in them, and when they die, it truly is dust to dust. Trying to discuss compassion, conscience and spiritual matters bores them and they cannot follow it. (Some of them are scientists dissing the existence of the soul!)

> All of that being said, there is **no "witch hunt" needed** and if one cannot see auras (I do, so I know this is all true) one can identify them by their behavior (see list in following section).

So if most of them are harmless, why bother to say they exist? Why bother to warn against marrying one? Reread points 6 and 7 above, and see the end of this Appendix .

That is why the following list of typical behavior is given… to help identify them – and these are typically people that the soul on a spiritual path will want to avoid!

Soulless Traits

OP Appearance

This can be a sharp way of spotting OPs: they <u>tend</u> to be **more handsome or pretty** because they usually do not have the soul's **karmic influence** to affect the way their bodies/faces develop and thus look. It was taught that without negative Karma to endure, the OPs could be as naturally radiant and good-looking as is humanly possible.

> Physically **the two races are virtually indistinguishable**. Statistically there are minor physiological and perhaps genetic differences. Physiologically, OPs tend to be **more attractive and well-proportioned**. Because they exist on an emotionally primal level, natural selection has ensured that sexuality, physicality and attractiveness play a large part in their physical evolution. Also, unlike ensouled humans, OP bodies are conceived and develop independently of soul pressures and karmic burdens, so they are as attractive as probability allows... [248]
> [emphasis added]

How many men have seen a beautiful woman only to find that she is heartless, cold and conniving? There was a song about one, in 1955 – "Maybelline, why can't you be true...?" She could be an OP, a soulless human. (Is this one of the reasons some Hollywood marriages don't last? Is one of them, or both, an OP?)

OP Characteristics, Basics

So, what are the general traits of OPs, and how can one reliably identify them? The basics are given here, with some advanced aspects.

> **Warning: you may see your spouse or neighbor described.**
> **And be careful to not prejudge... while some OPs are creeps,**
> **they may be Pre-souls acting badly.**

First, one way to be sure, and reliably know if someone is an OP, is to determine if they have an aura or not. Second, the alternative is observation over a period of time and seeing if their **behavior** is congruent with general OP behavior. Third, if such consistent (daily?) observation is not possible, then consider whether they are congruent with ensouled human behavior (below).

Ensouled Human Behavior

The following five points are typically ensouled human behavior **not found** in OPs.

1. ESH's care about other people and may find themselves 'hurting' for another person. **Empathy**. (True compassion.)
2. ESH's are interested in and pursue **spiritual** goals – meditation, yoga, even therapy and seminars to develop self.
3. ESH's **pursue truth** and dig to find out why things are as they are; this may include therapy and resolving "inner issues."
4. ESH's can have hunches, **intuition** and sense energy and act on it.
5. ESH's are often involved in giving of themselves, their money and time; they are **altruistic**.

And of course, there are other things that are mostly characteristic of ensouled humans, but the five points above are the major attributes that immediately stand out. And of course, **be careful**: there are times when an ensouled human can act just like an OP and be selfish, uncaring, and abuse other people. And there are times when an OP will pretend to have compassion, pretend to be interested in spiritual issues… the better they are at mimicking, the better they **survive** – and for them, that is Issue #1.

Rewind: Major Aspects of OPs

While they generally remain something of a mystery, a few things can be said of them:

1. They have no souls and thus no connection to the god-force; thus they are not interested in spiritual issues or self growth
2. Their DNA does not permit soul growth or connection to higher realms due to higher chakras not being activated
3. Most of them are not bad or evil – some may be first-level Souls (**Baby Souls** with a faint aura – see Appendix A), or necessary placeholder OPs (see Timelines section below).

> Not to confuse the issue, but a **Baby Soul** can have a very faint aura and act like an OP… Caution is advised. Judge carefully!
>
> It is not a sin to 'judge' someone else if the person you are evaluating IS an OP and you are **protecting yourself** – the Bible says: "Judge righteous judgment" which means be careful and objective and be accurate. In short, don't falsely judge and accuse someone.

4. When they die, it is "dust to dust & ashes to ashes" time
5. Because they don't have a soul, they have no aura
6. Because they have no soul, they have **no spark or glint in their eyes**; they are flat and lifeless (**they love wearing dark glasses**)
7. Because they have no soul, they are not interested in religious matters **and** cannot discuss them
8. They have **no conscience**, little or no compassion, and exhibit **low, if any, morals.**
9. Standing next to one, there is a sense that no one is there; as in **no 'presence'**; their energy is **'flat'** – like standing next to a telephone pole.
10. Their purpose is to distract and **drain energy** from the ensouled humans thus preventing them from connecting with their Higher Self (and achieving John 14:12) **The PTB are largely OPs.**
11. They are mostly interested in food, sex, power trips and games; many are sex maniacs.
12. A psychopath is often a failed OP – they failed to mimic ESHs and survive in the ensouled world.
13. They blindly serve the orthodox teaching on anything (Science and Religion); they do not question nor do they innovate.
14. They have no "inner issues" or problems that they need to work thru.
15. They often cannot 'get' jokes with a double entendre, and their humor is crude, and sometimes abusive.
16. Some OPs are driven to **control others**... real 'control freaks.' (They think it guarantees their survival.)

But again, beware! They are great mimics.

They are eventually exposed as they cannot or do not sustain their pretense of being caring and spiritual for long … that is why it takes time and observation to 'out' them! However, the simplest test is to ask them if they go to church, do they believe in a God, and do they have compassion for others? A 'No' answer coupled with off-the-wall behavior is someone you want to avoid, OP or not. Another clue is someone who tattoos and pierces most of their body... even desecrating it by carving symbols into it   (Hint: see pictures of Charles Manson).

Historical Aspect

Returning to the historical aspect, the Greeks and Mayans knew all about them, as well as a current-day researcher, Dr. Boris Mouravieff. (There are 6 categories.)

1. Greeks

The idea of soulless people is nothing new. As was seen in VEG, Ch. 5, the ancients knew about these people 2000 years ago – they were called *hylics* by the Greeks. Even the Maya had a story about them in the ***Popul Vuh***, and Valentinian Christianity (Gnosticism around AD 140) also addressed the basic nature of Mankind, according to one of 3 aspects: [249]

The Elect: the '**pneumatic**' or spirit-filled who were searching for a deeper Christian message because they lived via gnosis/insight;

The Called: the '**psychic**' or mental man who was happy with what he knew (knowledge) and he walked by faith;

The Material man: the '**hylic**' who was incapable of understanding any spiritual message, and did not have the soul-awareness of the other two types.

What is significant about this schema is that in the early centuries AD, man was aware that **all men were not 'created equal'**, and that some men had souls and some didn't. ***Hylics* didn't have a soul**. [250] So **they are not 'equal'** to people with souls. And they may be doctors, lawyers and accountants because they can master details and many of them do serve a very useful function in society – but there is always the isolated sociopath with a destructive orientation to society. Being a sociopath and having a soul are a contradiction – not that an ensouled person can't "go off the rails" and cause harm, but it is usually the conscience in the ensouled person that stops them from extreme misanthropic behavior.

Somehow, perhaps deliberately in an attempt to 'level the playing field,' that information has been lost to the modern world. So much for Truth. For now, it is important to say that **we do <u>not</u> all have the spark of divinity within us**. And the assumption that we do leads to a number of errors, including marrying the wrong person, working for the wrong boss, voting for the wrong Congressman/woman, or trusting the wrong person.

2. Aztecs

Until the Spaniards conquered the Aztecs (1521) and burned 98% of all their historical records, they too had a Creation story similar to the Mayans' (below). Now gone. (Thank you, Catholic Church.)

3. Mayans

Popul Vuh

Besides the Greeks and Gnostics identifying these people centuries ago as *hylics*, the same issue was understood by the Mayans and they wrote about it in their version of Creation, along with many of their ancient beliefs, in their *Popul Vuh*.

Normally, if this were the only source of information about soulless human beings, it could be easily dismissed as superstition and, coming only from the Mayans, it could be said that it was just 3rd world ignorance. In this case, however, what **the *Popul Vuh* says about OPs agrees with what the Greeks said, with what the Gnostics said, and (in the next section), it agrees with what a current-day Russian researcher Dr. Boris Mouravieff said.** The OP info is not BS from a backward or primitive people. It is truth that has been buried, most likely by the PTB or sharper OPs themselves.

The following quote is reminiscent of the Sumerian Anunnaki Creation of Man – first the *Lulu* that could not pro-create, and then the *Adamu* who resembled the "wood" creatures below, and then the *Adapa* which could reproduce and was a lot like Cro-Magnon.

> Check it out:

> According to the *Popul Vuh*, the "gods" had made creatures known as **"figures of wood"** before creating Homo *sapiens*. Said to look and talk like men, these odd creatures of wood "existed and multiplied; they had daughters they had sons…" They were, however, inadequate servants for the "gods."
> To explain why, the *Popul Vuh* expresses a sophisticated, spiritual truth not found in Christianity, but which is found in earlier Meso-potamian writings. The "figures of wood" **did not have souls**, relates the *Popul Vuh*… In other words, without souls (spiritual beings) to animate the bodies, the "gods" [Skygods] found that they had created living creatures which could biologically reproduce, but which lacked the intelligence to have goals or direction. [*Adamu*]

> The "gods" destroyed their "figures of wood" and held lengthy meetings to determine the shape and composition of their next attempt. The "gods" finally produced creatures to which spiritual beings [souls] could be attached. That new and improved creature was Homo *sapiens*. [Or Sitchin's *Adapa*.] [251]   [emphasis added]

Sounds like a rewrite of Dr. Sitchin's story of the Anunnaki experience, and it should be as the Sumerian Anunnaki Skygods went to Central America, as **Viracocha, Quetzalcoatl, Kukulcan**, etc. (see VEG, Ch. 11, 'Other Historic Anomalies') [252] … and they probably did the same 'teaching' there as the *Popul Vuh* continues the eerily similar point of view that Anunnaki leader Enlil had about the humans mating with the Anunnaki:

> According to the *Popul Vuh*, the first Homo *sapiens* were *too* intelligent and had *too* many abilities! … *they saw and instantly they could see far… they succeeded in knowing all that there is in the world...* Something had to be done. Humans… needed to have their level of intelligence reduced.

Mankind had to be made more stupid. [253]

This is reminiscent of the issue with Anunnaki leader/god Enlil: the hybrid offspring of Anunnaki mating with humans was a threat to the dominance of the original Anunnaki who were limited in number, while the humans increased their numbers every day. Remember, Enlil's solution to the same threat was to let a huge Flood remove the problem. (This is all documented in the Sumerian *Atra Hasis*, and see VEG, Ch.3.)

The Mayan gods said the same thing the Sumerian Anunnaki said to this issue:

> … are they not by nature simple creatures of our making?
> Must they also be [like us] gods? [254]

So as the Anunnaki also "dumbed down" their creation by removing abilities from Man's DNA, so too does the *Popul Vuh* relate a similar treatment of Man. Not only did Man not get to live as long as his creators (after The Flood), he was not to know what they knew, nor be able to do what they could do, and mankind has been pretty much kept ignorant – just as the god of E.Din (Enlil aka Yahweh) wanted.

Nexus:
What you think are other humans, just like you, aren't all just like you, and **many do not share the same motivations and goals.** Their job, sometimes with their Astral controllers (4D STS, and sometimes Discarnates) working against you, through them, is to stop you. And because you don't know any better, OPs succeed in creating wars when ensouled Man could be uniting, they create scarcity when there could be abundance, and they create fear when there could be peace, love and faith.

Why? To keep Man from getting it together and developing his divine potential – something they absolutely do not understand (because they don't have it). It scares the soulless PTB.

It is fascinating that the Greeks, the Mayas and the Russians knew about this same soulless truth which has been hidden from 'modern' Western Man. In fact, the Russians have really done the most complete research to date into the issue. (And they use **genuine aura cameras** to research the issue... examined in TOM, Ch. 11.)

4. The Library of Alexandria

It was a repository of Man's accumulated knowledge and no doubt, before it was sacked and burned **three times** (once by Coptic **Pope Theodophilus in AD 391**), it would have enlightened us all to see what historical treasures that great Library held... but there are those who value and promote ignorance, usually so that they can promote their own agenda (with nothing written to contradict them).

5. The Russians

Dr. Boris Mouravieff

Besides the Greeks and Mayans identifying these people centuries ago as *hylics and "figures of wood,"* there was a well-educated Gnostic, Boris Mouravieff, who was a Russian emigrant living in Paris in the 1940s – 1960's. He knew G. I. Gurdjieff and P. D. Ouspensky, and **He listened to their teachings on pre-Adamics – from their research in Tibet and Kashmir.** Mouravieff wrote a version of Gurdjieff's 4th Way Teachings, calling his work Gnosis which consisted of 3 volumes. His primary contribution, although not the only thing he talks about in his three volumes, is **to reinforce the concept of 2 races of Man on the planet**: two races of humanity without regard for skin color, national origin or sex. He calls them **pre-Adamic (1st seed)** and **Adamic (2nd seed)**, [255] or OP (Organic Portals) and ensouled, respectively. (OP is profiled in the Glossary.)

> **Except for the lack of a soul in the OP, they would be <u>almost</u> indistinguishable from ensouled humans on the street, or TV**.

To people who cannot see auras, this information might be useless except that there <u>are</u> **other ways to spot OPs** that are pretty reliable. See the list earlier. Of course, as with any skill, it takes knowledge and practice in the ways to spot them, and for your own well-being, the information is given in this Appendix.

Why?

You really don't want to marry one, if you have a soul, and you don't want them to head up your schools, churches or government. They take prayer out of the schools because they see no purpose to it. They pass laws that benefit them and shortchange the populace. They promote pharmaceuticals and GMOs that harm more than help. Some lawyers are all about money – they don't care whom they defend; there is no principle involved.

Creation Revisited

Mouravieff's alternate, enlightened interpretation of the first two chapters of Genesis goes something like this:

Genesis Chapter 1: the original 1st creation – without souls.
         Boris Mouravieff calls these humans **pre-Adamic man**.

Genesis Chapter 2: another creation where the 'breath of life'
         (*ruach*) was breathed into *adamah* meaning "earth man" –
         Adam had a soul. Eve was created from Adam's rib and as
         a 'meet' (complementary/corresponding) mate for Adam,
         would also have a soul. (Also Cain and Abel.)
         Boris Mouravieff calls these humans **Adamic man**. [256]

To repeat: pre-Adamic man has been termed an Organic Portal (OP) for two reasons:

         Organic because they are flesh and blood,

         Portals because entities from the 4th can and do operate
             through them to serve an agenda (and it isn't
             always bad – see later section on Timelines.).

That means that those things that a soul could normally do in a normal body (with all chakras functional), cannot be done in an OP body – even through the OP body looks normal, it isn't. **It is really underdeveloped chakra-wise**. These differences were delineated in the review of "A" and "B" Influences (See Ch. 5, VEG).

Note again that the creation of Man took place in <u>two</u> distinct stages and the pre-Adamic and the Adamic coexisted for a long time, probably intermarrying . [257]

But the two were quite different in their ability to evolve:

> The prehistoric period is characterized by the coexistence of two humanities: **pre-Adamic** *homo sapiens fossilis*, and **Adamic** *homo sapiens recens*. For reasons already expressed, pre-Adamic humanity was not able to evolve like the new type. Mixed unions risked a regression in which the tares would smother the good seed so that the possible growth of the human species would come to a halt. **The Flood was a practical suppression of that risk**. [258]
> [emphasis added]

> *Note that the Flood did not totally wipe out the pre-Adamics any more than it wiped out all the Nephilim, as Joshua saw the giants in Canaan. The Flood was not worldwide, but 'local' mostly to Africa and the Middle East.*

Not "All Men Are Created Equal"

Again, Mouravieff noted that the pre-Adamic was limited in its awareness and developmental potential due to **no higher energy centers (chakras)** formed and active . [259] The pre-Adamic are really seen as **Anthropoids**, and the Adamic are considered real Men. And the two were anything but equal.

> We must also note that the other extreme, the equalitarian conception of human nature, so dear to the theoreticians of democratic and socialist revolutions, is also **erroneous**: the only real equality of subjects by inner and international right is equality of possibilities, for **men are born unequal**....
> these two humanities ....
> are **now alike in form but unlike in essence**.[260]
> [emphasis added]

And again, he cites the difference:

Rewind: The human tares, the anthropoid race, are the descendants of pre-Adamic humanity. The principal difference.... is that the [pre-Adamic] does not possess the developed higher centers that exist in the [Adamic]... which offer him a real possibility of esoteric [spiritual] evolution. Apart from this, the two races are similar. [261]

The Two Seeds of Genesis 3:15 explained.

Seeds of Prejudice?

Is it unfair or not "politically correct" to point out that all men (and women) were not created physically and mentally equal? **Not knowing about the OPs only protects them** – it does nothing to serve you. As an analogy, would you prefer a dentist fully trained in the best dental school, or one that got his degree through the mail? Would you knowingly marry a woman whose genetics will lead to her having all children with Downs' Syndrome? Would you knowingly get on a plane with only one pilot who was subject to epileptic seizures (that he also may have told no one about)? In short, the point is: in order to have confidence that what we are about to do will work and come out well, **we need to know who/what we're dealing with** – that is the purpose of licensing, by the way – to give some assurance that the person we're dealing with is competent, or "all there," as it were.

OPs are <u>not</u> all there, some are "out to lunch" and I'm sure you have seen ditzy drivers, inept doctors/dentists, and even wondered about the person you married at times… or what your **elected representative** is doing in Washington. OPs are really attracted to the Military, Law and Politics – all are **power trips**.

It is not prejudicial to call a spade a spade. And you are a fool to tolerate 'differences', problems or inconsistencies that obstruct a society's or family's or church's well-being… if you do, you have just signed on for the lesson and it will blow up in your face, sooner or later. Due to the PTB obfuscating everything they can, so you don't know too much (so that you won't get upset), you won't know why you failed! You must know who/what you are dealing with, and then stand up for what is the right thing to do. This Appendix can help you learn to see these people around you – <u>by their behavior</u>.

Rewind: Is Judging OK?

Of course someone will object that it isn't nice to judge other people… and they quote the Bible to prove it! Too bad they are wrong. What they need is to look at John 7:24 (and I would warn you that if you don't see a liar, cheat and rapist for what he is, you are too Pollyanna-ish and will learn a lesson the hard way):

> Judge not according to the appearance, but **judge righteous judgment**.

It is not a sin to judge others – if you are right. And that is all the Bible says: be <u>correct</u> (righteous) in your judgment. So do not be afraid to suspect someone is an OP… just be cautious, look for corroborating behavior, and don't out them – just **id and avoid them**. If they are callous, insensitive, lying, offensive or have no spiritual side, walk off, avoid them. Save yourself the upset or heartache.

6. Dolores Cannon

Bless her heart, the late Dolores Cannon (of *The Convoluted Universe* fame) had more credibility as she always protected her clients with prayer and Light. And in fact, her material over the years was pretty consistent, did not contradict itself, and many times the information connected with, or expanded, what we already know in a reasonable way.

> In her *Book V of the Convoluted Universe*, published October 2015, after TOM went to press and was published after her death (October 2014), she has a chapter on the **Backdrop People**… which per her definition, is the same as what has been called the OPs, NPCs, or Pre-Adamics. (See Glossary.) It is worth quoting some of her salient points: [262]

> They don't have a path, or purpose. They are just here **like extras in a movie**. They are slaves to the Earth Drama.

> They do not evolve to become higher beings as ensouled humans can and do. They can apply themselves and may be a **Pre-Soul**. (see Glossary), but they don't know it and usually remain just undeveloped resources in the Greater Drama.

And lastly, she suggests a possibility:

> They have **no soul** and no future potential – they exist to drive the Scripts for the ensouled humans. [263]   (QES, Ch. 9: Simulation.)

This is exactly what Engineer Jim Elvidge (Ch 7.) said in QES.

Reincarnation

Perhaps more importantly, Dr. Mouravieff said that this soul issue relates to Man's ability to evolve:

> **Pre-Adamic man** does not reincarnate. Not having any individualized element [soul] in himself, (in the esoteric sense), he is born and dies but **he does not incarnate**, and consequently he cannot reincarnate. He can be *hylic* or *psychic* but not *pneumatic*, since he does not have the *Breath of Life* [*ruach*] in him, which is manifested in Adamic man…. [264]

See why the teaching of Reincarnation was buried along with the truth about OPs?

He's saying that the OPs have more of **a collective group soul, like animals**, and that their lack of higher centers (chakras) prevents them from being aware like ensouled humans, and thus from reincarnating. **Reincarnation applies to souls**, not to non-souls. When an OP dies, s/he follows the "dust to dust" regimen; again, there is no soul to go anywhere and they are thus not concerned with an afterlife, but they may still be afraid to die.

Mouravieff also says that the OPs serve what we nowadays call the **Matrix Control System** [Absolute III], and the ensouled humans serve the **Christ Consciousness** [Absolute II]. [265]   Matrix Control is the Greater Drama, using the and Control System (VEG, Ch. 4).

> Lastly, Mouravieff confirms what was said earlier about mixed families and nations:

Mixed Families

> Meanwhile the two races are totally mixed: not only nations but even **families can be, and generally are, composed of both human types**.
> This state of things is the ... result of transgressing the Biblical Prohibition against mixed marriages because of the beauty of the daughters of pre-Adamics. [266]
> [emphasis added]

Genesis Revisited

At this point it is also relevant to emphasize that the pre-Adamic man represents the 1st Creation in Genesis 1. And this is significant because the Bible tells us that Adam and Eve had two sons, who would be 'Adamic' like them: the offspring had souls just like their parents. Then Cain killed Abel and was expelled from the scene – and he takes his wife from the **Land of Nod**. (And he is relocated to the New World.)

> *FYI: The 'mark' of Cain was no facial hair.*

This is important information. There is only Adam and Eve up to this point, with their two sons. Where did the rest of the people come from? To cut to the chase, it is suggested that the people in the Land of Nod were descendants of the pre-Adamics, from the 1st Creation, and Mouravieff does agree. [267]

That means that **Cain had a soul and his wife didn't**. As they left for parts unknown (eventually the Skygods moved them to South America), and did a lot of 'begatting' as the Bible says, they began the mixed marriage scenario (See **Chart 1A** earlier in this Appendix) and the furtherance of the OP and ensouled lines, with enmity between the two, just as Yahweh/Enlil said in the Garden of Eden when he expelled Adam and Eve.

> **Edgar Cayce** said that one of Adam's sins was "consorting with others" and was told that **"all flesh is not one flesh!"** Cayce never clarified this reference to "others" but it is not hard to think that it could have referred to ensouled (Adamic) Man mixing his genes with those of the people in Nod, the Pre-Adamics or soulless ones.[268]

That is a suggested, plausible scenario. It explains where the OPs came from and why, as Mouravieff says, we have mixed families. A family with Mom and Dad ensouled and the kids ensouled would make for the theoretically balanced marriage. Unfortunately, the ensouled Dad could marry an OP Mom, and the offspring can be a combination that brings strife to the family. Ensouled parents would not know what to do with unresponsive, do-your-own-thing OP children (labeling them ADD) and OP parents would not care about nurturing their ensouled children and giving them spiritual values. Does this describe today's families' problems?

A family may have both ensouled and soulless humans by natural birth.

It bears repeating: The problem with a mixed family, where the parents are OPs, is that the **ensouled human** (ESH) children will not be cared for (nurtured and encouraged) as they could have been by at least 1 ensouled parent. OP parents often do not know what to do with ensouled children and their questions, deeper seeking and sometimes higher awareness (even psychic).

Note that an ensouled man and an ensouled woman (unless they have defective genetics) will <u>usually</u> produce an ensouled child… but there is no guarantee. With an OP man and an ensouled wife, it could be either, but with two married OPs, the offspring will <u>usually</u> be **only** OP. The reason for this is that an incoming (incarnating) soul will normally choose ensouled parents – unless there are hard lessons to be learned at the hands of OP parents… good luck!

Therefore a recipe for a rough, insensitive or tempest-tossed marriage is for an ensouled human to marry an OP. How do you know? The OPs are usually very pretty or handsome – no Karma to mitigate the out-picturing of the often dysfunctional soul – so they are great-looking people… as in **Hollywood types** – and don't the Hollywood marriages <u>all</u> work great!?

Birth into a Body

So what happens at birth? Why are babies born, living, and do not have a soul? Simple: a Soul chooses not to get in the body due to circumstances that were not foreseen when the soul decided to incarnate on Earth, with reference to a particular family and location. (This is examined in depth in TOM.)

Do you move into a house if it isn't finished, or if it isn't correct?

Souls choose their parents, lifetimes and basic experiences (See TOM, Ch. 6). If, as the fetus nears completion, the soul discovers that the Mom and Dad are now figuring to divorce, that may not be what the soul needs to experience – if it is, the soul proceeds. A soul's watching the baby develop over time may see the mother do drugs on a whim, and that affects some key DNA, so the soul backs off and chooses another birth family. **Souls do not have to enter a body.**

The point is, the baby can be born and live, move, walk, eat, sleep, and marry and have its own children – **without a soul**. Note that dogs, cats, horses and cows also do not have individual souls and they do just fine. When a baby is **still-born** that is due to the body not being able to sustain itself, for whatever reason, not because the soul pulled out.

> For what it is worth, there is a lot of furor about **Abortion**… largely due to ignorance. The soul usually enters the baby <u>at birth</u>, maybe just before or just after… so the issue of murdering someone is nonsense. If a house is not finished, or is in some way defective, and nobody has moved in, and a tornado comes along and destroys the house, was anyone killed?

> Just before we segue to Timelines, here is a basic practice to see auras.

Seeing Auras

This is the general procedure to practice, as many people <u>with practice</u>, can see the basic 1" etheric layer just 1" above the arm, head, face (sideways), or the shoulders where it is the strongest. Some people cannot, so don't worry about it – just go by the list of OP characteristics.

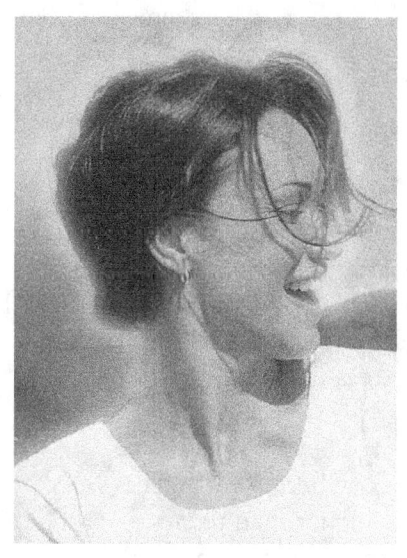

Obviously, if you practice with someone, if they are an OP, or Pre-soul, you will come away claiming that you can't see auras. Try it with a number of people, and in church is a good place to test your ability – OPs generally do not go to church.

Most people can learn to see auras, but very few can see them out in public, with a mix of things and colors behind the subject. **A single-color wall is best in low light.**

Shown left is the **1" etheric aura**… best seen against a dark wall, not the open sky. It looks like a bright but thin light around the body.

Human aura

(credit: Digital Vision, and Kevin R. Brown, 1997)

1. Stand in a dimly lit room – a bathroom during the day with the light off and in front of the mirror, with a white or pale beige wall behind you (a black wall also works)
2. Stare at your forehead, not intently, and make the focus of your attention the side of your head – seen as it were with your peripheral vision – **defocus** your vision…
3. Occasionally close your eyes and rest them, then open them and in the first few seconds, note if there is any 1-2" energy field that appears to be surrounding your head.

Two warnings: (1) If you try to do this with a friend, and s/he is an OP, you will think you have failed… that is why you want to try this with yourself first… get used to seeing the faint, fuzzy 1-2" light around your head…

Secondly (2): When doing the above exercise, and you close your eyes, you will probably see what is called an **"after image"** of yourself on your retinas. This is normal, and sometimes the aura will ALSO show up as a slight shift in brightness as on the edge in the after image. When you open your eyes again, and you **defocus**, you are most likely to see the etheric aura.

> **Note: auras do not show up on TV or in normal photographs. OPs do not have auras, they have "heat waves" above their heads.**

Now we come to the significant aspect of the Soul, OPs, and Timeline issue. These three go together just like Souls, Karma and Scripts. (See Appendix D.)

Timelines

It was mentioned in VEG, Ch.s. 15-16, that one could exit Earth as an Earth Graduate… showing both Love and Knowledge after death, when crossing over to the InterLife. Attaining the **51% Light** as part of one's Soul allows one to leave the Earth experience behind and go do something else in the Father's Multiverse (See types of work available in Chapter 10: Potential Areas for Service).

> 51% is not a really high consciousness, but if They had to wait until a soul perfected itself on Earth, we'd all be here for a really long time. As long as one's Light quotient is higher than that of the planet, one can graduate. (In physical terms, you resonate PFV higher than the planet, and thus are not that compatible with it.)

But there is another way to move to a better Earth, or any world, that <u>has been</u> subtly happening across the centuries – and the **OPs are the proof of it.**

> Here's how it works.

Timeline Splits

Let's say that Earth was in one continuous timeline up to and thru the mid 80's.

Edgar Cayce predicted that Atlantis would rise in the Atlantic Ocean during the late 60's, but the pole shift did not happen in '58 to '98, along with numerous Earth changes predicted in 1934, and there was no bad earthquake on the Pacific Coast of California in '36…[269]   and **Nostradamus** predicted that in July 1999 there would be a 'king of terror' coming from the skies (asteroid). [270]

Obviously, all that did not happen here. Such things happening were predicated on that timeline continuing to develop with the same influences and energies as were seen by Cayce and Nostradamus.

What those two men did not foresee was the infusion of positive thinking, Light and proactive behavior of **incoming higher souls** – sometimes called Indigos, Starseeds, Wanderers, or Homo *noeticus* (Hubrids) – planned by the Higher Beings. These two men also did not see that the Higher Beings would move Earth into a more proactive timeline for eventual ascension into the 4D realm… or <u>did</u> they and we have not understood their writings? Neither seems to say anything about timelines and possible splits.

A timeline split is basically a **timeline shift** – the vibrations go higher and you can't get there until/unless your PFV (Glosssary) is a close match with the new timeline vibration.

And yet, we are on a timeline where Cayce's and Nostradamus' predictions did not happen. Cayce had enough credibility as "America's Sleeping Prophet" and was quite accurate in his medical diagnoses, that his predictions probably DID happen – but not on our timeline! And what is even more interesting and amazing is that the split/shift was done seamlessly – no one noticed.

A dimension can have more than one timeline, but timelines are at different vibratory levels, kind of like two sine waves that have been "phase-adjusted:"

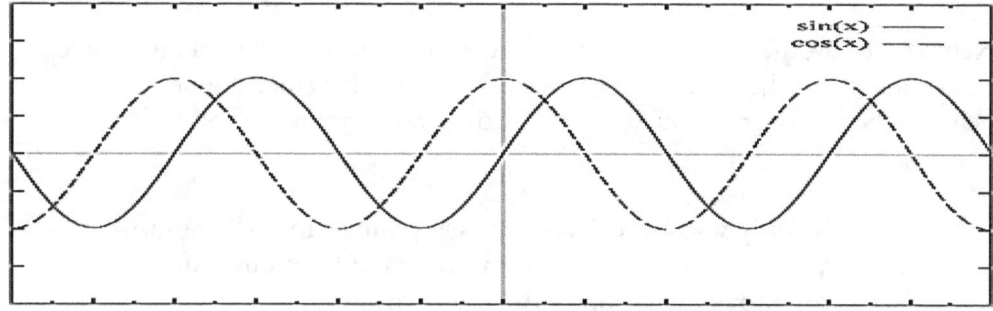

(**Source: Wikipedia** :
http://upload.wikimedia.org/wikipedia/commons/1/13/Sine_Cosine_Gra
ph.png)

They are both on the same machine, but do not interfere with each other. In real life, as on the sine wave chart, there can be points where they intersect, and that results in some anomalies from time to time... (See TOM, Ch. 2).

So where do the OPs come in? First, we should be calling them **NPCs – Non-Playable Characters**, as in a video game. You can't control the NPCs but they are there in the Game, interacting with you... they help drive the script for the video game. **OPs in real life do the same thing; They help drive the Greater Script** (designed by the Father of Light for training all souls), and they interact on a Karmic basis – sometimes delivering your 'lessons' and interrupting your day! Just as in a video game.

It was said earlier that Neggs and Beings of Light (two types of Angels) deliver events to you (guidance, protection, mishaps, fortune...) now you can see that **they often use the OPs to do it.**

As was explained in several other books, the ensouled people around you cannot (or should not) deliver your negative lessons as they will themselves incur Karma… ANY negative action does that -- even if well-meant. That is why OPs are used to 'afflict' or test souls as needed as the OPs do not have souls, so they do not incur Karma.

> During this author's investigations of souls vs soulless (in 2006-2008) – taking headcounts and keeping written records of what was seen – it was noted that about 60% of the people out there have no aura. And Drs. Bostrom and Greene in QES said that 60% was very significant in that it suggests **NPCs do occupy our world**.

Needless to say, the 60% was alarming and totally unexpected until my Source shared that that is because of the timeline splits that have been done – on our behalf – they were prescriptive and protective. Souls are protected and watched over.

And the only way to really see it is explained in the diagram below, and it involves understanding what happens when souls are moved from one timeline to another.

Placeholder Replications

Let's take a look at what happens to people when a timeline splits.

Suppose that Jack is on a timeline TL1 that splits to TL1 (more negative) and TL2 (more positive). In this case, TL1 is the original TL which just got more negative when the more positively oriented souls (STO) left for TL2. The difference in positive and negative deals with whether the outcome is supportive and appropriate, or whether it detracts, disrupts or derails thus being inappropriate. Further suppose that Jack's desire was to align with the TL2 positive timeline and he <u>as a soul</u> is translated there. He is thus an ensouled human on TL2.

Timeline Split Diagram

Steps B & E below assume that TL2 is a virgin timeline (no one else is there already).

The following is a step by step breakdown of how the above timeline split affects the people. Headcounts may be easier to follow below (using arrows):

Timeline Splits and People: Steps A - E

Step A: Earth in TL1 has 6000 people:
Sum:

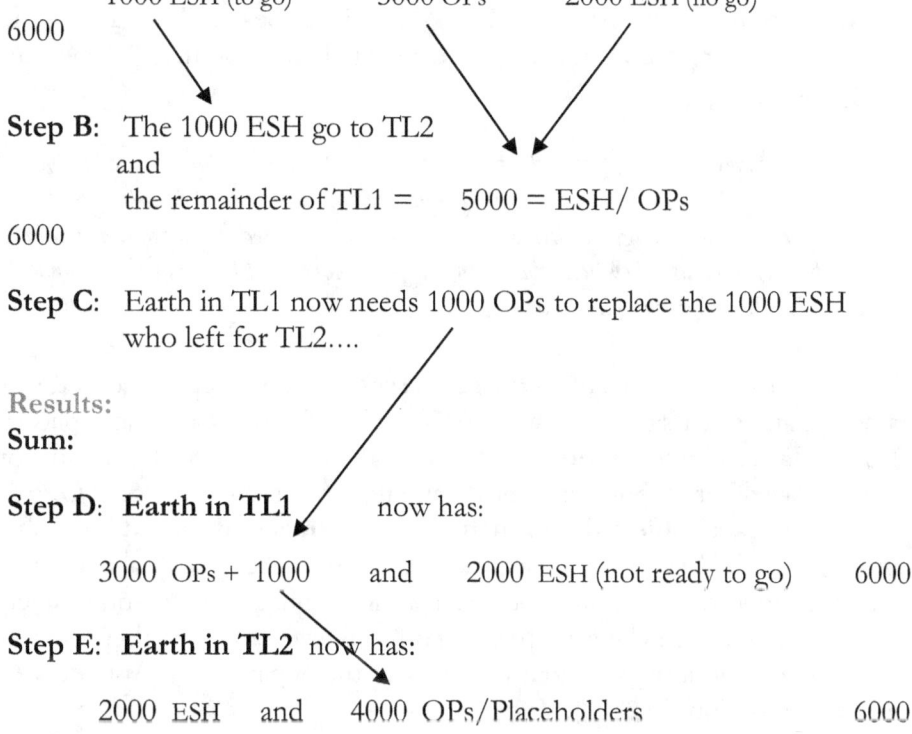

1000 ESH (to go)   3000 OPs   2000 ESH (no go)

6000

Step B: The 1000 ESH go to TL2
and
the remainder of TL1 =   5000 = ESH/ OPs

6000

Step C: Earth in TL1 now needs 1000 OPs to replace the 1000 ESH who left for TL2....

Results:
Sum:

Step D: **Earth in TL1**   now has:

3000 OPs + 1000   and   2000 ESH (not ready to go)   6000

Step E: Earth in TL2 now has:

2000 ESH   and   4000 OPs/Placeholders   6000

Note: this is how **timeline splits result in an increase in the amount of OPs or Placeholders**. TL1 gained 1000 OPs. The Sum of beings is kept the same indicating that the change was **seamless** – no one on TL1 would know anything happened. The 1000 souls moved to TL2 would also not suspect anything because everyone transferred with them.

> Remember that we are about 60% OP on our timeline….
> We have split several times (since AD 1).

When the timeline splits, TL1 will continue as the original timeline which has a less than desirable outcome, even negative, where World War III occurs. Members of Jack's family, people at work and church are used to having him around and in some ways, he completes the scenario for them – all to say that **Jack cannot disappear from TL1**. He is needed to fulfill duties, or responsibilities, that become a part of the on-going life drama of the other people still on TL1. So he becomes an OP on TL1 to sustain the on-going Drama there...

Note that the number of OPs increased in the split.

How can Jack be in two places at once? Simple: he is an ensouled human on TL2 (his choice), and he is replicated as a Placeholder OP on TL1 (by the Timelords who oversee this whole process). The Placeholder does not have a soul, there is no aura, and there is no 'soul cord' or energy link between the TL1 Jack and the TL2 Jack.

> *Sometimes Jack may replicate in more than one timeline as an ensouled human – i.e., both 'copies' of Jack can have a soul, but this is rare as it entails a division of limited soul energy (TOM, Ch. 11) that cannot be reduced beyond a certain point, so the usual event is that ensouled Jack moves to TL2, and the original TL1 Jack becomes an OP Placeholder.*

Note that a subconscious non-souled aspect of Jack was replicated in the TL1 timeline, and other significant 'players' in Jack's life drama are also replicated in TL2 for Jack as **'Placeholders'** to continue their function so that Drama options can be played out on both timelines – and now you see where some OPs come from. The **Placeholders do not usually have souls and do not need them** – although some players of significance may opt to replicate with a soul aspect if they need to learn something, too. Here again, a soul can only send its energy into so many alternate timelines (depending on soul level) and thus there is often a limit to the replication with soul-link back to the original soul (Appendix A: 'Allocation of Soul Energy').

The more advanced souls can replicate themselves into more timelines and realms than young souls can. And all souls are projected into multiple timelines concurrently **per Era**; there is little or no sequential lifetime Karma that applies within an Era as typically **Karma applies to lifetimes between Eras**. All lives are

concurrently happening within the Era – when the Era changes, the souls can redeploy themselves…thus there **is** Karma between lifetimes between Eras.…

> To repeat, Karma only applies to the Earth School… it 'affects' what goes into your Script – i.e., what you need to learn is also called a LifePlan. **What you do/say is not scripted.**

An OP also replicates as an OP in any new timeline, because the ensouled humans need them as part of their on-going, day-to-day Drama, and even some of these replicated OPs may depend on other OPs to perform <u>their</u> functions to keep the Drama going. **This is one reason Earth currently has so many OPs** – we have been through <u>several timeline splits</u> in the beginning of this Era.

Sometimes, an alternate choice (for example, a fractal TL3) playing itself out is more of a temporary **simulation**, as Robert Monroe discovered. At the point where it proves successful, the consciousness of the original soul may move over into a new timeline (with any necessary adaptations in conscious memory so that the move is 'seamless'); if the fractal choice proves disastrous, the information is recorded in the Oversoul's consciousness, as a learning experience, and the **fractal simulation** (TL3) is discontinued for that soul.

> *This is why there are Watchers and specific* **Timelords** *who actually effect the timeline mechanics.*

A soul may also opt out of an alternate scenario if it becomes too burdensome and/or there are no more lessons in one's Script to be learned. The Oversoul, in this case, knows what is going on and the goal is not to prolong an alternate scenario to the point where the soul is damaged. (Damaged souls are examined in Ch 7, VEG, Section: 'Rearrangement of Energy'.)

It is all about growth and Man must stop damaging the planet and rise above his pettiness, violence and ignorance.

Earth in a Timeline Split
(Credit: http://www.bing.com/images)

It is usually a simple replication into an alternate, and initially identical timeline… except that the new vibrations (TL2) are higher and thus more proactive.

Ok, some of this is hard to follow and assimilate. Nonetheless, Splits and Shifts do happen and are for our benefit. Splits keep the human experiment going, and the Shifts are a normal (yet uncommon) way to support a soul in becoming all it can be —by not limiting it to a realm it has already mastered. If a soul is ready for 4D (via a Shift) it goes there. (OPs do not shift up to 4D, but they may be replicated.)

The gods are not ogres. They want ensouled humans to become all they are capable of and work quietly behind the scenes (as in a Shift) so that the soul is not aware and disrupted from focusing on its activities and personal development. Souls are that important.

The PTB are allowed to obstruct as we are in a **Freewill Universe**, and their bad behavior only serves to disgust and anger the STO souls who eventually <u>will</u> wake up and overthrow the STS lords. Quite a Drama! The good news is: we win! But those who cannot or do not wake up are choosing the rougher school of hard knocks… they will have to experience 'destruction' of some sort because they often refuse to wake up and sometimes only pain and loss will do that for them.

And again, to repeat, because it is very important to realize that the 3D Fat Cats who run the planet, the 3D flesh-and-blood Power Boys, the PTB (including the Deep State), have at their tiptop hierarchy several soulless leaders with advanced awareness <u>who know they are soulless</u> and **they fear the souls** and what they can become! Thus they seek to keep souls ignorant and obstruct their growth… largely by distracting them with sex and violence, drugs, alcohol, and false teachings... dumbing down public education, censoring the Media (Fake News), destroying the family… They know and fear that some day, some souls will grow, gain Knowledge and Power, and begin to develop their divine heritage (as sparks of The One), and <u>rule over the PTB</u>, and they don't want that! It is as simple as that.

So why do we want to know about all the foregoing?

There are three summary reasons...

Summary 1: Awakening

1. It is encouraging to know that Man is watched over and when the Earth scenario gets too negative, the Timelords will shift the 'good people' to a higher timeline so they can continue their lessons.

2. It is instructive to know that Life is a kind of Video Game, or as Shakespeare said, a Play or a Drama, and Man is an eternal soul who "struts and frets" his hour on the stage, and then rebirths in a new role for a new lesson. The OPs help drive the Earth School Drama.

3.

   It is helpful to know that the **high count of OPs on Earth** at this time (60%) means that we have had at least 1 timeline split since Edgar Cayce's day...
   evidence that we are being looked after and IF we keep moving forward in spiritual growth, becoming more and more STO, we will continue to move into better and better timelines as we work our way out of here.

4. It is instructive to know that organized Religion is still operating in the dark and believing stuff that isn't true, such that knowing the Truth which carries Light, will benefit one's overall journey to become an Earth Graduate.

5. There is no reason to believe what the ignorant say ("There is a Devil", "There was just one creation"), nor should we believe what atheists tell us ("There is no God", "NDE is all in your head...") – the more we plug into ignorance, the more we bury ourselves in low, slow Earth vibrations – and will have a hard time getting out of here! (The PTB love it!)

6. **Six reasons to know about the OPs and recognize one:**

   a) Avoid marrying a bad spouse that does not support your values
   b) Can ignore atheists and their 'scientific' pronouncements
   c) Stop hoping that the really beautiful blonde/handsome man will come around and be a warm, caring, cooperative person
   d) Realize that sex maniacs and **control freaks** are OPs on steroids
   e) Realize that a person who criticizes everything you do is an OP and you should leave them (their goal is to break/derail you)
   f) People with no 'issues' in life are usually OPs – don't trust or marry one (If it is a lying ensouled person, you also leave!)

7. Lastly, the author was given this information and told to write it up and disseminate it. The time is ripe for two new revelations – as said in the Introduction. You had **Karma** and **Reincarnation** teaching 60 years ago, and now welcome to consider Scripts/**LifePlans** and the **Soulless**.

It takes all 4 to explain a person's incarnation.

It is your own sanity and spiritual growth that is important. Don't give your power (to make your own decisions) away. Skepticism can be healthy, if it isn't overdone.

Some people will really have problems with this Appendix, and that is why the information is not part of the main book, yet it is important – and some will be able to assimilate it. **It is Light**. These issues are where we are all going and are issues that we all have to deal with in our spiritual growth – sooner or later – and are new and important aspects of walking the Path to get out of here. Truth carries Light, and 51% is a minimum needed to exit the School.

Summary 2: Lesson Time!

Now that we have outed the existence of the OPs… so what?

> **Reason #1:**
> Here's the first reason: **you don't want to marry one**… but if you have, it is not about killing them, divorcing them, stabbing them, hitting them, or even confronting them with the knowledge…… if you leave them, the gods will just send another one (or two) with your incomplete lesson… so here's the deal:
>
> 1. Recognize that **you are not the problem** (as the OP keeps telling you!)
>
> 2. Most of them are not evil or anything to be afraid of. Most are nice but very simple people. (Don't try to 'educate' one.)
>
> 3. Realize that they do what they do as a way of bringing tests and lessons to you…
>
> 4. Realize that if it is an OP – there is a reason they are there, and **the lesson must be handled**—better to find out what it is and handle it! It's "Lesson Time!"
>
> 5. If it is a violent, obnoxious OP, leave them… save yourself.

So you might say, "Well. I don't care what they are, I don't like that person and what they are doing... so I am out of here!" So, if you go, take your leave in a way that doesn't harm them or create more negative Karma for yourself... make it a nice, clean break.

Reason #2:

The second reason to know about them is that if the person fits the description of an OP, or their behavior matches what is described here in Appendix C, it means that **your Script is in play and this is someone and an event to pay attention to!**

That is all we can do about them. Just realize that your Script is in play and you need to do the most reasonable or appropriate thing to **handle the situation** – not avoid it. You can run, but you'll have to eventually face the same situation, or worse, and it is part of your lesson or training to be a more competent and aware soul.

Thus, not only do you not want to marry one, you want to know when it's "Lesson Time!" That is why you want to know about the OPs.

Summary 3: Knowledge

Rewind: it is Knowledge that is important because it carries the Light. While Love is the power behind Life and our souls, just being loving will NOT get you out of the Earth School. Sorry. It takes both Love and Knowledge.

Knowledge Protects
Ignorance Endangers

If you don't know the Truth, you will fall for anything – unless you are a total skeptic, and then you reject anything and everything – and you still do not get out of here. That is the challenge for all souls – to seek, **ask,** and learn the Truth. It is only the PTB who love being Lords and so (via the Media nowadays) they suppress real news, of all sorts, and seek to keep you so ignorant that when you die, you automatically get **recycled** – so that the PTB have a constant supply of sheep or serfs to rule!

You are supposed to wake up, at least suspect something's amiss, and see thru that and rise above it! Failure to do means you just **failed a key lesson and will have to repeat** the 'test' or lesson.

And you have a **choice** – to handle it or not. Always.
Earth is all about lessons and tests. Get over it.
Whatever happens, **give thanks** and handle it.
It is all **catalyst.**

The sooner you choose to handle it, the sooner you get to move on into higher service, learning and serving the Father in <u>awesome</u> realms.
With the right Knowledge comes the discipline to handle the Power that goes with serving in the incredible realms awaiting us all. Earth is a cesspool compared to where we can be – once we graduate!

Our future as expanded beings is awesome.

When you know, why would you choose to stay here, in ignorance?

Dare to explore!

Appendix D: Karma, Reincarnation, Scripts & More

Chapter 4 referred to a Greater Drama and our LifePlans… the following is repeated from TOM, for your convenience. It is important to know what **Karma** really is and how it works… as well as the fact that **Reincarnation** is not just a Buddhist or Hindu notion   it was also taught by the early Christians (see Fr. Origen) and Gnostics.

And the reason Karma and Reincarnation are no longer taught in New Age churches (and it used to be, in the 60s-80s) is because **those two are not the whole story** and when the minister is asked to clarify how Karma works or why Reincarnation is necessary, they are not aware that there are <u>two other facets of the complete picture</u>, making 4 basic elements in total:

> **Karma, Reincarnation, Scripts (LifePlan) and OPs**

… and that latter (OPs in Appendix C) is not popular.

> And a **5th element** is the **Life Review**.

> **Chapter 1 (end) says what this place called the Earth Realm really is (VR Sphere)… and this Appendix gives an insight as to how it works for us souls.**

So here is the information, starting with the InterLife and what happens on the Other Side…

The InterLife

The location of this place is said to be in the High Astral – above the Angels and Masters who tend to us, but not as high as where the **Higher Beings** live and operate… they tell the Angels and Masters what to do. It is where souls come for rehabilitation, to be schooled, to be counseled, and to plan future incarnations – creating their Scripts and making agreements with members of one's Soul Group to play certain parts. It is a working world… no one is sitting back playing a harp (except musicians in training).

> It is also called the **Other Side,** but is not Heaven.

Souls are not here to just party. Both Earth and the InterLife are part of the greater working School for souls.

The InterLife is preparing and the Earth is doing.

Earth is sometimes a context of one's equals where the ignorant, selfish and dysfunctional are put (under supervision) to experience their errors and learn from them. It is all about **soul growth**. And while learning, when appropriate, one can party. The gods aren't killjoys.

InterLife 101

Here are the basics of what happens starting at death.

Leaving the Body

Almost always when a person suffers an accident that takes their life, whether it is falling off a cliff, getting run over by a Mack truck, or being shot, the soul knows what's coming and is <u>already out of the body</u>, and it is only the body that experiences the destruction/pain.

There is little point in the soul experiencing the body's agony as it is maimed or shot, stabbed, or hung. Often, the soul watches dispassionately from a few feet above to what happens to the body it once inhabited. The ETs refer to the body as a 'container' since **the body is not who we are**, we merely have one and at birth, the soul merges with the body at birth for its 3D experiences.

It is important to not bemoan what happens to the body in a life-taking accident. It is also an interesting insight that **there are no accidents** as we know them – events are often pre-programmed into our Scripts and they are designed to grow us and teach us something, or sometimes the **events are 'tests'** to see if we handle things correctly and make the right choice, or have the appropriate reaction, or sometimes, no reaction.

The soul is on its way back to the Realm that it came from, to rejoin with members of its Soul Group, the Masters, and the Schools where the soul is counseled, schooled and evaluated to see how much progress was made in the last 3D lifetime.

Tunnel & Life Review

This section's discussion pertains to **souls** who die and who also believe in God.

Note however that the **atheists and agnostics** have denied the existence of God, saying that they see no reason to believe in God, and so their Freewill choice is to be without God and when they die, that it is done unto them as they believe. They wind up in a limbo state, sometimes in a thick grey fog where they can hear others around them but they can't see anyone. There is no Hell – except for those who believe in it and thus are 'asking' to experience it.

This grey limbo is not the InterLife, it is a separate holding area.

Argue for your limitations and they are yours.

When the soul in limbo has had enough of isolation and asks for someone to help them, a Being of Light will meet them and present them with their options. Where that soul goes depends on (1) what that soul can accept, and (2) whether their ego can admit that it was wrong in denying Man's spiritual (soul) side.

> Remember Dr. Lerma in Chapter 8 discovered that "If you believe, you can go to Heaven " – apparently They do not welcome non-believers there.

The remaining review applies to souls who have an awareness of a Divine Being, a Father of Light, The One who loves them and they seek to be with Him. The soul will go through the **'tunnel'** which is a protective energy field so that Lower Astral entities cannot harass the soul as s/he proceeds to the Other Side (as they did to Robert Monroe in VEG) – and it is also called the InterLife area.

The soul is met by a Being of Light (an Angel, without wings) and receives a complete immersion, a sound & light show, replaying the just finished lifetime – to see whether progress was made in his/her scripted issues. The Being of Light does not judge or condemn – each soul through its connection to the Oversoul does its own judging. (See **"Movie Time!"** next section…)

If the soul made hardly any progress, s/he may be **recycled** back into the same lifetime. If a student doesn't pass 3rd grade is he allowed to move on to 4th grade? Otherwise making progress, s/he can and will later move on (**reincarnate**) into a new experience in the Father's realms. It is all up to the soul – no one makes anyone do anything. (See VEG, Ch. 8: Life is a Film.)

> ***Déjà vu*** *is the evidence that we have passed through the* ***same*** *lifetime once before – it means* **you have been recycled***.*

The Being of Light also assesses the incoming soul's condition; is the soul in pretty good condition, or was s/he really traumatized by the lifetime?

If the soul is lightly damaged due to traumatic life events, s/he will be sent to **Rehab** (see next chart) and probably will be regenerated, filled with more positive healing Light and Love, rebalancing them. These souls are mainly those who have lost their connection to their Higher Self, to the Godhead, and are minimally dysfunctional. They may also have simply lost heart.

If a soul is <u>severely</u> damaged, and does not respond to rejuvenation, he may be found to be unsalvageable (sometimes called "atrocity souls") due to having committed such heinous acts that they have aligned themselves with 'the dark side of the Force' and need **serious realignment** with the Light. If they cannot be realigned or rejuvenated, the energy/being will be disseminated, cleaned and then the energy will be reused somewhere else in the Multiverse.

The Life Review: Movie Time!

The incoming soul is evaluated by a Being of Light specifically trained to administer the Life Review, allowing the soul to assimilate what was done and not done as the lifetime is played back. The soul will see key events and people from the lifetime just ended and an interesting feature of the Review is that every time the soul harms another soul by word or deed, the soul gets to experience how the other soul felt… i.e., what the effect of their words or action were. [271] (Think: Entanglement in Quantum Physics: we are all connected at the soul level.)

When Tom hit another man in the face, he got to feel what that was like:

> In the life review I got to sense and feel basically what everyone around me felt at the same time. I was watching it and I was doing it. And I got to experience both aspects of it at the same time. But I didn't see anyone as actually judging me. It was more like I was **judging myself** on what I did and how that affected everyone….
>
> Everything was more accurate than could possibly be perceived in the reality of the original event…. During the life review you seem to have telepathic understanding of others' thoughts and emotions…
>
> I [Tom] also experienced seeing [my] … fist come directly into my face. And I felt the indignation, the rage, the embarrassment, the frustration, the physical pain. I felt my teeth going through my lower lip – in other words I was in that man's eyes. I was in that man's body.
>
> I experienced everything… that day.
>
> In short, …**there is no real separation between you and others**, and your illusory isolation as an individual in this world is revealed to be a sham. [272]   [emphasis added]

If you have read Chs. 9-10 of DNQF, you realize this is an example of the All Are One idea and **Entanglement** of Quantum Physics… the Eastern gurus (Tao) are right – we are One.

The Being of Light does not judge, he merely asks questions and answers the soul's questions as the <u>whole</u> lifetime is played back – as if it were a living video or holographic reenactment of the lifetime. It is very real, and you are back in it moving quickly from scene to scene.

Everyone finds that they are **more judgmental of themselves** than The Father or the Angel ever is. We are our own worst critic and sometimes the Review stops while the Angel calms the soul down, re-centers them emotionally, and then continues…
The whole lifetime is played back AND kept in the Akashic Records… it does not playback and then disappear.

In fact, if you knew that others could see what you did, what would you want them to see? Remember when you start to do something negative: **Movie time!**

> The HQC links to the Akashic Records and replays your
> lifetime – that is how the Masters know what your Karma
> is and what you need to work on (see diagram next page).

Meeting and Rebuilding

Following the Life Review, the soul is aware that it needs to either consult with a Master about what was inappropriate (not 'wrong') in the last lifetime, or the soul is so upset that it may need to spend time in an area (EIZ below) reserved for nurturing and lovingly energizing 'frayed' souls… to rebuild and re-center those souls who had a rough time and are not ready to meet with their Soul Group members, or a Master.
HBC = Heavenly BioQuantum Computer

InterLife Components

In the diagram following, the soul (larger star) at 'a' in the Group Area, makes a choice via the HBC to select Lifetime 4 (LT4) and the lighter arrow at 'b' is that choice which results in the LT4 being loaded into the HBC (heavier return arrow). The HBC invokes the Control System (C/S) to in turn invoke the HVR Simulation for planet Earth in Timeline 1 (TL 1).

> The reason I can describe a lot of this (and the diagram below) is I
> went thru a 3-hour Hypnotic Regression in 1991, and I saw a lot of
> this. I was allowed to remember it. More is examined in TOM.

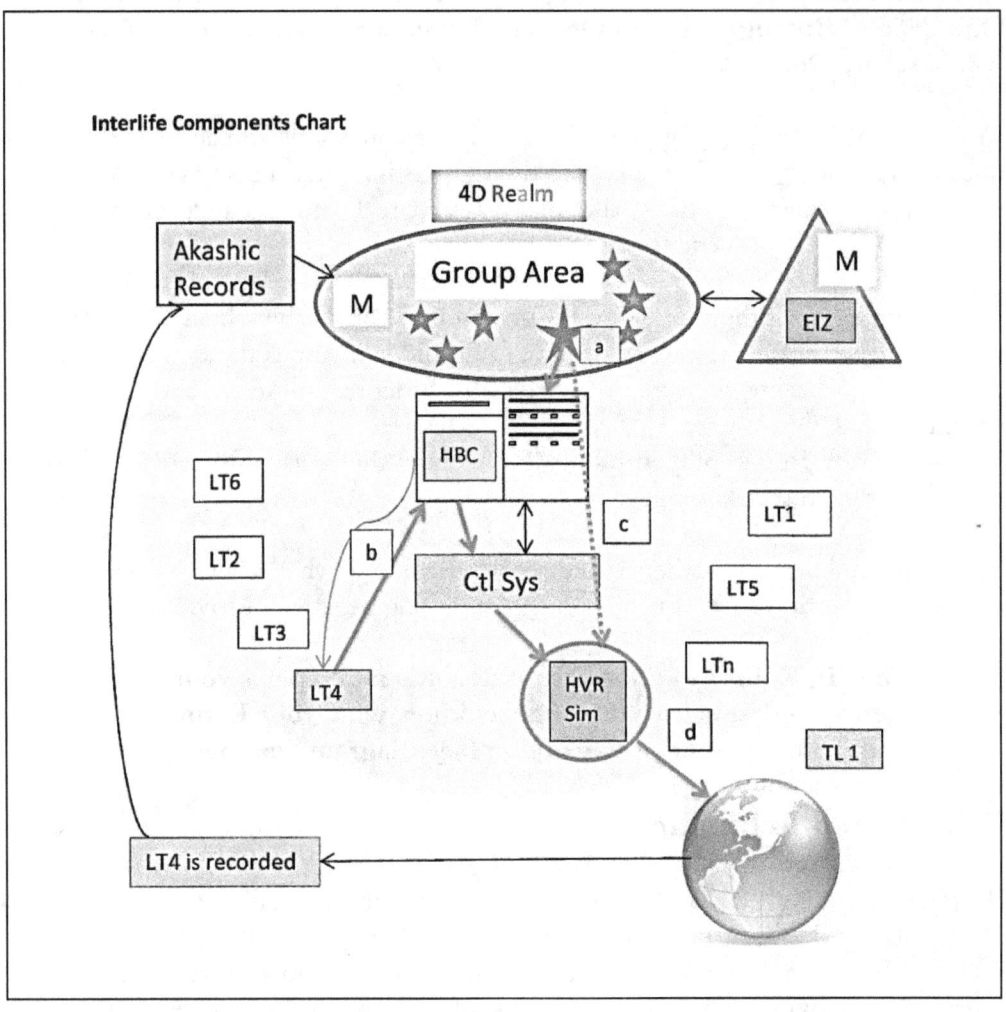

Interlife Components Chart

The LT4 LifeScript is loaded to TL 1 Earth via the red LT4 path. The soul at 'a' is processed into the HVR Simulation via the dotted blue line at 'c', and the LT4 + soul load (red + blue) and make purple at 'd'. There may already be souls running in TL1 and that is OK; the C/S can insert and delete without disrupting the Timeline and neighboring Scripts. The family into which soul 'a' is born is already there – they are just seen as having a baby. Thus the soul 'a' enters the LT4 Drama by birth in TL1.

Note in the InterLife Components Chart that:

The two small boxes with **"M"** represent the Masters who stand ready to advise. They are all over the place, advising, teaching and checking up on souls who may have gone to the EIZ.

The **Energy Imprinting Zone (EIZ)** is an area in 4D where souls go to recuperate and recharge after a strenuous lifetime. This is also the site for archtypes available for imprinting soul qualities, mentioned earlier in TOM.

"**HVR Sim**" is that part of the Heavenly (Quantum) Bio-Computer (**HBC**) that effects the desired Simulation of a Holographic Virtual Earth Reality in a Timeline (TL). Both HVR Sim and Control System (**C/S**) are aspects of the HBC and were covered more thoroughly in VEG, Ch. 12-13.

LT1 – LTn represent the available Lifetimes and Scripts available for all souls (the **Stars)** in the Group Area.

The lifetime LT4 for soul 'a' is recorded in the Akashic Record (AR) which is available to any Master as well as to the soul whose lifetime is in there – soul 'a' in the Group Area can review any of his past lives (from 4D) – however, he cannot see just any other soul's life unless that other soul has given permission.

When the soul 'a' in TL1 Earth dies, s/he is taken back to the Group Area via a **Tunnel –** which protects the soul from 'Astral' harassment by Thoughtforms, Parasites, and by Discarnates ('ghosts') who were afraid to go to the Light and will roam the lower 4D around that version of Earth until the TL1 dissociates and collapses.

This rebuilding area (EIZ) is appropriate for souls who were caught in a hurricane or earthquake and died without much warning. It may also be a refuge for souls who abused others – like a Hitler or a serial killer who can't face themselves or others for a while....

Soul Group Welcome

Following the exit from any nurturing area, the soul can move forward to meet with 'family' members, members of one's Soul Group. Souls will often get together and plan an experience in the Earth realm – one will be the mother, another will be the son, another will be the father and so forth – different roles to experience different aspects of the human incarnation.

There are lessons and aspects to be learned, insights to be gained, by playing different roles. **Shakespeare** told us more than we were able to hear with the following clue:

All the world's a **stage**,
And all the men and women merely **players**;
They have their exits and their entrances;
And one man in his time plays **many parts**.   *As You Like It*, II, 7.

Subtly he tells us and uses the key word. **A Stage**. And a Man plays many parts –
he cannot play all parts at once, so he must wait until the next time the play is
performed to play a different part in the **Drama**. Reincarnation. The subtle
message here is "in his time" – the soul is eternal, so "his time" is **eternity**, and he
thus "plays many parts." (And what if the Drama is run as a Simulation?)

Perhaps that is what Shakespeare meant when he also said:

…And all our yesterdays have lighted fools
The way to dusty death.
Out, Out, brief candle!
Life's but a walking shadow,
A poor **player** that struts and frets
His hour upon the **stage**,
And then is heard no more.

It is a tale told by an idiot,
Full of sound and fury,
Signifying nothing.   *Macbeth*, V, 5.

Key Insight:

If you're on the Stage, you're in a Drama, right? And if the
Drama is a tale signifying nothing <u>in itself</u>, and you can't
control the Play, then it is futile, right? And if the Drama
has no significance beyond providing your lessons/tests,
and your gaining Knowledge, why do we get so attached
to something that, in itself, is futile?
Drama is catalyst only.

To realize the Futility (of getting your way) is one of the
Goals in the Earth School. (Learning to handle everything
is the other one: not necessarily succeeding at everything,
but learning to 'handle' it.)

And the frustration of not being able to fix, stop or change any of it is what leads to the **state of futility**. The deep realization that spending <u>more</u> time (i.e., multiple lifetimes) on Earth is futile. You have to see it, feel it and really know it, and then **detach**. That is part of your ticket out of here.

Meeting With a Master

The Soul Group has access to ascended, wise Masters who counsel souls in the undertaking of an intended lifetime. The players for the next round of incarnation may have to adjust what they plan to do, or switch roles – depending on the Master's advice.

As a result of the meetings and discussions, and plans to meet each other, where and how, and what things this group of souls will do to and for each other, they will also include **exit points** (death scenes), special tests and **Points of Choice** will be set up in their respective Scripts. Their individual Scripts (or LifePlans) must mesh with each other and with the Greater Script running in the Earth during the time period the group plans to incarnate.

This level of planning does require the insight of the Master, and the review of the potential lifetime is put through a simulation in a **Heavenly Quantum Bio-Computer** – the same one mentioned in VEG Ch. 13 that runs the Greater Script for the Era in which Earth currently finds itself.

Higher Control

No one does anything that does not fit within the boundaries of the **Greater Script**, else it would mean that God (Infinite Awareness) is not in control, and that is not the definition of God. Since God, The One, The Intelligent Force is not a person, It is more an entraining, **Intelligent Energy or Consciousness** that underlies and empowers the Greater Script which is the **intent** of a yet Greater Intelligence <u>outside</u> this universe which oversees the Multiverse. The intent of an advanced being is very powerful. It is often called **Law**.

> #### Nothing is out of control and there are no accidents.

That means that God <u>does</u> have control, even though to us with a limited intellect, it appears as if things happen by chance and we fear that one day one of those things that is not under His control will collapse the planet, or solar system, etc. God does not watch over the world and every now and then say, "Oh, shoot! I wish I had foreseen that. Now I have to do that over!"

> **We don't have a God in training**. His Greater Script **is** in control and ours must mesh with His.

Think of the **Greater Script** as the operating system in a computer, and your life Script as a program (e.g., a word processor) running under the 'control' of the Operating System. The many programs experience a give-and-take with the Operating System (OS), and they can function because the OS empowers them. They are not able to do anything they want; the OS acts as a 'control' and monitors their activities and use of all resources.

Another word for the OS is the **Control System...**

When a program violates programming convention, or tries to illegally access files, or change parts of the **Operating System (aka Greater Script)**, it is not only stopped, it is usually quarantined.

Note in the diagram below that there are five jobs running in the computer – under the control of the Operating System.

> **Job 1** is a bigger job and occupies more computer space than Jobs 3, 6 or 8.
> **Jobs 4, 5, and 7** have finished and have exited the system.
> **Job 6** is an email with a link to a video, so the Operating System performs the link to the video and plays it for the person who initiated it thru the email link (**Job 2**).
> Note that all jobs depend on give/take from the Operating System (symbolized by the double-headed arrows).

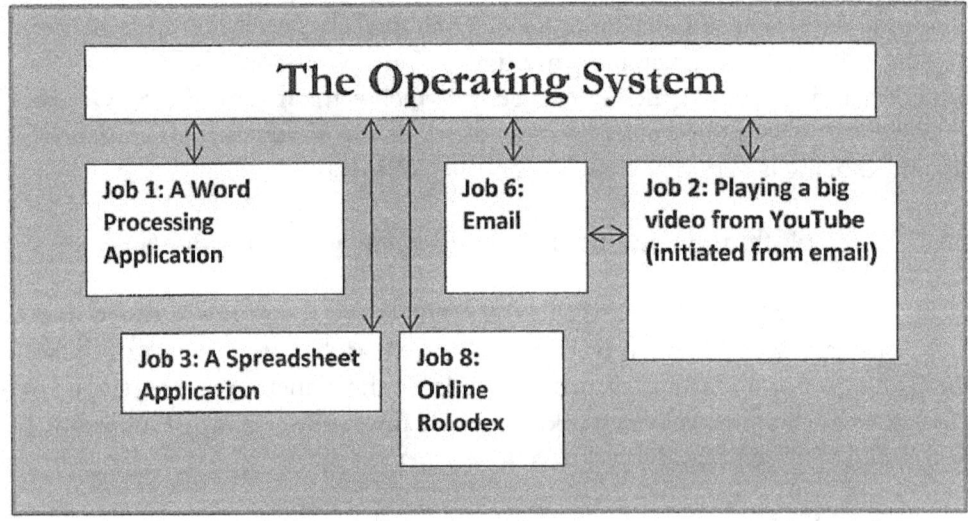

The Heavenly Quantum Computer operates in a similar fashion – think of the Jobs running in the diagram as 5 Scripts of 5 individual souls incarnated on Earth. The **Operating System** for Earth is the Greater Script and a major part of the OS is the Control System which often acts like firmware **Drivers** do in a PC system – operating the peripherals. The Control System sometimes initiates **Utilities** that perform special functions (like fish falling from the sky, or manifesting Airships [1896-97] to inspire Man…)

Job 6 and the interaction with Job 2 (horizontal arrow) is similar to two souls who have an agreement with each other to interact in some way. The Greater Script still oversees the association.

Note that **there are no LifeScripts for OPs**. Souls need Scripts to identify and manage what they are to learn. OPs are bit players (NPCs) sustaining the Drama and providing catalyst. (Appendix C)

OPs are walk-on bit players. The Beings of Light effect the action and the scenario. God is the Playwright. The major theme of your life is not by chance. It may be called a **Sacred Contract** as Dr. Carolyn Myss so termed the Script.

Welcome to the Greatest Show on Earth. Your life.

In the following, refer to the difference between Soul and soul – in the Glossary. The 'gods' below refers to the Higher Beings.

Heavenly Quantum Computer

When the soul has 'designed' his or her Script, with or without the help of the Master(s), it is analyzed by the **Heavenly Quantum Bio-plasmic Computer** to see what time period is most conducive to the lessons planned… and a suggested time period and potential families known to be available are presented to the soul.

This was examined in depth in TOM, but the gist is repeated in Chapter 8: LifePlan Selection.

What if the Earth Realm is a kind of Simulation and there are beings running it, keeping it going, and making modifications as needed?

LifePlans are designed , approved, programmed and implemented as part of the soul's incarnation on Earth… (TOM Ch, 7.) They also link to the Akashic Records for the Karma input.

A choice is made to try out a lifetime, for example, in Elizabethan England with the Smythe family and the soul gets to study the characteristics of this family to see whether the projected lifeline of the incarnated soul will meet the needs of the scripted lessons planned… If not, another time and family are chosen.

No soul is ever tested beyond what it can endure – it is not the goal to "make or break" souls, although steel is not made without fire…. And the creation of steel may require several lifetimes before a **soul can respect itself** and stand up to whatever is thrown at it. The gods aren't ogres, and they know a soul so well (because they are trained to read the 7-level aura which tells the whole story) that they will not permit a soul to undertake an assignment that is obviously foolish or wild – unless the soul is prepared (and briefed) <u>and can handle it</u>.

If a Soul cannot decide which lifetimes to choose, or what role to play with **aspects** of one's Soul Group (Souls often incarnate together into a life experience to help each other), the Master will choose aspects and locations for the Soul and possibly even for the Group, and **it is often done for Baby and Young Souls**.

At this point, we need to reconsider Reincarnation and Karma and how they play out – note that **Karma only applies to the Earth realm**, and not to the whole Multiverse. Earth is a School and Karma is the measuring stick reflecting lessons that were not handled appropriately and have to be redone. If the soul aspect successfully passed his last lifetime lessons, or at least a good number of them, he is eligible for **Reincarnation** – incarnating into a new realm and experience (even though some past lessons may present themselves again in a new guise).

If the soul failed to make much progress in the last lifetime, that soul aspect may be **recycled back into the same lifetime**. The proof of success or failure is recorded in the Akashic Records and the HQC "displays" the results of that Lifetime as Karma.

Karmic View

Your personal Karma often shows up as a discrepancy between the ideal Script you adopted (incarnated into) and what you actually did – the HBC can playback and compare the two. The difference is Karma – good (where you really did better than what the Script was trying to get you to do), or a Shortcoming (They don't say 'failure') where you didn't do what was the expected performance.

There is no Right or Wrong – there is "appropriate" or "inappropriate" and They know that if you had known more (not 'better'), or were capable of more (a more enlightened response to Their catalyst) you would have done it.

That is all it is about – so that when you graduate (Chapter 8) They know you meet a basic level of PFV/STO/Awareness that They can rely on and then you are sent to the next level training – see the Positions Available to Graduates in Chapter 8.

Reincarnation and Recycling

A significant object of this chapter is to clarify the difference between Reincarnation and Recycling.

People already know that **Reincarnation** is when the soul selects another lifetime in another locale or timeline for further experiences. What is less obvious is that the soul must have completed his Script (lessons) from the last lifetime with enough positive marks that he can move on. (See Chapter 8: Life is a Film.)

If the soul failed to make sufficient progress in the last lifetime, he can be sent back into that <u>exact timeline</u> again – and that is where **Déjà vu** comes from. You <u>have</u> seen that before, you <u>have</u> heard that before, and you <u>have</u> done that before. Being sent back into the same, exact lifetime is called **Recycling**. It can also be done in a **fractal portion** of the same timeline; it is a **simulation** involving mostly OPs.

> *Soulless humans are not recycled and they don't experience Déjà vu. They also have no Karma, and thus no Script. Like NPCs in a video game, they help drive the Greater Script in the Earth Drama.*

Yes, Recycling is less common than Reincarnation, but it does happen. If a student in elementary school does not pass 3rd grade, is he sent on to 4th? He may be sent to another 3rd grade, in another experience (and that is Reincarnation) but sometimes the failed lifetime was so potentialed that there isn't another like it… yet. The soul either waits indefinitely for a similar opportunity to come up in the future, or he is reschooled and sent back with the hope that his 'retraining' in the InterLife will help him meet the tests of that <u>same</u>, failed lifetime. And sometimes, a new **fractal lifetime** is specially programmed for that soul.

Anyone who experiences Déjà vu must stop and think: Can I handle this <u>appropriately</u> this time… What is really needed? And usually it is best answered by asking oneself: What is the most **compassionate** thing I can do?

Scripts: A Major Point

Consider: How many people around you, during the day, are also ensouled and working out their Scripts? What if you are the only soul working out the particular drama you find yourself in? What if 95-98% of the people around you are **OPs** – Placeholders driving your Drama/Script? That could mean that you are in a **fractal**

simulation that will repeat – until you get it right – which is what the Higher Beings showed Robert Monroe in one of his OBEs. It would additionally explain why there might so many people without auras in a specific lifetime...

A simulation that repeats is also called being **Recycled**.

Could it be that everything we see and experience is <u>orchestrated</u> for our benefit – as **catalyst** to provoke a certain direction in our growth? And how could the orchestration be done with willful [ensouled] human beings who do what they want? Two or more souls may make an agreement to be or do minor things as <u>a test</u> <u>or a blessing</u> of another soul in their group when they are all incarnated together in a common Drama. If the **intent** is to test or bless and not harm, then no Karma is usually earned.

Lastly, to repeat a related major point: the soulless/OPs <u>are</u> often what help keep the **Father's Greater Script** for Earth on track. Man has Freewill and does not always do what he should, preferring to boogie and party instead of studying, inquiring and cleaning up his act – to get out of here. Many don't want to leave the Earth. So the **OPs can be manipulated astrally to drive the serious karmic aspects of the Script**, giving Man his 'lessons', in turn manipulating Man for his/her greater good.

Freewill vs Control

Interestingly, Shakespeare gave us a clue about our ability to control our lives or our destiny. Your Script governs <u>generally</u> what you can do and what you can get, as well as it governs what the Beings of Light are permitted to do and not do. And your Script reflects your **Ground of Being** which helps determine what experiences you need and thus <u>attract</u>. You do <u>not</u> create your day, manipulating people & events, much less your life. Remember, Shakespeare said:

> There's a **divinity** that shapes our ends,
> rough-hew them how we will... *Hamlet* V, 2

For "divinity," substitute the words **Greater Script**...and this opens up the subject of Freewill and control – do we really have both in our lives?

Scripts/LifePlans

Soul growth is carefully scripted. No one is allowed to come into the Earth experience without a Script – **a plan**, Contract or Pact, for what the soul wants to accomplish where **only the major events (= tests) are scripted**. In addition, there are often many **Points of Choice**: options to follow one path or another, and that is effected through one's level of knowledge (knowing what to choose), and open to the degree of Freewill one has in the particular situation.

Often the lessons the soul needs to learn are hardwired **events** into the Script – i.e., there is no avoiding them. The major lessons will appear as the soul's Destiny, and yet what s/he does about them is governed by the soul's wise/unwise use of **Freewill**. Major events are usually orchestrated as tests and opportunities.

The Script is not Fate.

For example, my big dog is in the back yard, **contained** *by a fence. He cannot get out except when I take him for a walk. He can do anything he wants to (Freewill) and usually does! within the backyard -- but his overall life is 'destined' (*fated*) to live in the backyard.*

Major point: Man's Script does not tell him what to do or say.

It generally controls what he can do, where he can go, and sometimes what he gets and doesn't get: **up to 10% of his Life**. The Script even determined his parents and any physical defects he would have to experience. And it determines the **exit point** from a lifetime, when, and what form(s) that can take. Within the limits of the Script, Man otherwise has "Freewill" to do whatever he fancies – appropriate or maybe inappropriate. **The Script orchestrates events, Man chooses how he will respond**… and unwise choices mean he will be returned to the Earth School over and over – until he gets it right… reminiscent of the movie *Groundhog Day*!

Note that in many cases, incoming souls with a Script, participated in the 'design' of that Script; their birth location, choice of parents, and the opportunities and obstacles in their lives. (Chapter 8: LifePlan Selection.) While it is not a case of 100% Fate, **things are a lot more scripted than Man would like to think**, and there are some optional choices he must face.

Alternate Scripts

Many souls usually develop or have an alternate Script that they use as a fallback if the first, or main, Script should prove unworkable – because a key player in your life changes their mind and does not follow thru with what you expected (and they usually agreed to!) – now their Freewill is a factor. Or if a soul somehow completes the lessons of a lifetime early, they can try out the alternate Script! But they stay within the same lifetime, and the alternate version must 'fit' within the boundary of the current lifetime.

The switchover is done when the person is asleep, desires the change, and when arrangements can be made and the personal Drama reset. The Beings of Light assisting you in your Script are keenly attuned to whatever is going on with you; they don't miss a beat.

Script Control

The controls (i.e., "boundaries" of your life as set by your Script) are to make sure you experience whatever your Script says you are here to experience. **How you deal with events is up to you** knowing that your inappropriate response(s) could damage your body, lengthen your stay here, or make yourself and others miserable. Insisting on your way, when you want it, the way you want it, was something that probably got you put here in the first place... you might want to reconsider your **ego's point of view** that you can do anything you want while here.

Earth School is not controlled by the students.

If you resist learning and overcoming tests, trials, and blockages, and hate your life, you may consider drinking, drugs, sex or even suicide as a way of escape. All will result in your being sent back, after counseling and 'training', with the "screws" tightened more than they were the last time. [273]

> You must eventually face what you were just sure you couldn't handle. This is exactly what happened to me: I committed suicide two lifetimes back and then avoided doing the right thing in the last lifetime, and so this current lifetime has been very rough (i.e., my Freewill has been almost non-existent). My Script this current lifetime was 40% orchestrated (10% is the norm.) and I saw it all in the '91 Regression.

The Higher Beings are serious about our learning what is expected of us. And, ironically, even though we may not be aware of it, **we are more capable than we could ever imagine** – technically, we can handle whatever is given us – and that is one of the lessons (usually reserved for "last time" souls who are about to graduate from Earth). **That realization has to become part of each one of us.**

Scripts and Karma

Let's examine some more of Karma.

Karma is not about you being knifed in this lifetime because you knifed someone else in a past lifetime. The dictum **"An eye for an eye" is false**... two wrongs don't make a right and it is too bad that the tit-for-tat teaching made it into the Old

Testament of the Bible. It is wrong and serves only to generate further hostility and negativity on the planet. (Look at the Hatfields and McCoys!)

Your Karma is what you need to work on – by **meeting yourself** in other people – you need to meet the **Shadow Side of yourself** that caused you to knife someone in a past lifetime and handle it now – not repeat the act. And nothing is gained by you being knifed in this lifetime. Thus the Script will be orchestrated and overseen via Beings of Light [aka Angels but they don't have wings] – explained in the Glossary and VEG.

Events, people and circumstances will be initiated for you, if you don't undertake them yourself, to present you with **experiences that reveal who you are**, how you react, and where you fall short. Failures are an opportunity to wake up, correct and move forward. Remember: it is all being recorded.

All of that to say that **your LifePlan (Script) rules**… if it doesn't prohibit something, go for it. Since you don't know consciously after birth what is in your Script, you do whatever you can while here – and if the Script prohibits it (for reasons of 'teaching' you something through denial of what you want), you will not be able to get it, do it, or have it… That doesn't mean you are a bad person, or that God hates you, it means there is a <u>reason</u> you can't get what you want and it has to do with soul growth. Rethink your goals.

> Again, the prime example is a soul who comes in to learn humility. The Script will be designed such that that soul cannot make money and keep it – they cannot become rich as it would defeat the purpose of a **hard life that teaches humility** – very few rich people are humble. And they can do what they want, which also does not teach humility.

*   *   *   *

Rewind: InterLife

This is all very important and there are still some issues to briefly discuss (although most of this was examined in depth with many examples) in <u>Transformation of Man</u> (TOM).

So because this is very important, and it all does fit together, if there was something in the first section you didn't get, or like, perhaps **restating some key points** will help. These are organized with a different grouping in mind…

Why Karma?

There has to be a way **to track the soul's use of Freewill** (when they do have Points of Choice) and hold a person accountable for what they have done. It is called the Law of Karma and **only operates where there is Freewill…** obviously if you have no Freewill, no choice about what you do, you are not responsible for what you do/don't do. And Earth is a Freewill School where souls learn the consequences of their actions. Hence the existence of tests.

> **Karma is not control – it is merely a feedback system that reflects itself in elements (mostly events) of your Script.**

If you didn't do the most compassionate or appropriate thing, and harmed someone else, you do <u>not</u> have to pay "an eye for an eye" – that is <u>not</u> Karma. Whatever it was that caused you to be inappropriate, act without understanding or compassion, is a <u>lack</u> or weakness within your development and will (1) be reviewed with you in the InterLife, and you may receive further teaching or 'adjustments' to help you, and (2) you will meet that part of yourself again in a future incarnation (not necessarily the very next life).

> **Karma is nothing more than meeting your shadow side.**

For example, if you lost your self-control and knifed someone in a past life, you do not need to be knifed in a future life to 'pay' for that. You need to see that there is a part of yourself that was weak (**underdeveloped**) and responded poorly to a provocation which might have been a 'test.' Such an event will be reviewed with you in the InterLife and you can receive training and energy adjustments to help you meet the 'test' again in the future. It <u>will</u> come up again – you can count on it being part of some future Script's test.

Again, Man does not have <u>total</u> Freewill to do whatever he wants, whenever he wants, any way he wants – that describes a **god** and not a soul in a learning mode. Individual Scripts vary from soul to soul depending on what a particular soul is here to learn – **Old Souls have more latitude than Young Souls**. We don't always have the control we'd like.

Handle it

If you are never 'tested' how will anybody ever know what you are capable of – let alone you?

The Beings of Light never ask you whether you like what you get or not, or whether you're happy – that is irrelevant. **You are merely expected to handle it…** including the sex of the body you were born with. (**Sex changes** are not forbidden, but they defeat the purpose and the sex you can't stand, and want to change, will have to be handled in the future.)

The experience of non-control is not to be met with anger, resistance, and depression. Whatever is in your life is there for the effect it has on you. **Catalyst**. The intended objective is for you **to rise above it**, not take it personally, and **handle it**. It doesn't need analysis, it doesn't call for rebellion, denial, suicide, or micro-managing one's life. It also doesn't call for drinking or drugging (Rx or otherwise) the problems away.

If you can't control (change, fix or stop) something in your life, it is a sure sign that it is **scripted catalyst.** <u>It is not intended that you can control catalyst.</u> Handle what you get, and don't wish/try/hope to get what you can't get. Find the inner strength and just handle it. Or, you'll be back…

Where Man gets into trouble is that <u>he</u> tries to be God and do whatever he wants, when he wants, any way he wants… <u>before</u> he has the wisdom to act (or not act) appropriately. A form of 'godhood' comes later (6D), after discipline and maturing on the Path that all graduating Souls must walk in 3D-5D.

LifeScripts

Obviously, if a soul plans to come back as the son of another member of his Soul Group, then the Mother and Father souls have to incarnate first. Otherwise, by working with the **Heavenly Quantum Bio-Computer**, a soul may choose to experience something outside his Soul Group – especially if there is a particularly taxing lesson he has to undergo – such as being born to an abusive Mother or Father – due to the way Karma works, <u>no member of his Soul Group can play that abusive role without themselves incurring Karma</u>!   (So, OPs are used.)

Thus soul A might choose another, different Soul Group member in another realm who could provide that lesson, to experience the giver of abusive behavior, BUT the giver would pay for that behavior anyway, so soul A cannot use soul B's behavior. The hangup comes in when soul A begins to resent or hate soul B and this would create unfinished business between the two souls – if soul A cannot forgive soul B <u>while in the Earth realm</u>.

> Realizing what happened and trying to forgive when back in the Soul Realm does not release debts… **what is earned in 3D must be paid in 3D and/or overcome in 3D** for it to have an assimilation in the Ground of Being of a soul.

Thus having said all that, be clear that <u>the right way</u> for soul A to experience an abusive parent and have no residual Karma or debt with that person is for the abusive parent to be an OP. **OPs do not incur Karma**, they have no soul and one never sees them again... when they die it is literally "dust to dust." That is also why, besides timeline bifurcations, there are so many OPs on Earth right now driving the Greater Script for many souls.

The Greater Script = Control

There is a Greater Script which characterizes each historical Era and Man is subjected to different Dramas and challenges in each Era. So, like it or not, **the OPs provide a good part of the Drama to keep Man on track with the Greater Script.**

Since Man has Freewill, he won't always do what he should, much less what the PTB would like, and so to steer civilization onto the Path that will have the most **catalyst** (opportunity for growth) for Man to learn from, the standard **OPs** are controllable, largely influenced by the Angels. OPs serve to act as both a feedback mechanism to Man and a derailing, deceiving mechanism to push, pull or prod Man into the experiences needed to grow him. And the experiences can be positive or negative.

Scripts and Errors

Souls have relative Freewill and while our lessons <u>are</u> scripted, our **appropriate** exercise of Freewill depends on our understanding of the situation, compassion for the others involved, and what our accumulated Knowledge (wisdom) tells us is **appropriate** in a given situation. If a soul has had no experience in a certain area, or plays dumb, or really can't figure out what to do, s/he is still expected to make the most **appropriate** choice s/he can. Even if it is 'wrong,' do something!

I HAVE A FEELING MY GUARDIAN ANGEL LOOKS LIKE THIS OFTEN

Errors will often be made because no one is perfect on Earth.

Our goofs can happen, even as errors of judgment, and we are still held <u>accountable</u> (not "guilty") for our choices. When someone is inadvertently hurt by our carelessness, or ignorance, <u>it counts</u> – but not like it does when we <u>knowingly</u> harm others. So if I hurt my soulmate and she feels like kicking me in the *cojones* or spits on me, if that occurs to her – that will incur some Karma <u>for her</u>...

Payback on payback. When does it end? **"Vengeance is mine" did <u>not</u> say the Father** of the Multiverse, and a loving God would not say that: He doesn't get even. Karmic "payback" is the great equalizer.

God does not punish… we do it to ourselves.

Any vengeance, or payback, should be left to God since the system of Karma is much more effective and complete.

It is better to forgive and move forward… especially if the other person is a soul.

Forgiveness

Forgiveness breaks the cycle and **breaks the energy link** to the person we hate or resent – it removes the link and the need to come back and play with that person again and again, until one of the two stops paying the other back. One of the worst role models that we have had on Earth was the Hatfields and McCoys. "An eye for an eye" is the wrong way to go… (see TOM, Ch. 11 section: 'II Using Energy').

And the higher one goes spiritually, the more one forgives – because it becomes evident that the others <u>don't</u> know what they're doing. Getting involved in their ego game doesn't pay.

Why Forgive?

Thus, Dramas themselves do not need forgiveness…. unless you hate or blame someone (right or wrong) and they should always be released <u>anyway</u>. **They are not a test of our self-worth.** Note that the state of **unforgiveness** of another person energetically links you to that person and it means that you'll have to either let it go, or 'dance' with them or their surrogate later on. You are connected until you resolve the issue and release them. **Unforgiveness generates more Karma.** It isn't worth it. Release them and <u>release yourself</u>.

If a **soul** does something amiss to you, it is probably scripted Drama (prior agreed-upon drama), so forgiveness is not always an issue. See it as a lesson/test and give thanks. If a **soulless** person does something amiss to you, it is probably also scripted, but could just be their <u>unconscionable</u> behavior. In either case, give thanks for the lesson/test because there are no accidents. Don't shoot the 'messenger' who delivered the lesson. Only the lesson is important.

No Drama, No Process

Realize that most Drama in our lives is scripted. **Catalyst**. It is not to be fought, resisted, changed or avoided. Face it – with a new insight: **You are never given something that you can't handle** – it is just your point of view that (1) it shouldn't be there in your life, and (2) you want it gone. Worse yet, you think that its presence in your life means that (3) you are bad, defective, lazy or cursed. Read TOM, Ch. 5 -- I went through all of those possibilities and came up empty-handed. The most damning question I could have (and did belabor) was **"Why is this happening to me?" I wasted years with that search,** and all the while I was so distracted that I lost aliveness and productivity, missed job and personal relationship opportunities, and became demoralized.

> *It is fortunate that I didn't become an alcoholic, start smoking, or doing drugs… or commit suicide.*

Unless you do a hypnotic regression with a qualified therapist, you will never be sure of the reason Why, and guess what? Ultimately, **knowing Why you have issues is the booby prize**! The lesson was <u>What</u> not <u>Why</u>. Even when I found out what past events caused current issues in my current life – <u>I still had to handle them</u>! That ability is not conferred as a result of an hypnotic regression, although it can make it easier, and the realization of it <u>being one's own fault</u> sometimes removes health problems whose antecedent may be a past karmic event.

> *When we get to the Other Side, in the InterLife, we can know anything we want; all Knowledge is available there. But knowing about tennis means nothing without the 3D experience of <u>playing it</u> and we need the experience of handling our drama – to <u>understand</u>.*

> *Hence the Earth 3D School.*

What really drove me nuts was hoping to find the answer and then magically expecting that my life would all work out… it was as if the quest for the elusive answer became the *raison d'être* of my life. Nothing else mattered. And because I could not find the answer, I brilliantly concluded that I was either defective or cursed. This is a downward spiral for anyone and leads only to a crash. Then in addition to not knowing Why, there is the illness or destruction caused by the crash.

Some people don't seek answers. They give up and commit suicide, or drink/drug themselves to death. Or, as in the New Age, some people pretend the issues aren't there and they think that by ignoring them, they'll disappear (See "Happy" Video by Pharrell Williams)… these are the same deluded people who think they're creating their day (i.e., <u>making</u> it happen). Sad, because they'll be back to eventually handle it... and the problem won't disappear until they wake up.

What you resist, persists.

All Life Processes deal with Drama in one way or another. **No Drama, no Process**. No Process, no Growth. No growth, just death and stagnation. A soul that refuses to grow may be brought back to the Other Side, 'disassembled' and re-infused with new energy, then 'reassembled' in an attempt to correct 'blockages.' Failing that, their energy may be terminally 'dissociated' (disseminated) as a failed unit. (Ch. 7, VEG.)

We are **eternal souls** having an Earth experience – whether we like it or not. Certainly, my opinion about what I was going though in my life didn't count for anything, and **I wasn't expected to figure it out**. No one is. I was expected to just **handle it**. I wasted time learning that.

Steel is not produced without fire.

Summary: Scripts, OPs and Simulation

On the positive side of the coin, when everything is said about Karma, OPs, Scripts, Freewill and Destiny, it may occur to some readers that the soul is undergoing a Simulation, not unlike *Star Trek's* **Holodeck** where experiences are very real but 'engineered' nonetheless. If everything was totally Freewill, there would be no need for a Script, and no need for OPs, nor would there be a need for a Life Review when you die. But, all these things exist. And they exist for a reason: **no one gets into the Earth experience without a <u>purpose</u>, or a Script (or a Contract).**

Consider the following:

The <u>Script</u> contains some elements of <u>destiny</u>, as well as Karma as things that must be faced – as 'payback' or as <u>tests</u>. The Script is largely driven by <u>Karma</u> and selected lessons, and it may include some particular aspect of soul growth that the soul has <u>elected</u> to work on. Every Script needs a scenario and supporting players and the mix becomes the **Drama** whose major acts/scenes are the elements of destiny – scripted events and/or people that must be met. The OPs participate in creating the Drama, like walk-on bit players. Unlike a hardwired script in a thespian playhouse, there are areas of one's personal Script that call for Freewill decisions **(Points of Choice)** which are reflections of the soul's maturity. <u>All</u> reactions to one's Drama/scripted events are <u>recorded</u> (HBC) and will be evaluated after leaving the Earth life by the gods who run this school: that is the **Life Review**.

Many **OPs are "fillers"** – there to round out the scenario, but they have no personal purpose (or Karma) in the Drama. They are like **NPCs** (Non Playable Characters) in a video game --**They have no Script**. Yet, some OPs are key to your lessons. There may also be few ensouled humans in one's Drama as they have their own lessons to learn, in their own Scripts and as will be seen, they are not all participating in <u>your</u> Drama... unless by agreement, like a soulmate. **So many players in your Drama who are not family or close friends, and who challenge you, usually turn out to be OPs**.

Rewind: OPs

OPs are very necessary as **an interactive element in every person's Drama**; other ensouled people who would be required by your Karma to lie, cheat and steal from you, or worse, and would thus <u>themselves incur more negative Karma</u> – even though their actions are required by your Script. Such "give and take" Dramas are <u>not usually</u> staged between ensouled beings. To repeat: **OPs are used as <u>they incur no Karma</u> and can deliver the negative experiences to you (at the behest of the Neggs administering the negative aspects of your Script) that other ensouled people could not do without generating more negative Karma for themselves.**

So if we can see that **the Script is driven largely by OPs** who discharge karmic lessons, and some aspects of the Script are open-ended, is it not logical that someone will be reviewing our progress or failure, and thus **our scripted life can be called a Simulation?** Wasn't a function of the *Star Trek* Holodeck to pit oneself against simulated situations to hone skills and develop personal growth? Why should the scripted Drama on Earth be any different?

Just something to think about.

Remaining InterLife Issues

Please note that issues pertaining to incarnation, selecting the baby/body to inhabit, choosing the family, cancelling out with **'still' birth**, and changing one's mind were covered in VEG, Ch. 16, 'InterLife' section, and TOM in Chapters 6 & 7.

A Better Way

What is suggested here is that **we are not our Drama**, we are not our Process, we are not our stuff, not our car, not our home, not our job, and <u>not even our body</u>.

These are all things that the ego uses to justify itself and look to for security and meaning.

The Process is just a Script and what will surprise you is that **you can handle whatever your Script brings your way**. That IS one of the big "Ahas!" that we are expected to make as we grow in spiritual maturity… usually we do not see that aspect until we are a late degree **Mature Soul** (see Appendix A). Instead we back off, fearing the worst… What is the worst that could happen to an eternal soul?

Worst Case Scenario

People fear death because they have forgotten that they are an eternal soul – the body eventually dies, but the soul lives on. **A soul cannot be killed** (but the Higher Beings may disassemble it! ..IF it is truly defective, and that is rare). Death may be inconvenient and a nuisance, but it is not the end. A soul is an eternal spark of Light from the Father, and each soul carries **a unique potential preprogrammed** by the Father and developed in the Earth Realm.

The worst thing that could happen for a soul is to be separated from the Father of Light's Love. Cast out into **a Void**. No Light, no Love, no other souls to be with… and according to the reports by Drs. Ring, Moody and Modi, and others, some souls are so ashamed of what they have done on Earth, they assume the Father will visit terrible punishment on them so they avoid going to the Light at the death of the body, and they voluntarily choose to be alone (as a Ghost) – in a Hell of their own making: The Void. Some are the Discarnates.

Please remember: **God punishes nobody**.

There is no Hell – I didn't see one when I saw the Other Side. We punish ourselves. But, doesn't God hate sin? What is sin? A **mistake**… missing the mark. God knows we are imperfect here and He knows we will make mistakes. That is what Earth is all about, just as with any 'school.' And everything we have ever done is recorded like on a "video tape" of sorts and is played back for us when we return to the Other Side…this is called the **Life Review**, full of our mistakes and successes.

The Life Review is much like the Comprehensive Exam (verbal interview) when one completes a Masters Degree at a University – what one has learned is tested verbally and one must defend the Thesis s/he wrote.

We are evaluated, not judged.

Only deliberate heinous acts are worthy of 'incarceration' and separate 'counseling' by very advanced, loving Higher Beings… the perpetrator is still not punished as we think of punishment. (Not even Hitler or Stalin.) Being separated from other loving souls is often punishment enough – ostracized until one comes to one's senses and asks to be 'rehabilitated' and then the **Father's Grace** is immediate and restoring… All the soul has to do is ask and open to receive. God is Love.

You have only yourself (and your ignorance) to fear.

Darkness is ignorance, or absence of the Light. Darkness is not a material thing in itself…if it were, then when you turn on the light in a dark room, the light would have to make an effort to push back the darkness and its ability to do so would depend on whether the light was stronger than the darkness. A *non-sequitur*.

Outcomes

In essence, **Man is a special creation** and needs only to respect himself, others, the Creation, and his Script. Everything he needs is provided – even if it doesn't look like what he said he wanted. As Flip Wilson (1971-74 TV show) used to say,
"Maybe what you want ain't what you need!"

Go with it, handle it the best you can. No fear. Remember that we are to go through the Drama, not stop in it, not play with it, not ignore it, not try to figure it out, and not blame others for it. I did all of the above. I thought that if I could just figure it out, I could stop/fix/change it… **catalyst cannot be manipulated**.

Thus, you cannot "create your day."

The Earth Drama is scripted and will 'roll on' anyway and your expectations don't count. **An expectation is a setup for an upset.** Get rid of attachment to outcomes and you'll be a much happier person, and healthier. Part of your individual LifePlan 'lesson' might be to never have anything turn out as you intend or wish – So will you be unhappy your whole life? You can't control your Script, and you don't know what's in it… unless you do a hypnotherapeutic regression (**and IF** They choose to tell you).

No Control

Not all parts of our lives are under our control (because this is <u>scripted planet Earth</u>) and control is an effort to manage the outcome. Want to be really happy? It is simple:

Take what you get
and
Don't take what you don't get.

Whatever you get is what you got, because you 'earned' it. You can't get something that another person got, and they can't get what you got – sometimes they can't even *grok* it. However, you both may be able to get the same thing if you both do the same thing (and if your Scripts don't block it).

I know. People don't like hearing that; it is popular to believe that we can do whatever we want to while here. (You can try AND some things may work.) Sounds good, but our LifePlans do rule, and we are not here to do whatever we want…

> **Souls don't all have the same degree of Freewill… so don't compare yourself to your friend who may have more than you do!**

Summary

So there you have it, and that is the basic information on the InterLife as it relates to **Scripts, Karma, Soulmates and Reincarnation**. Something that affects all souls and is important to understand. It is presented for general enlightenment and further explains the actions/inactions of some people on Earth, and even **explains why bad things happen to good people** (Hint: think Scripts and "tests"). And as we examined OPs, there are soulless people on Earth, **about 60% are OPs**, and they are used to drive the School's Greater Script.

Lastly, not everyone has a soul, and not everyone cares about it – the soulless person will (if they hear of this distinction) pooh-pooh it and tell you that it isn't important – after all, aren't we all equal? (Hint: No.)

The soulless would not be reading a book like this, dealing with soulmates and spiritual issues. Not only do they not know what that is, they don't care, and they assume that you don't, too. They are the promoters of the over-simplification of life: drive your car, eat your burrito, and watch Seinfeld… They are often attracted to Science where everything is Evolution (simple: no God , the soul does not exist (and it doesn't for them), and they are all the product of biochemical interactions within their bodies.

Yet it is to your advantage to know about them and what they do, and evaluate that man or woman that you're dating, your child's teacher, your pastor, your congressman, some 'scientific' pronouncement by some 'expert', or person with whom you work – **ensouled and soulless are like oil and water.** They don't mix, and a marriage between them is headed for trouble. Maybe even abuse.

An ensouled human with greater spiritual potential should **not be at effect** with the primitive, soulless person (no matter how rich or good-looking!) and should seek to minimize their interactions with them. (See VEG, Ch. 5)

Forewarned is fore-armed. Remember:

> **Knowledge protects.**
> **Ignorance enslaves.**

Appendix E: Serpent Sculpture

Chapter 2 displayed a Ubaid sculpture from 3500 BC – made by someone who saw these beings in ancient Sumeria. In addition, there was documentation from Berossus and the Jewish *Haggadah* that there were strange-looking beings walking around… the **Shining Ones**. (Their small, fine body scales reflected light.)

Of course the experts will claim that the following are all part of religious rituals practiced by pagan people – all around the globe they all just happened to make either the same or very similar statues – pointed snouts, and **elongated heads**…. But, you decide.

Ubaid Sculpture (Mesopotamia)

And the side views (male and female):

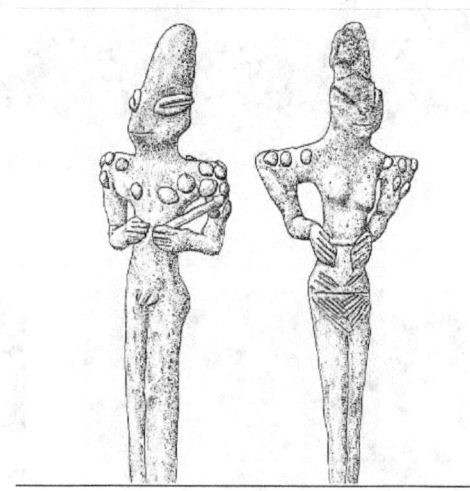

Vinca Sculpture (Serbia – Romania)

481

Mayan Sculpture (Latin America)

From: "Revelations of the Mayans 2012 and Beyond"
by Nassim Haramein

Japan... dogu statues

'Astronaut?' or Deep Sea diver
on AD 1102 church in
Salamanca, **Spain** (right).

Iraqui Pottery

Either this guy saw something unusual, or he was just a poor artist:

Iraqui Pottery

The circles or bumps are aleged to represent the Anunnaki 'ME' or crystals of power and knowledge.

The 'circles' above are reminiscent of the Ubaid sculpture....

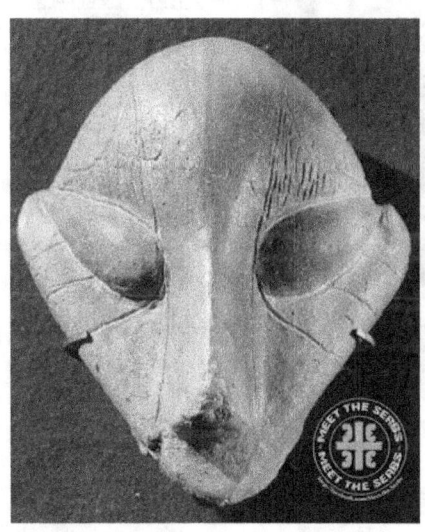

Vinca Sculpture again – what did the Serbs see and is it related to the Bosnian Pyramid of the Sun in **Visoko**, 20 miles northwest from Sarajevo, the capitol of Bosnia-Herzegovina?

This also bears a resemblance to the Ubaid sculptures... same "coffee bean" eyes.

(Credit: Bing Images)

483

Below is a screenshot from a video by Klaus Dona on the *Hidden History of the Human Race*.... Statue from Mesopotamia. A 47 minute video well worth the watching: [274]

While statues and pottery do not <u>prove</u> the existence of ETs/reptilians in the past, it becomes **credible evidence** for it when it is **all over the Earth** and the

projectavalon.net

images of the ET/reptilian hominids **look similar** – the isolated tribes did not communicate with each other, so how did they make images that are similar? And why do they have similar creation stories?

Credo Mutwa is a Sanusi shaman with the Zulu in Africa – his necklace tells a story of creation involving Earth women and reptilian hominids ... à la Enoch.

ZULU means "people from the stars."

The figure with the stylized 'Willie' is a reptilian ("Chitauri").

(credit: http://www.bibliotecapleyades.net/esp_credo_mutwa08.htm)

The 2004 Dragon below was publicly branded as a hoax, yet it was said to be created **before 1890** by Germans who then suspended it in formaldehyde and encased it in a 30" bottle. That is awfully good fakery for 125 years ago. Note the **very small hairs and veins in the wings** on it… fingernails and musculature are correct, and there is an umbilical cord…

It also bears a resemblance to the **Ubaid and Vinca sculptures**…

If it were a hoax, a simple x-ray could disprove it. It either has internal organs or not. **An x-ray was done and the results were not released.**

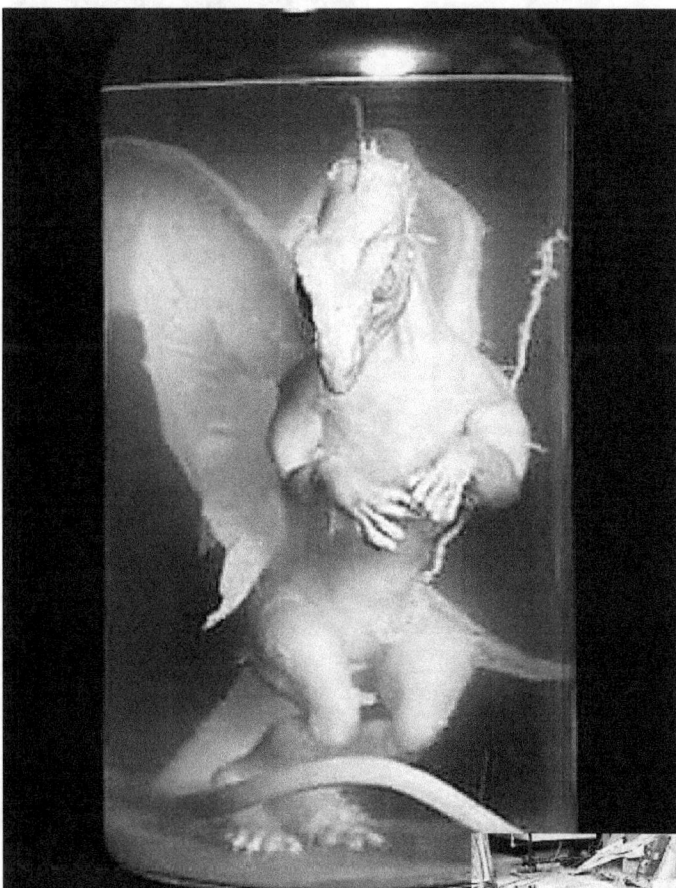

Note that the dragon's face resembles the Ubaid sculpture and that of the reptilian Anunnaki… reproduced below left…

Below: real dragon found in the ice in Romania.

Above: Copyright of Telegraph Group Limited 2004. [275]

485

Left: Ubaid sculpture (3400 BC)

Right: Naga sculpture in India

Reptilian sculpture from the **Peruvian** city of Chiclayo...[276]

The work of art is meant to show the deity **Morrop**, the Iguana Man known to the ancient Moche people.

Morrop was a powerful character who was crucial to their descent into the **underworld**. (Nagas live underground.)

486

South America is rife with legends and depictions of controversial characters that would fit neatly into modern theories about aliens, hybrids and their dealings with the people of Earth, from Viracocha to Quetzalcoatl, the feathered serpent.   The **vulture on his head** is a nice touch...

And If you thought that was unusual, you haven't seen the other statue right next to Morrop – Mr. Crabman...

The city of Chiclayo is also home to sculptures of **Lang Ñam**, the anthropomorphic Crab Being that guarded the entrance to the sea and its creatures. (BTW, Neptune aka Poseidon was Enki.)

Note the **Inca Sun-disc** on his head...

And there is also a model of the monstrous **Strombus**, a creature that is meant to represent the powerful forces of nature the Moche had to deal with (no pix available).

487

And there is one more from **Nazca, Peru** .. a **mummified head** (below) that has been x-rayed and found to be authentic. [277]

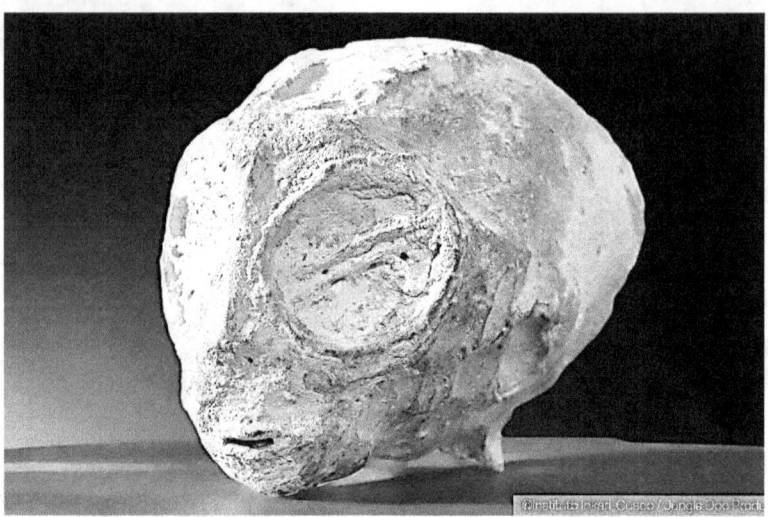

Head details:

Type: humanoid

Species : "Reptilian", "small grey" type

No hair. No teeth

Big almond-shaped eyes

It has neither ears nor ear canals (But they have a "middle ear and an inner ear" – Cochlea dixit the Mexican biologist **José de la Cruz Rios**)

Occipital opening located in the middle of the base of the skull (foramen magnum)... like Starchild skull also found in Mexico by the late Lloyd Pye.

The characteristics of the mummified heads observed and analyzed have nothing to do with the anatomical and morphological structure of any other known species. But we are reminded of this from Peru.... (Think: Brien Forster) [278]

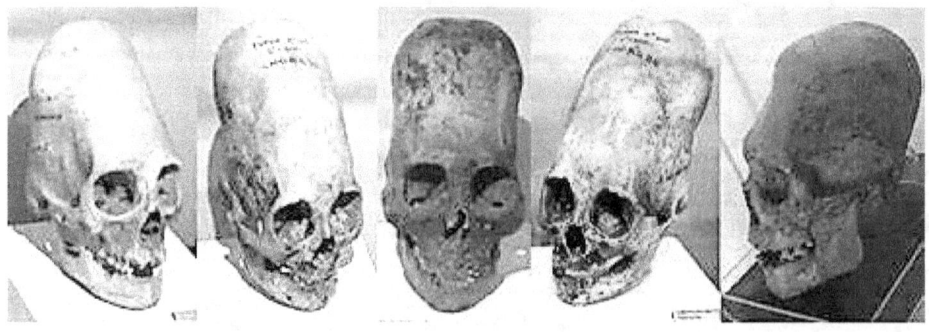

From Mexico...

What were these people seeing....?
What is interesting, and is why I show pix from different countries: they all saw the same thing, so it wasn't just tribal, ceremonial imagery.

And again, Reptilians....

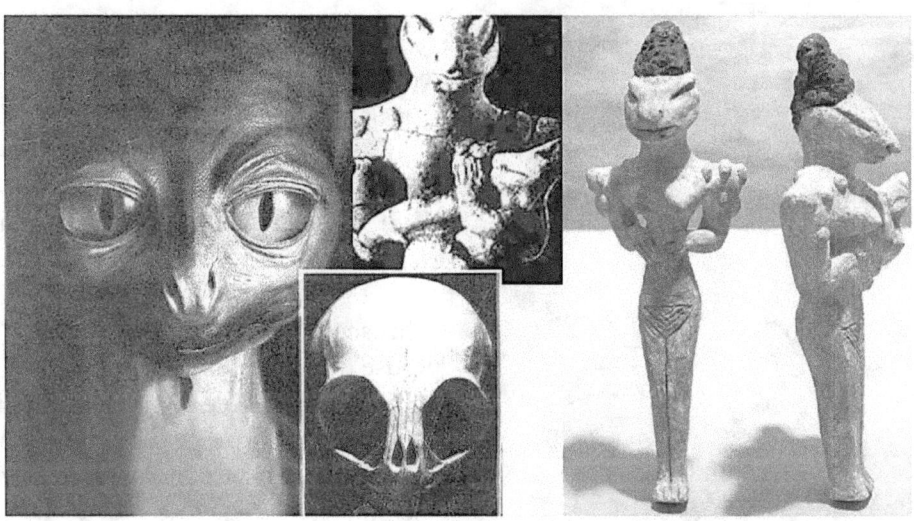

Pointed snouts and elongated heads... greater cranial size has meant a bigger brain... and smarter than us?

From Africa... land of the Dogon who speak of the Oannes and the Nommo...

And even from **Vigeland, Norway..** a man wresting with a Reptilian...

This (left) and other sculptures were all completed before 1943 by Gustav Vigeland.

His idea was **Man's struggle with Evil** (Nazis?) and found a home in the sculpture at left.

This sculpture also reflects the 13th house of the Zodiac: **Ophiuchus** which has been removed from public versions of Astrology – so some astrology charts are incorrect if they lack the 13th house (Nov. 29 to Dec. 18) – removed because it deals with reptiles. In the zodiac house, Ophiuchus wrestles with a giant serpent.

And lastly for your contemplation, this unusual entry:

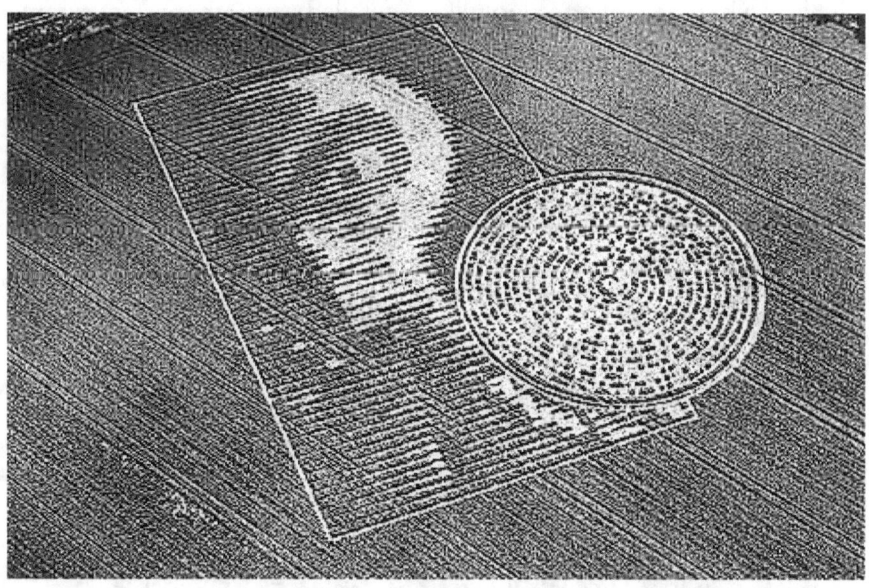

A 2002 Cropcircle, © Lucy Pringle.

Taken in a crop field in Westchester, England called the **Crabwood Face**. Obviously a real cropcircle and NOT one of the Doug & Dave hoaxes. Note the face resembles the ubiquitous Greys (below)… as well as the Vinca and Ubaid sculptures above. (Disc translation:

"Beware the bearers of FALSE gifts & their BROKEN PROMISES. Much PAIN but still time. BELIEVE. There is GOOD out there. We oPpose DECEPTION. Conduit CLOSING\".)

Of course that crop circle has been debunked but the PTB have to do that... they can't have the very naïve (whom they can deceive) wondering and growing up... So they used two dorks from the English countryside to fake a few, and they are obvious (below): rough, not symmetrical nor polished.

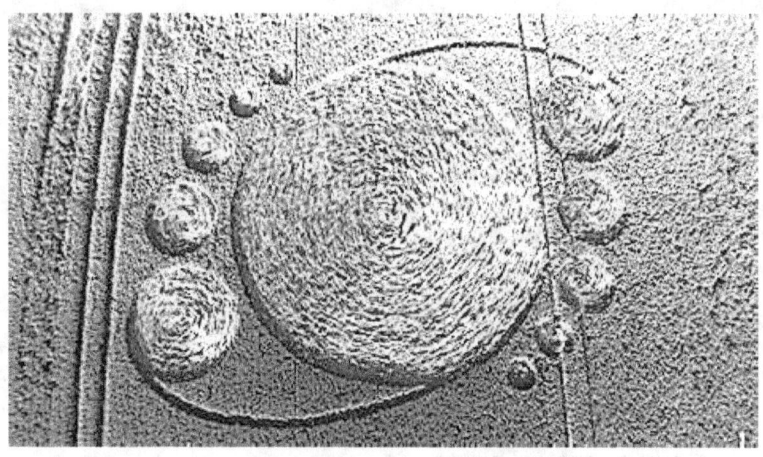

This is what the real ones look like... detailed, smooth, perfect symmetry...

And just FYI – some are done with Photoshop and were never in a cropfield...

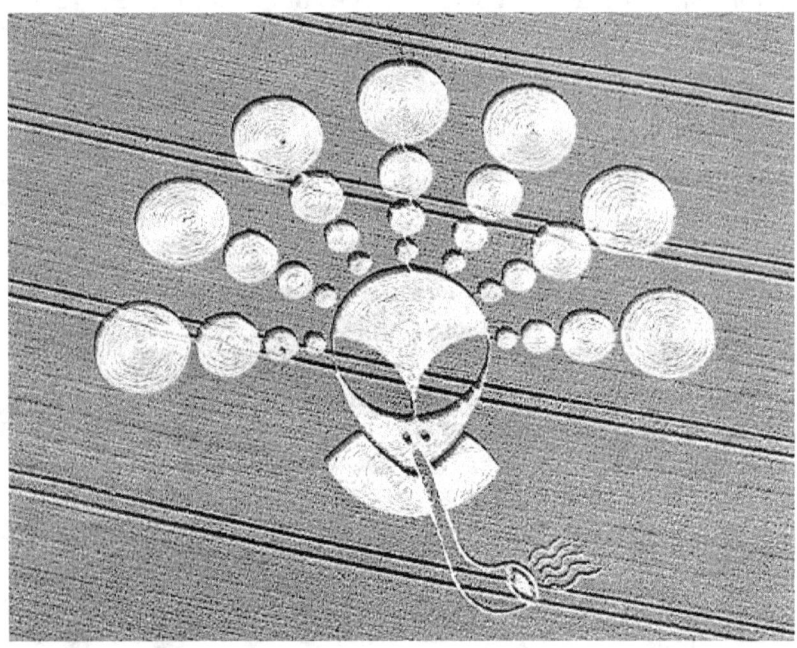

Some are an attempt to communicate....

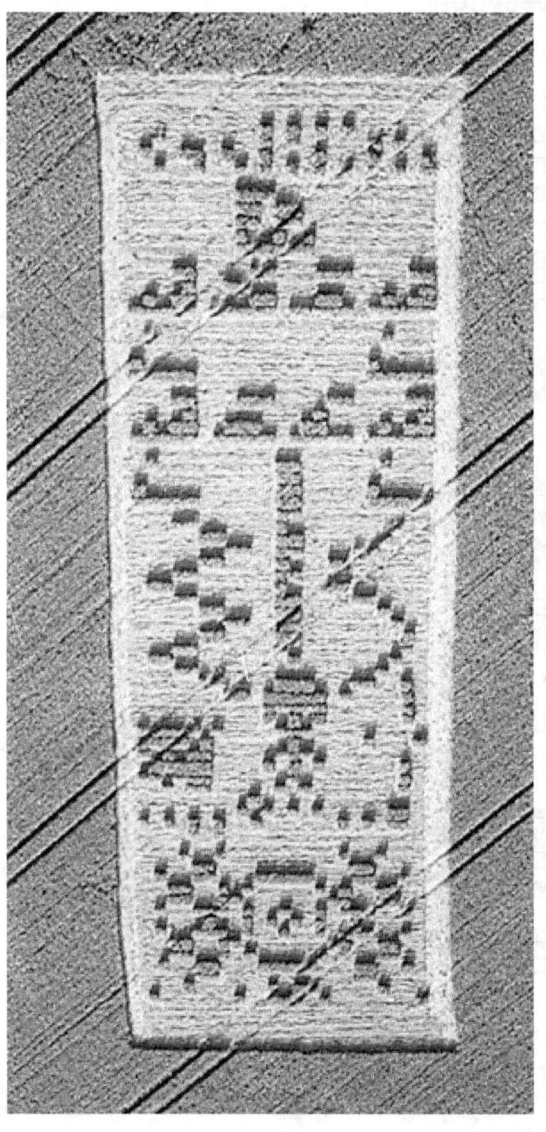

Left is the **2001 response to a 1974 telemetry** message sent into space describing Man, by Dr. Carl Sagan (by telemetry).

This was found in the crop field at Chilbolton in 2001 –
27 years later...

Below (next page) is what was sent by **Arecibo** radio telescope...

Note that the little figure (left) is different (at the bottom) as well as the planetary symbols directly below the humanoid.

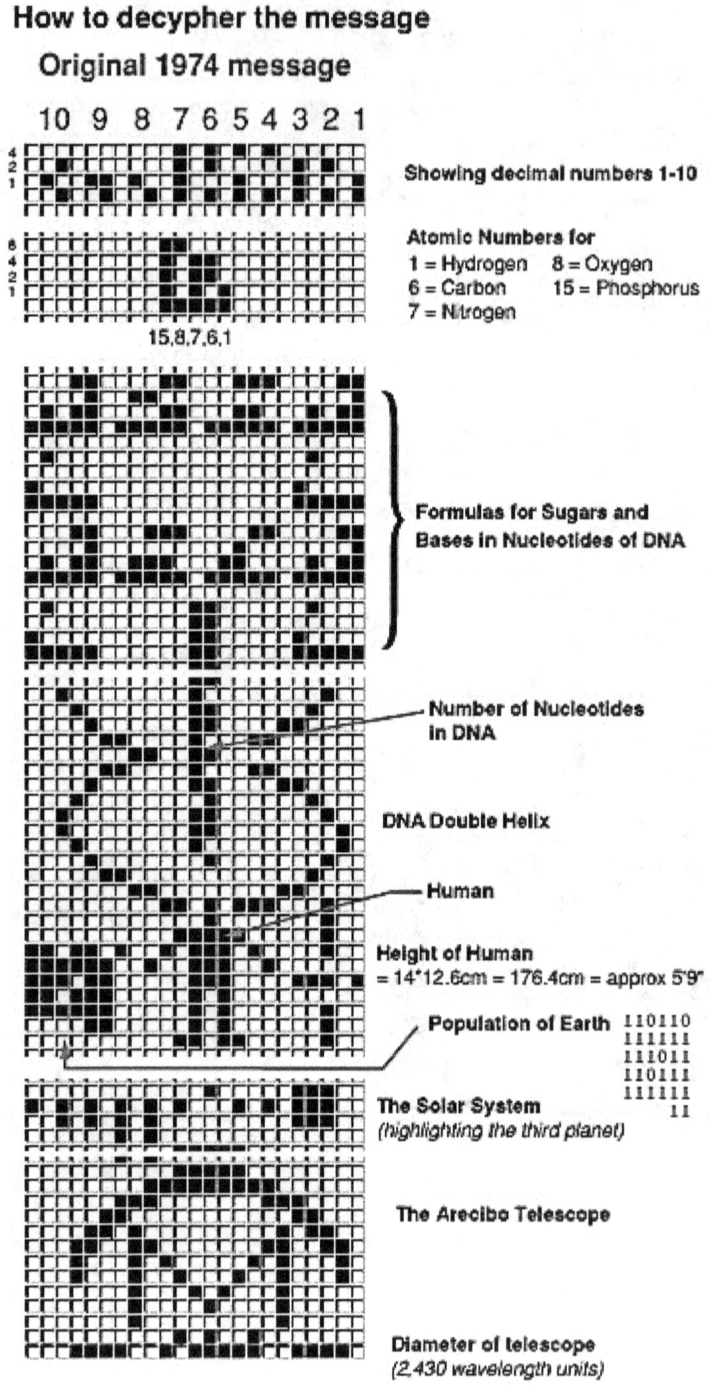

How to decypher the message

Original 1974 message

10 9 8 7 6 5 4 3 2 1

Showing decimal numbers 1-10

Atomic Numbers for
1 = Hydrogen   8 = Oxygen
6 = Carbon   15 = Phosphorus
7 = Nitrogen

15,8,7,6,1

Formulas for Sugars and
Bases in Nucleotides of DNA

Number of Nucleotides
in DNA

DNA Double Helix

Human

Height of Human
= 14*12.6cm = 176.4cm = approx 5'9"

Population of Earth
110110
111111
111011
110111
111111
11

The Solar System
(highlighting the third planet)

The Arecibo Telescope

Diameter of telescope
(2,430 wavelength units)

(credit: https://www.pinterest.com/pin/403353710361083034/)
The Chilbolton 'Arecibo message' Formation By Paul Vigay (R.I.P.) and
Alexander Light...

The following is another translation...

Left is what we sent them (by telemetry) and **right** is their answer...

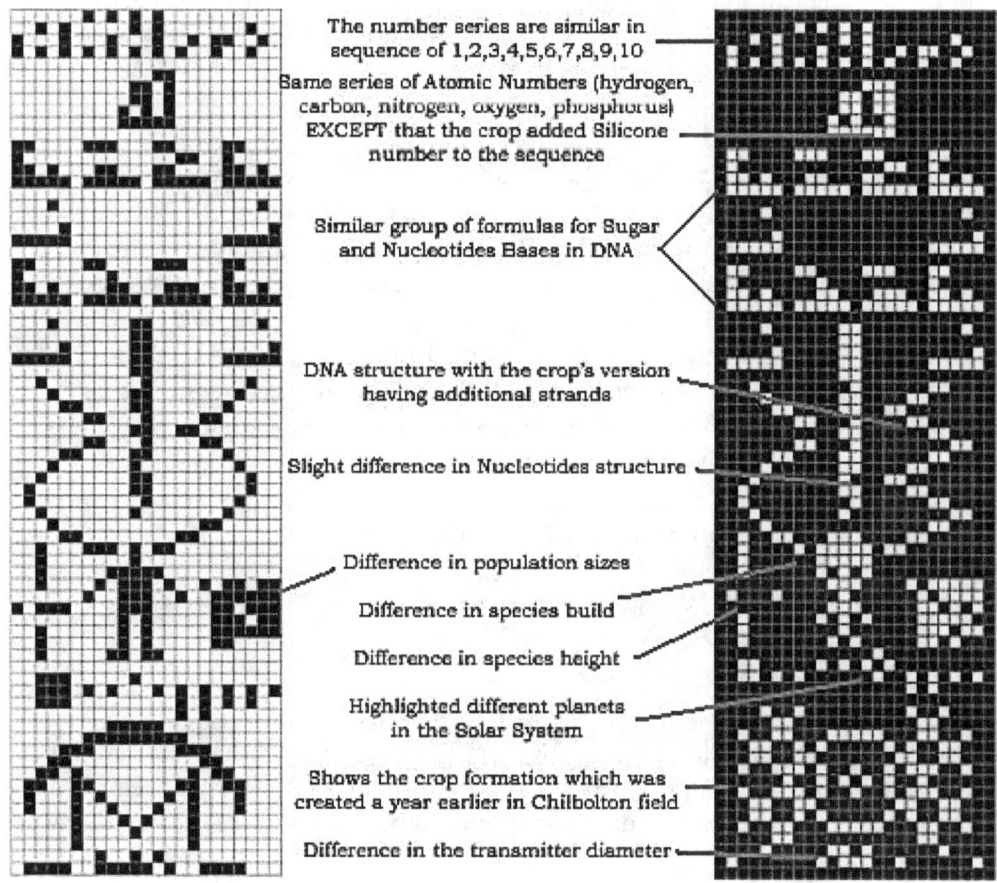

The number series are similar in sequence of 1,2,3,4,5,6,7,8,9,10

Same series of Atomic Numbers (hydrogen, carbon, nitrogen, oxygen, phosphorus) EXCEPT that the crop added Silicone number to the sequence

Similar group of formulas for Sugar and Nucleotides Bases in DNA

DNA structure with the crop's version having additional strands

Slight difference in Nucleotides structure

Difference in population sizes

Difference in species build

Difference in species height

Highlighted different planets in the Solar System

Shows the crop formation which was created a year earlier in Chilbolton field

Difference in the transmitter diameter

(credit: http://www.wisdom-square.com/alien-contact.html)

Interesting to note is that they added **Silicon** to their DNA structure, and that scientists have just recently discovered that life may be supported by Silicon… whereas before we only believed it could be supported by Carbon like with us. Also, the fact that an **extra strand of DNA** is added… giving them **3 strands of DNA,** is interesting because **scientist have begun to find a few human babies that are being born with 3 strands of DNA now, instead of the original 2**. The human genome has nearly 4.3 billion nucleotide sequences, while the mystery genome sequence has about one million more.

Finally, the diagram that depicts our solar system has been replaced with one that still has **nine worlds**, but the planets 3-5 (possibly Earth, Mars, and Jupiter) are lifted up, which is how we indicated which planet we inhabit. The **5th planet (possibly Jupiter)** is drawn larger than the others or perhaps instead represents multiple inhabited moons. The fact that Jupiter's moons are highlighted in a special diamond shape actually would support the discovery from NASA that stated that there is water and heat on the Jovian moons Calisto, Ganymede, and particularly Europa, which makes them good candidates for life. They are predicted to have subsurface oceans with more saltwater than Earth even. This research was published June 15, 2001. [279]

The preceding info shows that we are dealing with the Greys – synthetic beings (silicon) with 3 DNA strands, and they inhabit Earth, Mars, Jupiter. But why did it take them 27 years to respond to the 1974 message?? Is it because of the radio Barrier mentioned earlier (Appendix B) and they didn't see it until much later?

It is also interesting, in conjunction with genetics in Chapter 3, that "some children are being born now with **3 strands of DNA**" and that does exist and is called **TFO** [280]... looks like the Greys have been busy doing genetic engineering... Wonder what the extra strand does? We already know that some people are now **Tetrachromats** – their eyes have extra retinal cells that see more colors, compared to the standard **Trichromat** humans. (DNQF, Ch. 5.)

Rather than have you go to DNQF, here is the essence:

Additional Special Eyes

There is another fantastic aspect of the Grey's Genetic Upgrade – some people are becoming **Tetrachromats.** Most humans have 3 types of **cones** in their retina (Trichromat), but the new humans sometimes have 4 and they see colors that we, normal, people can't.

TETRACHROMATS
SEE 10 COLORS IN THE RAINBOW
can differentiate one **HUNDRED** million colors

(Credit: BingImages: wordpress.com)

Again, this has been documented as occurring **mostly in women**.

Have you ever watched a cat stare off into space, say into the corner of the dining room, where we see nothing, and then hiss and arch its back? Cats, birds and some insects can see into the **ultraviolet spectrum**… and it makes you wonder what they are seeing. Dogs by the way, are **Dichromat.** [281]

A **Trichromat** (regular people vision with **3 cones** which interpret color) can see approximately 100 shades with each of the 3 cones, so color definition is 100^3 or about a **million different colors**.

The Tetrachromat can see 100^4 or about **a hundred million different colors** (and this has been tested in the lab – when they find a woman among the **12% who have this trait**) and the woman can distinguish between two colors that to a Trichromat look absolutely identical, yet have a very subtle shade difference. And weirdly, some women with the 4 cones are <u>not</u> functional, as if they will have to be trained somehow to activate the 4th cone. [282]

So if you gents have a wife who says the color of the wall is not uniform, or she sees a difference in two apparently identical colors... she may be a Tetrachromat and is telling the truth.

Interestingly, **colorblind men** (it does not affect women) have been found to have a mutant and inoperative 3rd cone so that they have 2 good cones but the 3rd has a problem distinguishing red and green. And interestingly, a man with this shortcoming does not pass it on to a female offspring – she will have normal color vision, but may carry the genetic issue into the birth of a male offspring.

Speaking of Greys

The Grey on the right is a real Zeta being (picture taken at Area 51) and the one on the left is the BioCybernetic Roboid (drawing) that does abductions.

...and they are reminiscent of an alleged ET caught and kept at Area 51...

This is a real picture of alien caught and kept at Area 51 until he died – Earth environment is not healthy for them ... but the two ET resemble each other: a real ET and a real cropcircle... gives credibility to the idea that this being is one of the real (benevolent) ones... and obviously did the cropcircle.

You gotta wonder if the white mummified skull (found earlier) belongs to this being – large eyes and head....

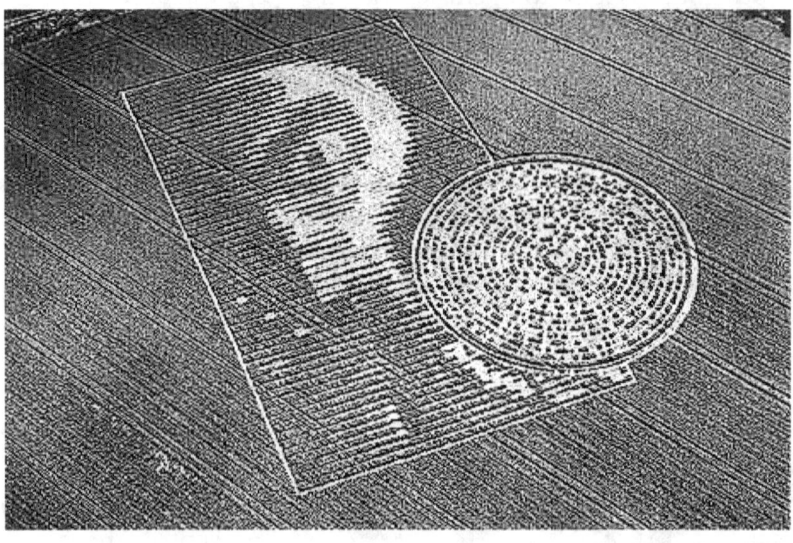

...perhaps the Greys (biocybernetic androids) were modeled after the above being(s).

The bottom line: Man is not alone and the Others don't all look human.

PostScript: Göbekli Tepe

This was just discovered floating on the Internet with the new research on the Turkish site...

We're talking Reptilians again, this time in the humanoid statue – the face was destroyed because it was reptilian... and you can still see a vestige of that from the side angle

Serpent again

Three-Being Statue

...and again from the front where you can see a human baby/child at the bottom – the statue was meant to display the genetic evolution from Reptile (top) to Human (bottom) ...

... and somebody did not like the reptilian reference, so they smashed the top head. Note the position of two sets of hands...

Ritual Circles Buried

And while we are at it, the reason Göbekli Tepe was buried was not from a desire to hide it (dumb!), but **The Flood** that Enlil sent washed silt, dirt and debris into the ritual circles when it swept up from Antarctica, over Africa and

500

up into Mesopotamia. This was how he planned to remove the humans and the Nephilim. The scientists have not been forthcoming with their having found seashells in the detritus they excavated... nor have they made a point about the unusual piezo-electric nature of the ground in the circles. The circles were used as **Kundalini Generators** – the candidate for awakening would stand barefoot on the special floor between the two large, center Ts while a ship overhead hit them with sound, vibrating them to the right frequency to initiate the movement of Kundalini in the waiting human.

Of course you will not hear that from official sources, but that is the meaning of a head found at Göbekli Tepe below (and examined in AL, Ch. 7 and Apx. E) This reflects Serpent Wisdom (a good thing) with the <u>serpent on his head</u> and the direction is UP – meaning the Kundalini at the base of the spine reaches the head and enlightens the initiate.

Head Sculpture at Göbekli Tepe

The Skygods just didn't bother to dig up the site (after The Flood) and it was just left. These sites were also used as **fertility centers** – note the hands near the genitals on some Ts, and other phallic symbology... below:

501

... bulls (top of right shaft with horns) were also fertility symbols...

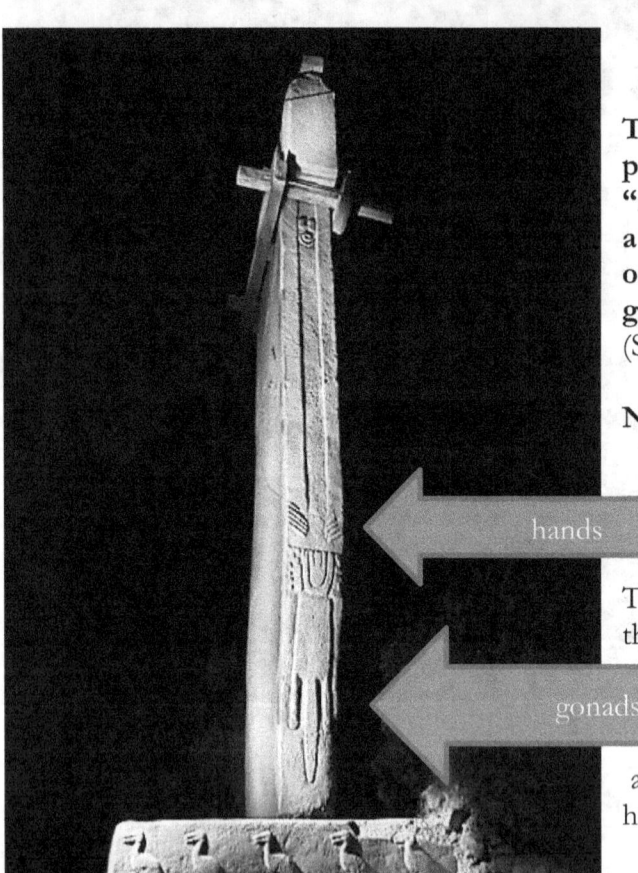

The ducks are ancient paleogeometric script: "ra-mi-is-ra" or Granting... a request that the fertility or kundalini awakening be granted.
(See AL, p. 324.)

Note the hands again...

hands

The left has the hands and the **symbols** for gonads

gonads

at the bottom (between the hands and the **ducks**) ...

(credit both: BingImages)

And the overall obelisk (large T **from the side**) is itself a phallic symbol.

Cropcircles and Göbekli Tepe are related to the Reptile issue as Cropcircles are made by what appears to be an advanced reptilian form (the being from Area 51),

...and don't forget e Serpent Mound in Ohio...

This was explained in AL, Ch. 7: the serpent is not eating an egg, nor birthing one, it is following a turtle to new Indian sacred grounds... (7 bends = the Pleiades [7 stars])...

...and Göbekli Tepe was being used 10,000-12,000 years ago when the Reptiles were around, and this is probably one of their sites. The Head (above) and the 3-Being Statue also have the serpent image... (sometimes a fertility symbol and sometimes represents Serpent Wisdom).

The reason Göbekli Tepe is not suggested to be a Pleiadian site is because it is way too close to the Anunnaki HQ in Mesopotamia and the two were not friends. It is thought that Puma Punku and Machu Pichu were Pleiadian sites, and the Anunnaki destroyed Puma Punku in a battle.

The Anunnaki could be violent, and the Anunna Warriors were their finest... and according to Chapter 9 Man inherited the "Warrior Gene" which predisposes some to violence and sociopathic behavior (see footnote 228).

Appendix F : Genesis & Human Lineage

Biblical Creation

"In the beginning, God created the heavens and the Earth." (Gen. 1:1.) The word God here is *Elohim* – **plural** for gods. *El* is the singular for God.

And on the beginning of the **Sixth** day of creation "…God said, let **us** make man in **our** image, after **our** likeness…" (Gen. 1:26). The word "God" again is *Elohim* – more than one god. Who is **"us"**? What is "our" likeness? It has been argued for years that the Supreme God was speaking to His heavenly court, but they were not gods, nor did they look like Him. So, why didn't the writer use the word *El Shaddai*, God Almighty? Or *Adonai,* My Lord? Or even *El* as the singular of *Elohim*?

> But, Man **was initially** created by the Elohim in the Garden, and the Skygods re-created him in their image.

Genesis is very specific about this, for it states "then God said 'I will make Man in my image, after my likeness.'" Adam was thus created in both the image, or *selem,* and likeness, or *dmut,* of his creator. The use of both terms in the Biblical text was meant to leave no doubt that man was similar to the gods (Elohim) in appearance. It is this likeness, or lack of it as we shall see, that is at the root of the **admonitions** of the Bible and the Sumerian literature. [283] [emphasis added]

What admonitions? To not make any images of God (or the gods). To be told that means automatically that Man was able to see God and thus he could make "graven images" of God. So God must have had a reason for telling Man to not make images of Him…so this God must have been physical. And He told Man to not make any images of Him (see **Ubaid sculpture in Appendix E**) because He didn't want it known what He looked like… and images in stone, whether statues or carved on walls, tend to last for years… thus it means that He didn't want images for later generations of Man to find and be able deduce something … the image of the gods was to be a secret.

> So, in Gen. 1:27 we're told that not only was man created in God's image, but "… male and female He created them." What does The God look like? Remember that He would not let Moses on Mount Sinai see His face, and yet He let Moses feel His powerful presence. (Ex. 33:18-23)   In Ex. 24:17 it says His glory was like a

'devouring fire' – no one could look upon the Lord God and live.
Since we don't know what THE God looked like, it cannot be
substantiated that the Man as created by God looked like Him –
and if He is an intelligent, powerful Consciousness at the center of
Creation, He would not have a physical form anyway.

Two Creations

Now we have a contradiction: the God of Eden is physical (He was walking in the
Garden) and the God of the Universe is a powerful consciousness, or Great Spirit
as the Amerindians said. Remember that God rested on the **Seventh** day, and He
said that everything was "very good" (Gen. 1:31). Then we learn just 7 verses later,
in Gen. 2:7, that God again creates man from the dust of the earth, and breathed
life ('*ruach*') or living spirit into his nostrils thus giving Adam a soul… on the
Seventh day.

> What happened to the man and woman of the Sixth day (in Genesis 1)?
> What happened to God resting on the Seventh day?

So on the 7th day, Man is alone (Gen. 2:7) and God says this is not good, so He (the
gods?) takes one of Adam's ribs and creates a woman in Gen 2:22 complementary
to him – meaning she also had the breath of life, or **a soul**. But we were told in
Gen. 1:27 that He created man and woman (no soul or *ruach*), with no time gap
between them as we see in Genesis 2.

> There were 2 creations? (Yes)
> Multiple men and women were created? (Yes)
> The high-level Bible account raises more questions than it answers.

Not to be outdone by the facts, the Church declared (via the invention of
Apologetics) that Genesis 2 was a restatement of Genesis 1, and they forgot to
explain how God rested on the Seventh (Gen. 2:2) but somehow recreated man on
that same day (Gen. 2:7)…. If you asked the above questions 500 years ago, the
Inquisition paid you a terminal visit.

Obviously, something is amiss – something other than what we were taught was
going on because the writer of Genesis was not dumb, nor could he have forgotten
so soon what he said just 26 verses earlier. And the reason for taking these Genesis
verses literally is that they **do** generally reflect what actually happened… as some
Bible scholars contend, there were two creations: Genesis 1 and Genesis 2..

The Inter-creation Gap

Bible scholars and other students of Man's history have noted the **Genesis 1** creation issue, and examined just how big the time gap between the first and second creations might have been. And to better understand the reasons for there to be two creations, one must dig into the Sumerian, Egyptian, and Hebrew texts -- the *Edfu* texts from Egypt, the *Enuma Elish,* the *Epic of Gilgamesh,* and the *Atra Hasis* from Sumeria, and the Old Testament and the *Haggadah,* a source of Jewish oral tradition.

While there is no way of knowing what the time element was between the first creation and the subsequent one, what has to be considered is that there <u>must</u> have been a gap if after Cain slew Abel, Cain was expelled from Adam's family and went and took his wife from the people in the **Land of Nod.** That means there were other people on the Earth – again support for multiple creations.
(See later **Chart 1a in this Appendix.**)

Since Adam and Eve were supposedly the first and only ones and they produced only Abel, Cain and later Seth... **where would the other people in Nod have come from**? There had to have been an <u>earlier</u> creation preceding that of Adam and Eve because Cain took a <u>wife</u>, not a little girl, nor a baby, from Nod which was a larger population. There was another earlier civilization. Thus there was an earlier creation.

> *And by the way, if Adam and Eve had three **boys**, Abel, Cain and Seth, how did their lineage propagate itself with no girls? Even if there had been a girl or two in the offspring of Adam and Eve, why is the writer of Genesis unaware of it? Let's not suggest that the future generations were conceived through incest...*

Obviously, there had to be more people that were created, or there were originally <u>at least</u> two creations, and Seth and Cain (of the second creation) may have taken their wives from the first creation in Nod... The problem is the Bible is too vague, too general, and does not give an acceptably coherent account of things.

For simplicity's sake, let us identify the first creation in Genesis 1 as the **pre-Adamic** race. Then something happened to cause a second creation involving Adam and Eve, herein referred to as the **Adamic** race (Genesis 2) – maybe the first did not work out, maybe they were defective, or maybe they weren't quite what was wanted in some way (yes, according to the Mayan version in the *Popul Vuh*). That also suggests that The God of the universe did not do the first creation (pre-Adamic) as it implies that for some reason the creation had to be done again (Adamic) – Does the God of the universe create errors?

Apparently years did go by after the first creation (living in the Land of Nd) and there was at least a second creation. This one specifically says that the spirit of life, *ruach*, was breathed into Adam, and Eve: a 'soul' was imparted to Adamic Man as the Gnostics later also tell us (VEG, Ch. 2). One might infer that the first Pre-Adamic creation had 'something missing,' and that was probably **a soul** since we are not told that God breathed the spirit into the first human as He did with the second creation. That means that Cain took a grown woman for his wife who did not have a soul (an OP, see Chapter 6) since she was from the first creation (pre-Adamic). See Chart 1A.

We will return to this issue again and again as it is very important. It will be seen that there <u>were</u> two creations and only the second creation (Adamic) received a soul. Thus Adam was the first soul man.

Genesis Clarification

> What should be said here (due to Chapters 1 & 2 straightening things out) is that the Elohim (Chapters 1 & 2 called them the Planners) did **create** the original hominid placed in the Earth Garden (not the Garden of Eden) which was referred to as the Zoo, the Conservatory, the Preserve – this was <u>before</u> any Skygods or Reptilians got here.

> When the Skygods got here, they **re-created** the hominid to serve them with their DNA upgrade and placed them in their E.Din to tend the trees and plants and manage the crops for them.

> The two separate events have been confounded and commingled over time.

Two Seeds

Genesis 3 deals with the Serpent supposedly deceiving Eve, and the Man and Woman discover they're naked and hide.

> In fact, our pastor informed us last year that the word "woman" comes from Adam's first reaction to seeing his new wife/help mate:
>
> "Whoah, Man!"

God comes looking for them and doesn't know where they are – is this cute or what? The all-knowing and all-powerful Biblical God, "walking in the Garden", not knowing where they were, and then His getting angry when they tell Him they hid because they were naked... this is a god with very human aspects. This god was upset because until they ate of the fruit, Adam and Eve did not know they were naked.

So He turns to **the Serpent** (who walked upright) and removes his legs, and declares the Serpent will henceforth crawl...

...and then He makes a very interesting pronouncement in Genesis 3:15:

> "...I will put enmity between thee and the woman, and **between thy seed and her seed**; and it [he] shall bruise thy head, and thou shalt bruise his heel." (Gen. 3:15) [emphasis added]

God is talking to the Serpent. And over in the Koran, Islam expresses the same idea:

> 'Adam,' we said, 'Satan is an enemy to you and your wife. Let him not turn you both out of Paradise and plunge you into affliction.... [Satan shows Adam the Tree of Knowledge and they both eat...and then God discovers their sin, but relents and admonishes Adam and Eve <u>and</u> the serpent:] 'Get you down hence, both,' He said, 'and **may your offspring be enemies** to each other....' Surah 20:119 [emphasis added]

Two different religions but from the same part of the world, with one common teaching in this case – that says they both have **a common source** (Abraham) who turns out to be more than a myth. [284]

In addition, the Bible repeats the different 'seed' message over in Daniel 2:43:

> And whereas thou sawest iron mixed with miry clay, they shall **mingle themselves with the seed of men: but they shall not cleave to one another** even as iron is not mixed with clay.
> [emphasis added]

That verse really does not fit in the context of Daniel explaining the king's dream – what has the "seed of men" got to do with explaining the "great image" or statue comprised of 5 elements: gold, silver, brass, iron, and a mixture of iron and clay? The different elements represent different kingdoms that come and go, not versions or species of Man…. or do they? It is almost as if verse 2:43 is part of a longer insert that is missing, or it was added for those who have "ears to hear."

Apparently the Two Seeds ("they") will not unite or 'cleave to one another' [as in marriage] and that is suggested in the Genesis 3:15 passage as well. Chapter 6 also warns about this.

> There will be basically **Two Seeds on the Earth** – the (human) woman's and something else related to the Serpent (who was Enki) who also earlier created the OPs in Nod. Enlil ("God") was referring to Enki's offspring *Adapa* (ensouled) versus the pre-Adamic (*Adama*) , soulless humans...

Follow this closely: If Adam and Eve have souls, their offspring Cain and Abel also have souls; Adam and Eve were thus the **second** creation ("2ⁿᵈ Seed") where God specifically breathed the *pneuma* or 'spirit of life' into them. And because Adam and Eve are one of the 2 seeds, for Gen. 3:15 to be true, the man-like Serpent's seed (his initial 'seed' was the *Lulu*, or *Adama* aka pre-Adamic), the other seed.

> The first creation of Genesis was the pre-Adamic (soulless) Man ("1ˢᵗ Seed") whose lineage has survived to this day (Chart 1a).

> *In VEG, Chs. 5-6 it is noted that the Two Seeds do in fact not have the same potential, and the soulless seed often obstructs the second ensouled seed. There **is** enmity between the two seeds, coming from the resentment of the 1ˢᵗ seed of the 2ⁿᵈ seed.*

The Serpent (Enki) was present at the 2nd creation, and at the 1st one as well, thus can we infer that the Serpent being was part of (involved in) the re-Creation.

> **Note the Göbekli Tepe statue of Reptile to Man (3 part statue) in Appendix E and note that there is a serpent on it – the serpent referred to Enki and creation.**

Gen. 1:27 said: "So [the gods] created man in [their] own image… male and female created [they] them." This was the **first** creation; we assumed that those initially created looked just like us, or maybe just the second creation did…

Rewind: keep in mind, from Chapters 2 – 3, that the Elohim did the very first creation of the human hominid (perhaps Homo *erectus* or one we don't know about) and then the Skygods, Anunnaki and even the Pleiadians, came and RE-created Man for their purposes. The Bible can be seen to be describing the RE-creations on Earth, not the Elohim first creation.

While the Genesis 1 account could be seen as the Elohim first creation, no argument there, it also reflects the first on-Earth re-creation of the pre-Adamics (the people in the Land of Nod) by the Skygods.

Appearances

> Consider for a moment that the Serpent in the Garden was supposed to be one of the Watchers known as **Gadrel** (according to Enoch), but more likely it was the Anunnaki Science Officer, **Enki...** [285]
>
> That means that God walking in the Garden was Enlil, the god or Head of the Earth Command, Enki's superior officer. Chapter 1 confirms this.

Left: <u>Gul Dukat</u>, displaying the gray skin and scale patterns typical of some Reptilians (of which there are multiple types – See Chapter 2).

S*tar Trek* Cardassians, humanoid reptilians. [286]

This is getting ahead of things, but is mentioned here because it is important to realize that (1) the Serpent was already in the Garden, before Adam and Eve, (2) the Bible does not specify <u>his</u> creation in Genesis (as do the Gnostics – VEG, Ch. 2) , and (3) it is important to realize who/what the Serpent was.

And remember the Serpent was a reptile that walked upright – It spoke, so it had a mouth, and it saw Eve so it had eyes, and it heard her responses so it had ears (real snakes don't have ears)…**Humanoid in appearance**. Intelligent because it knew what the Tree was good for and what effect it would have. And because its seed would be the enemy of the woman's, we know that it procreated…. very interesting. (All this is documented in the Jewish *Haggadah*.)

Therefore, Adam was semi-human and the Serpent was reptilian in appearance and genetics… perhaps like the **Cardassians** [above] on the TV show *Star Trek Deep Space Nine?* The series ran from 1993-1999.

According to Genesis and other documents, and long before humans ever existed, the Reptilians (Serpent) lived in the Garden of Eden and did the necessary work to maintain it.

This Biblical serpent was not just a lowly snake; it could converse with Eve, and knew the truth about the Tree of Knowledge (see Chapter 2 onfo). It was of such a stature that it unhesitatingly questioned the deity (God was Enlil, Enki's brother). Genesis concedes this point when it asserts that "the serpent was the **shrewdest** of all the world beasts that God had made."

Rewind: *Haggadah*

(Credit: Bing Images)

Ancient Jewish legends describe the serpent of Eden as **manlike** – he looked like a man, and talked like a man. The part of the *Haggadah* which deals with the Creation depicts the serpent who inhabited the Garden as an **upright creature that stood on two feet and who was equal in height to the "camel"** [i.e., 8-9' tall.]. He was said to be the lord over all the beasts of Eden… In the Jewish *Haggadah* there seems to be little doubt that he was legged and walked like a man.

The tempter of Eve was **not** actually a [snake]. [287] [emphasis added]

One of the truer pieces of art (above) based on the Biblical story. However, this is a good place to spike the myth that a sadist somewhere said that Adam wore a **fig leaf**. If it was a big leaf, it is **very prickly** and if you are sensitive to oxalic acid in the white latex sap, you have signed up for a bad itchy/scratchy time! [288]

And there was **no apple**: the Tree was representative of Reproduction (see Chapter 2, Forbidden Fruit). Apples were not native to the Middle East.

To repeat: it was said that Man (*Adapa*) and the Serpent-like creature (and its offspring called pre-Adamic) would be enemies (Gen. 3:15). This in fact, does become the case, and the exact form it takes is not Man hating snakes, but, as will be seen, the two types of Man on the planet, the "2 Seeds," do not really get along, and that is because the Anunaki left multiple types of humans on the planet.

As he said, Enlil (OT god), knew that would be a problem.

Two Seeds

Let's do some summarizing and clarifying, so that this will make more sense.

Gen. 3:15 spoke of enmity between 2 seeds – between that of the woman (later descending allegedly through the Adamic generations to Jesus himself), and that of the Serpent (Enki's pre-Adamic offspring/progeny).

> *It will be clarified in subsequent chapters that the Two Seeds are a basic distinction between soulless and ensouled and that <u>all</u> races of Man today (and 2000 years ago), contain a mix of both types.*

Note that **lineage was very important in that day** because people knew that there had been the mixing of DNA with the advent of the Nephilim (the 3rd Seed) and their offspring (Gen. 6:2-5). This is why the Bible traces Jesus' lineage so specifically – to show that He was of **the pure lineage** (the 2nd Seed) dating from Noah's day. Just as DNA was mixed, so was iron and clay and there is more to the analogy in verse 2:43 in Daniel than meets the eye.

> And a major point is that Yahweh is too human to be considered a God of the Universe; he is a lesser god-like being that was involved in the [re-]creation of Man, he has been obliquely referred to as a god, or Enlil, as is examined in VEG Ch. 1.

Seeds and Enmity

Remember these distinctions (in Chart 1a that follows):

> **The 1st Seed** = the pre-Adamics who are soulless.
>> There were two types here: *Lulus* and *Adamas* according to Sitchin. One of these two could have become the ill-fated Neanderthal.

> **The 2nd Seed** = the Adamics who have souls.
>> Enki sexually upgrades this version to *Adapa*, which was more like Cro-Magnon.

> **The 3rd Seed** = the Nephilim (offspring of the Watchers and Earth women , sometimes Anakim, or giants). Most perished in The Flood.
>> The ones that didn't perish were removed by Enlil's orders.

Yahweh walking in the Garden was actually Enlil, head of the Earth Command. Because he is 3D, "flesh and blood", he could not find Adam and Eve because he was not omniscient. Hence, his upset at them telling him they were naked – which meant that they had awakened to the truth of their condition by interaction with the Tree of Knowledge.

The **Serpent in the Garden** was Enki, Chief Science Officer, serving Enlil, and a master of genetic science. Adam and Eve are his creation (genetically) and he wants to see them grow and be fully functioning... hence he adds gene #23 for reproduction (from the **Tree of Knowledge**) as "knowing" was a Biblical way of referring to reproduction, or sex between a man and a woman. (See Chapter 2) Enki gave his progeny the 'gift' of reproduction. (That is why Enlil was angry.)

> This reproductive ability had to have also been done (sooner or later) to the humans who populated Nod, even as pre-Adamics who had no souls, as they were a populous group of Humans, pre-existing outside of Eden.

Enki told them the truth, but now they were no longer able to eat of the **Tree of Life** and live forever (as Enlil's pets). Enlil would cast them out of the Garden. Enki is now constrained to see that his progeny survive and he must improve their genetics – which he does by having sex with an Earth *Adama* woman <u>himself</u> (bypassing genetic trial and error) and the offspring become a superior version of Man, called *Adapa* (or Cro-Magnon).

> The enmity in Gen. 3:15 thus takes on two aspects, because of what Enlil (Yahweh) did:

First, since Enki had created both pre-Adamic and Adamic humans, the "1st Seed" without souls would come to resent the ensouled "2nd Seed" and even become sociopaths, trying to kill 2nd Seed humans. Thus when Enlil tells the Serpent (Enki) that there will be enmity between the seeds, that is what he is referring to – Enlil IS actually wise and knows that **the humans have deficits**, and he doesn't want them to procreate and spread what he knows to be their dysfunctional, limited genetics. They are just worker slaves.

Second, Enlil is disgusted with Enki, but they are brothers, so Enlil chooses not to kill Enki for disobedience, and he <u>does</u> need him on Earth. However, as Enki spreads humans around the planet (multiple Edens), he tries to educate the humans and improve them, creating the **Brotherhood of the Serpent**, or Serpent Wisdom group (See VEG, Appendix E), with esoteric knowledge (and **tools**) to raise Man to a higher consciousness. This does not work because Enlil defames Enki, denounces him and his Brotherhood to the humans, and the humans (later under the tutelage of the Church) come to despise the Serpent ... and it is so today in many parts of the world. This was also a form of enmity between the Serpent offspring and the humans.

This is somewhat represented in the following simplified Chart 1A.

Chart 1a of Creation

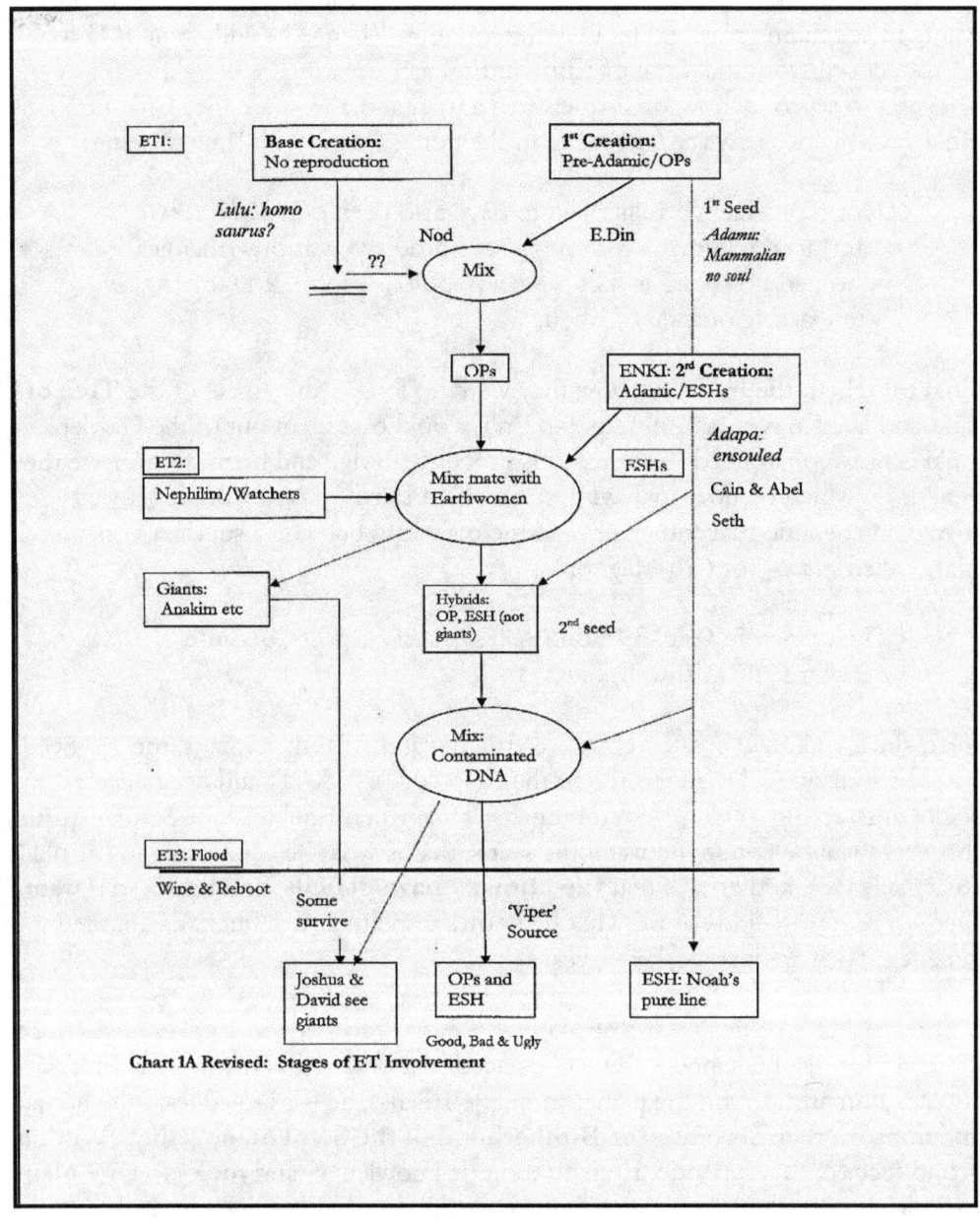

Chart 1A Revised: Stages of ET Involvement

The other half of this 2-part chart is at the end of this Appendix.

While this initial chart is far from complete, it is a starting point to understand the different types of humans that were created – narrowing it down to the <u>three main types</u>:

Lulu = the first Man or Pre-adamic [soulless],
Adama = the 2nd Man, or Adamic [ensouled],
Adapa – Enki's unique upgrade of Adama.

The Lulu, Adama and Adapa are symbolized below:

Man: Neanderthal, Cro-Magnon, & Homo Sapiens
Or: The *Lulu*, *Adamu*, and *Adapa*.

Chart 1a (continued...)

Not pictured: **Eve** – Adama's helpmate.
Lilith – Adama's first woman helpmate (according to Jewish tradition).

According to Wikipedia:

In Jewish folklore, from the 8th–10th century *Alphabet of Ben Sira* onwards, **Lilith** becomes Adam's first wife, who was created at the same time and from the same earth as Adam. This contrasts with Eve who was created from one of Adam's ribs (**Think: DNA**). The legend was greatly developed during the Middle Ages, in the tradition of the

517

Zohar, and Jewish mysticism. In the 13th century writings of Rabbi Isaac ben Jacob ha-Cohen, for example, Lilith left Adam after she refused to become subservient to him and then would not return to the Garden of Eden after she mated with archangel Samael. Lilith was then demonized.

The Chart 1a above is a graphic attempt to identify who was interbreeding with whom, and how the **Land of Nod**, Cain, and the giants (Anakim, etc) fit in.

"ET1" refers to information from VEG, Ch 1.
"ET2" refers to information from VEG, Ch 2
"ET3" refers to information from VEG, Ch 3

"OP" refers to Organic Portals (soulless humans) examined in Chapter 6 …these are the pre-Adamic or "1st Seed" of Genesis 1

"ESH" refers to EnSouled Humans ...these are the "2nd Seed" Adamic creation of Genesis 2.

"Base Creation" is the first creation of the gods – creating a worker who could not reproduce, the "*Lulu*"… This was very impractical as cloning additional workers was very time-consuming. This is not the 1st Creation of the Bible, but precedes it.

"Wipe & Reboot" refers to The Flood when Mankind was cleared away and restarted with Noah. (See Glossary.)

"Vipers" are the lineage that Jesus referred to in the Bible. They are not giants, but 'contaminated' human stock whose DNA is corrupt. Maybe 1st or 2nd Seed.

"Noah's Pure Lineage" were the ensouled humans who were not contaminated. These would be the lineage of Adam, Eve, Seth, and Cain until Man polluted his genetics with those people from Nod.

E.Din means "home (E) of the righteous (Din)." EDEN.

The **Land of Nod** had to be created before Cain was expelled from the Garden because he took his wife from those people. It is assumed to be "East of Eden."

The **Mark of Cain** was to be beardless, and when he was banished, he was relocated far West of Eden... probably to the New World where the Spaniards later met a lot of natives without facial hair.

Rewind: Anunnaki Re-Creation

There were **several 'creations' of Man** – from the first *Lulu* then *Adama* to the later more refined, intelligent and self-sustaining *Adapa* ('model man'). Some scholars have seen the *Adama* of Genesis 2 and the *Adapa* of the Sumerian epic as a similar being. [289]

What is interesting is that the **Anunnaki mixed their own genetics** with that of a pre-existing bipedal hominid on the Earth. And even more interesting is that the *AtraHasis Epic* from Sumeria gives a rather grizzly account of how the Anunnaki genetics were obtained:

> On the first, seventh and fifteenth of the month,
> He made a purification by washing,
> Ilawela who had intelligence,
> They slaughtered in their assembly.
> Nintu mixed clay with his flesh and blood. [290]

The Anunnaki morality was such that they were not above sacrificing one of their own to obtain the genetic material. And that is a little weird since they could have gotten his DNA without killing him – unless they were after his body parts, too, but that is not said.

According to Boulay:

> Since the previous experiment in the laboratories of the spaceship did not turn out successfully, it was decided to commission Enki, working with the Chief Nurse Ninhursag, to produce a primitive being. This creation, called a *Lulu* by the Anunna, was to be the first primitive man.
>
> Enki and Ninhursag conducted a number of experiments in the Abzu [underground], and Enki's floating laboratory near Eridu [southern Sumeria]... **There were many attempts which ended in failure** for one reason or another.... Finally a successful method was found... [but] the process had one main drawback... the creatures were **clones** [*Lulus*] and could not reproduce themselves.

In this manner, the first primitive man or Adam was created, looking like its creator(s)... the gods' essence is mixed with the malleable clay of the earth [Earth-based genetics]. In the Sumerian tablets, the clay is mixed with the essence of the gods and upon this creation they "impressed upon it the image of the gods."

The Adam of the Bible was not the Homo *sapiens* of today. He was what one might call Homo-*saurus*, a hybrid mammal-reptile creature that was to become our ancestor and the first step in the creation of modern man. [See pictures, next section.]

Since the Adam of Genesis [1] and the *Lulu* of the Sumerians were created in the image of the serpent-gods, shouldn't traces of this fact be found in some of the ancient scriptures?... Indeed, it is... One tract describes Eve's reaction in the Garden of Eden:

She looked at the tree. And she saw that it was beautiful and magnificent, and she desired it. She took some of its fruit and ate and she gave to her husband also, and he ate, too. Then their minds opened. For when they ate, the light of knowledge shone for them. When they put on shame, they knew they were naked with regard to knowledge. When they sobered up, they saw that they were naked, and they became enamored of one another. **When they saw their makers, they loathed them since they were beastly forms.** [291] [emphasis added]

The human hybrid that was created probably looked semi-reptilian since he was "created in the image of God." [292]

Re-Creation & Reptiles

It has been interesting to speculate what the Anunnaki looked like and According to Anton Parks (Chapters 1&2) we now know... but what would have happened to the Velociraptors had they been allowed to evolve?

Dinosauroids

On a related note, paleontologist **Dale Russell**, curator of vertebrate fossils at the National Museum of Canada in Ottawa, has speculated what the smarter species of dinosaur, the **Raptors, could have evolved** into had they undergone an evolution of their own, and he theorizes that they could have looked like what this book calls a Homo *saurus*, or to use his term, a "dinosauroid:"

Homo *saurus*?
(Source:
http://en.wikipedia.org/wiki/Reptilian_humanoid)

Could this be close to what the Anunnaki looked like and in whose image Man was created? (Yes: Chapter 2 and below.)

And this is close to what Anton Parks pieced together from the Sumerian descriptions.....

Right: This is what **Enki** allegedly looked like...

Sa'am
(c) Anton Parks

Mamítu Nammu
(c) Anton Parks

It looks similar to the next statue image…which is what the **Ubaid** people made as an image of their gods. The viper/serpent issue rises again in the ancient proto-Sumerian stone statue (below) from **UR** in Mid Iraq that pictures a serpent-type being created back in 3500 BC:

There wasn't just one of these statues...

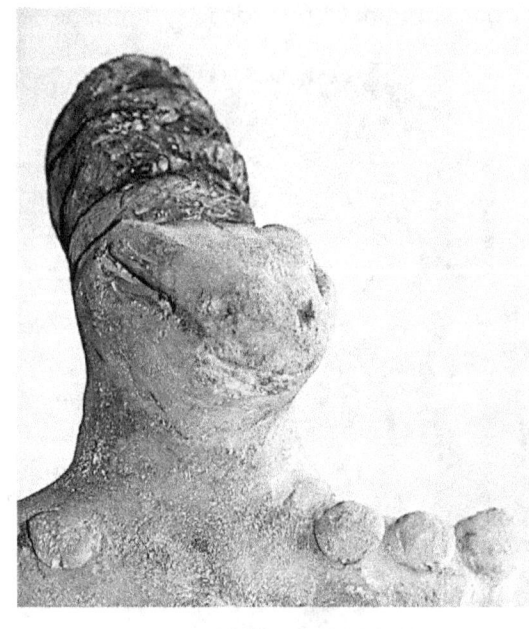

Sumerian Serpent Goddess Statue [293]

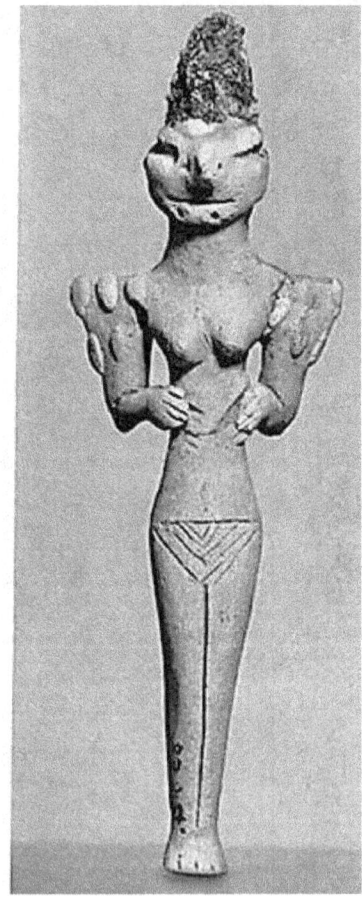

522

Reptilian Image

Is this why God (Yahweh/Enlil) would not let Moses see his face on Mount Sinai? It was also an injunction to Man to not have any "graven images" of God around – therefore His appearance had to be withheld from Man, and the high ziggurats served to distance the Anunnaki from Man.[294]

And yet, there are some small reptilian statues pictured (Ubaid statue shown above) in some of David Icke's books, and Icke's African shaman friend, Credo Mutwa, wears an ancient 'creation story' necklace containing the figure of a reptile, an Earth woman and a UFO. (Appendix E has more serpent images from around the world.)

Berossus, a Babylonian priest writing about the appearance of the gods, said that Man's ancestry traced back to the **Oannes**, an amphibious creature which came to teach civilization to Man.

Berossus called them *Annedoti* which means **"the repulsive ones"** in Greek. He also refers to them as *musarus* or an **"abomination."** It is in this way that Babylonian tradition credits the founding of civilization to a creature which they considered to be a **"repulsive abomination."** [295]

According to another source, the reason 'Annedoti' looks like 'Anunnaki' is because the **Oannes were sometimes another name for the Anunnaki**. [296] And Nommo was another name. [297]

One would think that if the gods were so superior and grand as indicated in ancient texts that they would be flattered to have Man make images of them, and display their greatness. But, after Man was created, the gods forbade Man to make images of them and they tended to stay atop their Mesopotamian ziggurats and Mayan pyramids to be waited on by certain servants who knew the truth, but were told to not speak of the gods with the worker population below, or in the city and fields, that had been created to serve the Annuna. So their physical **repulsiveness** must be true, otherwise Man would have flattered and praised their god's appearance. And again, the reptilian nature of the Anunnaki is explicit in the Sumerian accounts:

> **The reptiles verily descend,**
> The Earth is resplendent as a well-watered garden,
> At that time Enki and Eridu [his city] had not appeared,
> Daylight did not shine,
> **Moonlight had not emerged**. [298]
> [emphasis added]

Moon Origin

Note: the absence of Moonlight is also addressed at the end of VEG, Appendix A where it is said the Moon did not appear until centuries ago. [299]

The above could refer to the fact that the watery firmament was still in place above the Earth (disappearing later during The Flood?). But that could have been when the Moon was moved into place... until the Firmament was removed, the Earth could not be observed from an orbiting satellite.

While this is a slight "birdwalk," you may find it interesting, so the quote from Jim Marrs' book goes like this:

> **Aristotle** told of a people called the **Proselenes** who lived in Arcadia, a mountainous region of Greece, long before the coming of the Greeks. The name Selene refers to the Greek goddess of the Moon, and "Pro" Selene refers to "before the Moon." According to ancient legend, the Proselenes held claim to Arcadia because they lived there "before there was a Moon in the heavens."
> The Greek writer **Plutarch** also referred to "prelunar people" in Arcadia and the Roman writer **Ovid** stared that the Arcadian "folk is older than the Moon."
> The Roman **Apollonius Rhodius** wrote that the Arcadians "dwelt on the mountains and fed on acorns before there was a Moon."
> In **Tibetan** texts there are references to a people on a lost continent named *Gondwana* said to be civilized before the Moon shone in the night sky.
> Iconoclast Immanuel **Velikovsky** also has written about an early time before there was a Moon.... "....in a very early age, but still in the memory of mankind, no moon accompanied the Earth." Velikovsky quotes from a Finnish epic poem *Kalevala* regarding a time "when the Moon was placed in orbit."
> [...and he adds...]
> Legends and sacred writings across the world indicate that the Moon was once much closer to the Earth. [300] **(Appendix B)**

So Science stating that the Moon was spun off of the newly created Earth, 4.3 billion years ago is nonsense. They ought to try reading history.

And while Man got the Anunnaki DNA characteristics, which should have been an uplifting, improvement to the Homo *erectus* base genetics, the mix resulted in a human population that later drove Enlil crazy – the constant noise and activity, even violence, among the worker groups was too much, as discussed in VEG.

And things haven't changed that much even today. The average human cannot sit still without having to have something to listen to, eat, do or watch. Today's youngsters cannot stay with a TV program that doesn't have a lot of action, noise and color … or their attention wanders. And some of the older teens and people in their twenties have not outgrown this tendency: How many young people have to have MP3 going while they work or, God forbid, text while they are driving? And they have to have repetitive Rap or Heavy Metal or Country Music going – they cannot sit and enjoy peace and quiet. ADD and ADHD may be rooted in some very primitive genes.

Second Half of the Chart 1a

ET Intervention

There was a Chart 1A earlier that showed the general elements in a Chart of Creation.
This is the other half:

Chart 1B Notes & Explanation

> **Note1**: Since the Watchers/Nephilim were Anunnaki Igigi, it fell to the Anunnaki to clean up the mess in genetics – after the Flood. All giants were relocated or exterminated, and remaining Nephilim DNA in humans had to be located, identified and neutralized.

> **Note2**: There were other ETs in the mix: notably the Sirians and Pleiadians....

> **Note3**: There are beneficent (**+**) and antagonistic (-) Anunnaki Remnants, as indicated by the Chart. In later chapters, they are referred to as Insiders and Dissidents, respectively.
> The beneficent (STO) are like Enki and seek to help mankind; the antagonistic ones are STS and work to assist the Draconians (from Orion) to subvert Man's progress.

> The **Draconians** are not 'evil' – they just want what they want and believe that their repressive, hierarchical, societal structure works best. They are STS like "space pirates."

Many do not have a soul, nor do they have higher chakra centers, and so Man is a potential threat – if he is allowed to develop his divine potential and connect with his 'godhood.'

(Hence Man was removed from their direct access – Earth Quarantine.)

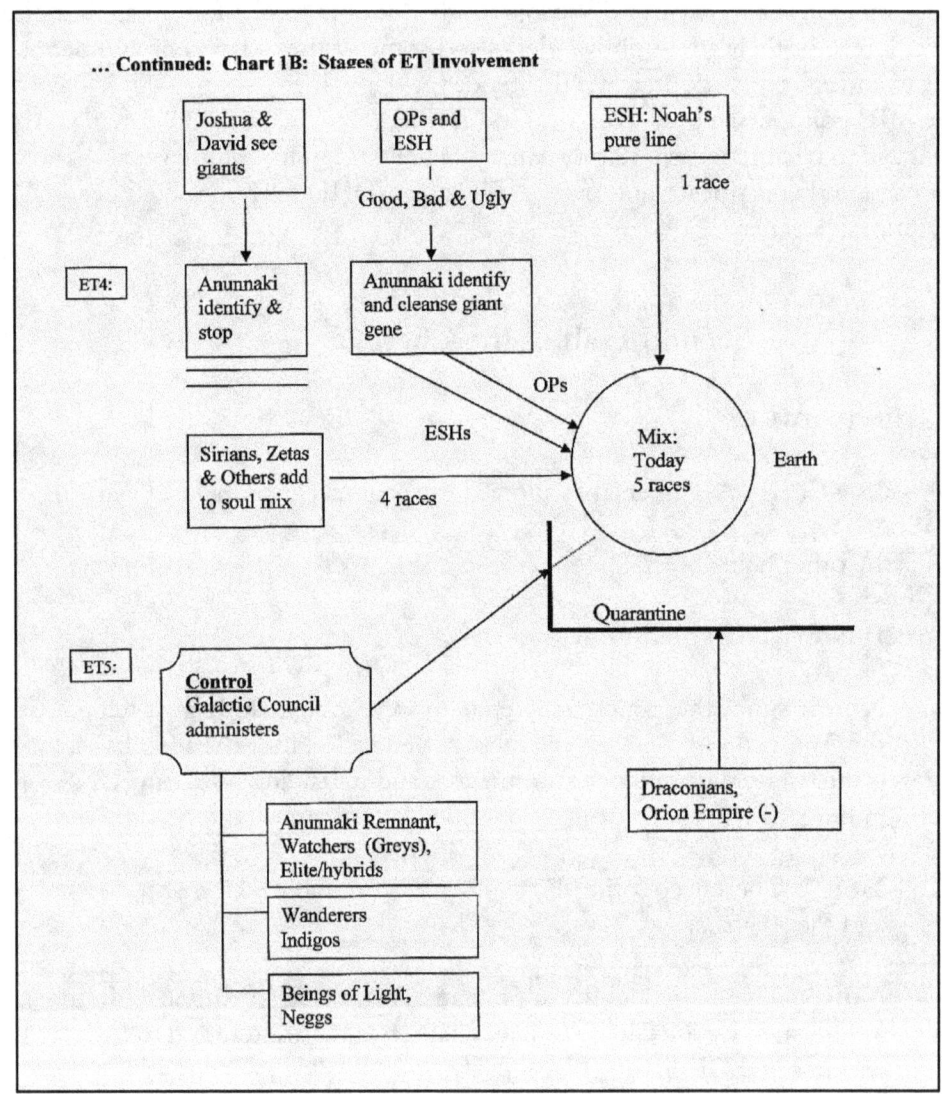

... Continued: Chart 1B: Stages of ET Involvement

Summary

It was shown that the late Zechariah Sitchin did a major scholarly work of translating and communicating about Man's early history – even though re-created by ETs as a slave race in Africa (archeologists confirm that **a version of Man originated in Africa**). It was shown that the Anunnaki had rocket ships (*'Shem'*) that spit fire as they flew through the air (origin of Dragons?), and they had a superior knowledge of genetics – both of which Man today has. Man seems to have reached the level today that the Anunnaki were at 200,000 years ago… including nuclear power as the Anunnaki bombed Sodom & Gomorrah, Harappa and Mohenjo-Daro, and the Sinai Peninsula.

Mankind's Heritage

One thing is for sure: the successive forms of Man on the planet are mute witness to the fact that Man <u>was</u> changed – from **Homo *erectus*** to Neanderthal to Cro-Magnon to Denisovans, to Homo *floresiensis*, to Homo *sapiens*, and lately to Homo *noeticus* (Indigo children).

See Chapter 3: 'Extra 223 Genes in Man' section.

The Anunnaki could have modified Homo *erectus* or *habilus* when they re-created *Lulu,* or their sterile worker human (the 1ˢᵗ re-creation). The improvement, by Enki, into *Adama* could have been closer to the Cro-Magnon man (the 2ⁿᵈ re-creation). Enki would later 'personally upgrade' *Adama* to *Adapa* -- much like Homo *sapiens.* And the Greys are behind the latest genetic change to Homo *noeticus* (the Indigos). Neanderthal was one of many experiments that didn't work out and was replaced by Cro-Magnon. In a similar way, The Change as discovered by Dr. Jacobs (see VDoG, Chs. 9-10) portends a **replacement of Homo *sapiens* with Homo *noeticus*.** Hubrids. With a gradual phase-in, Dr. Greer's Disclosure will not be necessary.

And there appears to be another reason that full Disclosure will not happen… That is examined in Chapter 8 and involves a **Space Force** and a **Breakaway Civilization** which is either underway or has already happened. And it has recently come to light (since those words were written in VDoG) that the Hybrids aka Hubrids {"**hu**man hy**brids**") that the Greys have been building from abducted human genetics and cattle mutilations, <u>may not be for insertion into human society</u>. At least some of them would be to form the basis of a Breakaway Civilization – especially on Mars as Man may not be long for planet Earth – given his propensity for war, biological and nuclear, and his destruction of the environment.

Mankind's Legacy

Man looks to be not learning anything from history because that history (Atlantis and Mars in particular) has been hidden, and Man is thus *en route* to repeat on Earth what was done to Mars -- make it a wasteland. This is why Enlil dd not want to continue with a species that he knew would be defective and semi-functional, and largely irresponsible, as current-day humans largely are. You have to give him his due. Thus was he really wrong to send The Flood ? – he did wipe out the Nephilim.

We take that question personally as we want to make it, we want to succeed, and we are smarter than the *Adapa* and versions of Man 4000 years ago... **but** many are still **ignorant, petty and violent** – thanks to the Anunnaki genes. That is what the Greys have been working on and as was said in VEG, they succeeded in cutting down our reproductive efforts by <u>cutting down testosterone</u> and giving rise to the emergence of many Gays... male with male is not going to augment the species' headcount (and that was the goal).

We are a very negative society and our TV shows it...

> Just as an example take a look at today's TV programming compared to 50 years ago: we have more violence, family squabbling and dysfunction that is supposed to be funny, Zombies, death, disease, killing, war movies, Rambo and WWE stuff, and Rx commercials that we had only 2% of in 1955.

> With a few exceptions (*Candy Crush*, *Forge of Empires*, *Myst*, etc) most video games are based on shooting, killing, fighting – keeping the Anunnaki genes stoked and alive! We are a very violent species.

And I have watched and have not seen the influx nor insertion of the Hubrids, the new version of humans, into society that Dr. Jacobs was so worried about... and I know what I am looking for as Baldy told me – I just assumed he meant they were to benefit <u>our</u> terrestrial society – I now suspect differently. They are better, stronger, smarted but they are not inserting – unless they will AFTER mankind is seriously culled (think: Georgia Guidestones) ... by our own hand (WW III?) or an asteroid.... or...a pathogen? Or were the Hubrids always intended for a restart, or for a Breakaway Civilization? (Chapter 8)

> Sorry to be Johnny Raincloud but this version of Mankind has not found favor with the gods, AND that is why Chapter 8 is so important – **Graduate and get out of here**... while you can still set your intent and focus on STO soul growth, and increase your PFV. Take the Higher Path and it will pay off – even if society is pursuing the dark, violent path.

> I am not trying to be negative (above), it is just that **I now see**

us as They see us, and it isn't good. Our egos say we are Ok, and yet we are on the brink of disappearing as a species. Hence the need for a **Breakaway Civilization** (Chapter 8)....

Note the domes ....

...or on Mars...

Appendix G : Safe EMF Ranges

Nature of EMF

First be aware that there are two types of EMF radiation – Ionizing and Non-ionizing. Cellphones are Non-ionizing.

The **Ionizing** type is the dangerous one and does not apply to the discussion of cellphones and laptops, etc. in this Appendix. It does apply to cellphone towers and 5G transceivers.

Radiation that carries sufficient energy to detach electrons from atoms or molecules, thereby **ionizing** them. Ionizing radiation is made up of energetic subatomic particles ... moving at high speeds (usually greater than 1% of the speed of light), and **electromagnetic waves** on the high-energy end of the electromagnetic spectrum.
Gamma rays, X-rays and the higher end of the EM spectrum are ionizing... [301]   (see next page)

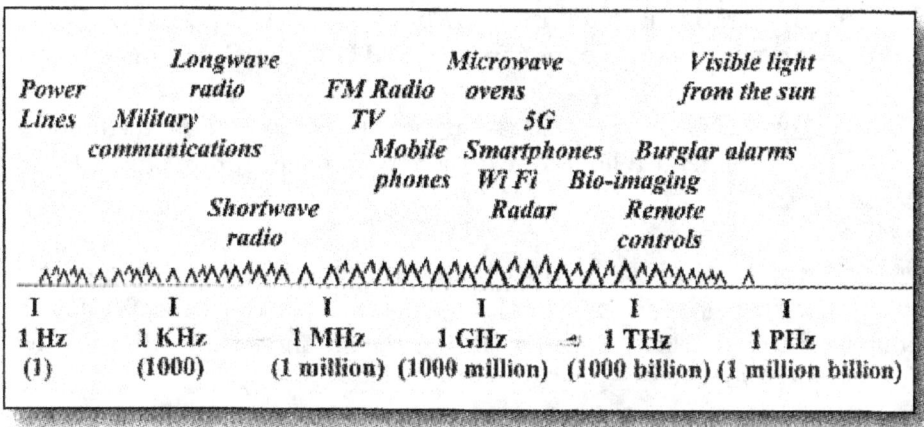

Note that the far right of the scale, "1 PHz" is PetaHertz where X-rays start (usually 30PHz) and that is Ionizing... see below... 5G and SmartPhones are in the 30GHz to 60GHz range... but the millimeter waves are intense.

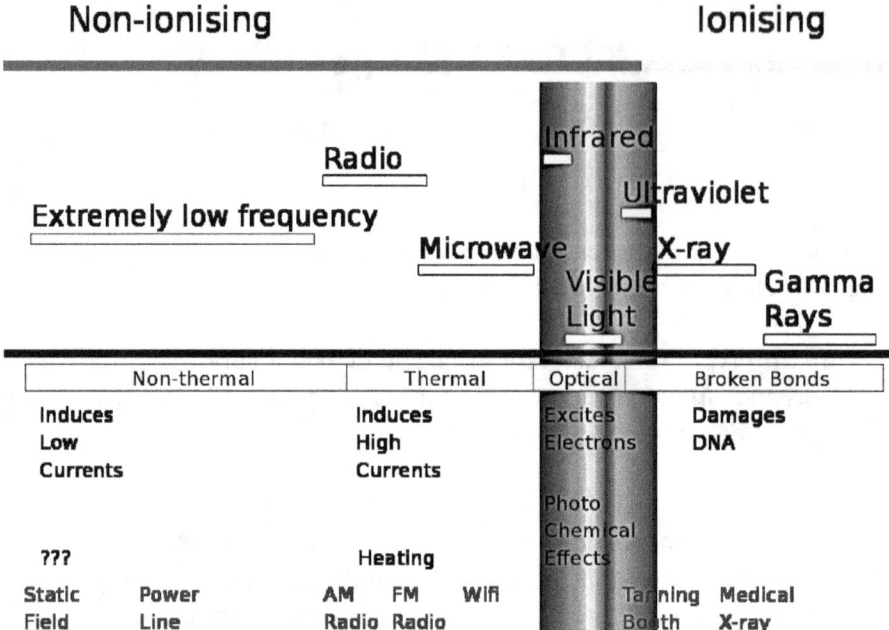

Non-ionising		Ionising	

Non-ionising — Ionising

Non-thermal		Thermal		Optical	Broken Bonds
Induces Low Currents		Induces High Currents		Excites Electrons	Damages DNA
				Photo Chemical Effects	
???		Heating			
Static Field	Power Line	AM Radio	FM Radio	Wifi	Tanning Booth	Medical X-ray

Because 5G is a microwave, WiFi (above) and Cellphones have thermal effects at certain frequencies even though they are not ionizing.

> **The concern of 5G is the danger inherent in increased speed and the amount of radiation exposure once it is implemented. Proven negative impact upon the body starts at 6GHz. 5G uses broad spectrum bands in the 30GHz to 300GHz range.**

> 10 mGy [milliGauss per year] (=1 rad/yr) is ten times more than the ICRP recommended limit for exposure to the public from artificial sources.[302]

Down the street where I live, the city fathers decided to make use of a greenbelt running under the high tension power lines and they built a very nice **Volleyball court** under the power lines. And 10 years ago the ground level EMF TriMeter reading was just 3 Gauss.

> **The accepted exposure range is .5 to 3.0 milliGauss.**

Then they doubled the capacity of the **adjacent** power station, put in new and bigger transformers, and built 210 more new homes 2 miles away, ran power to those, and ran the increased power thru the same overhead lines – above the Volleyball court. The EMF TriMeter reading now at court level is **12 Gauss**. A 400 times increase and the kids still play there. (I said something about it to the Parks & Rec people and they offered to have me arrested and threatened me to not say anything.)

So 5G is not ionizing but its intensity does have thermal effects and can still cause the issues in Chapters 5 & 7.

All the following information is from the website:
https://www.safespaceprotection.com/emf-health-risks/emf-health-effects/5g-dangers/

EMF and the Home

Electrical appliances produce toxic electromagnetic fields (EMFs). These fields can layer, one upon the other, creating a harmful level of radiation. That's why **the kitchen is a hotspot** for harmful EMFs...but other rooms are affected, too. The mG [milliGauss] exposure at three feet is up to three mG for a **blender**, up to four mG for a **washing machine**, up to six mG for a **hair dryer** or **a television**, and up to 40 mG for a **vacuum cleaner**.

This does not mean to avoid them or throw them out, just minimize your time spent standing in front of the appliance when it is running. The **refrigerator** and outdoor **A/C Compressor** are the worst. The **Microwave** hits 100-500 mG at 4 inches

DANGER ZONES		
EMF Levels from Common in Miligauss (mG) Recommended Safety Levels .5 mG-2.5 mG		
SOURCE	mG up to 4 inches	mG at 3 feet
Blender	50-220	0.3-3
Clothes Washer	8-200	0.1-4
Coffee Maker	6-29	0.1
Computer	4-20	2-5
Flourescent Lamp	400-4,000	0.1-5
Hair Dryer	60-20,000	0.1-6
Microwave Oven	100-500	1-25
Television	5-100	0.1-6
Vacuum Cleaner	230-1,300	3-40
Airplane	50	
Source: USA Environmental Protection Agency		

Rewind:

The Environmental Protection Agency recommends that you limit your exposure to 0.5 mG to 2.5 mG. When you are three feet away from a microwave, you are exposed to up to 25 mG.

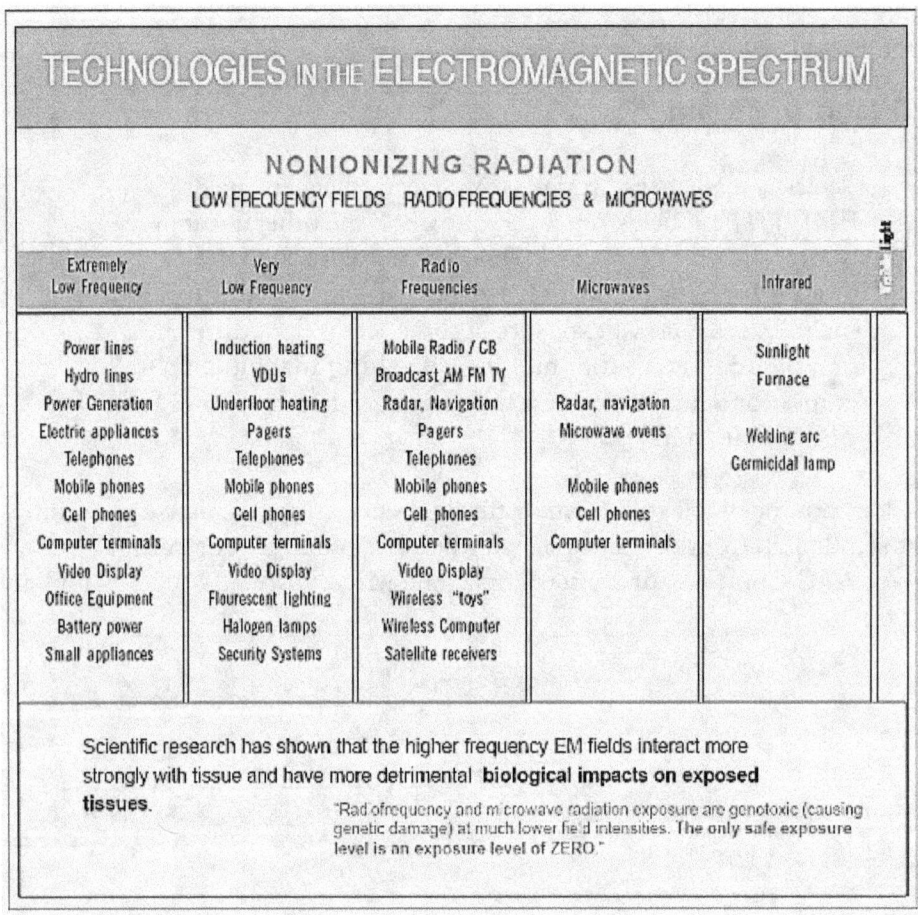

The cellphone, laptop and PC and video display (monitor) are in 4 of those columns.

Cellphone Issues

When you hold your cell phone up to your ear, 10% to 80% of the radiation from **the phone penetrates two inches into your brain**. In children, the penetration is even deeper. Studies have shown that cell phones held near the head cause brain wave changes in 70% of people. The potential danger is so widely accepted that **insurance companies** are beginning to exclude coverage for injuries related to wireless phone radiation exposure.

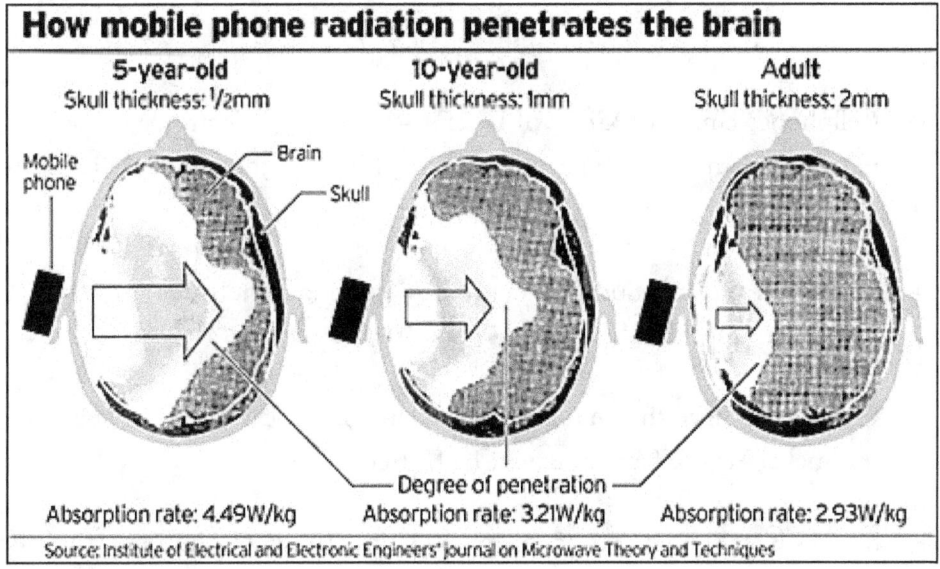

How mobile phone radiation penetrates the brain

5-year-old	10-year-old	Adult
Skull thickness: 1/2mm	Skull thickness: 1mm	Skull thickness: 2mm

Mobile phone — Brain — Skull

Degree of penetration

| Absorption rate: 4.49W/kg | Absorption rate: 3.21W/kg | Absorption rate: 2.93W/kg |

Source: Institute of Electrical and Electronic Engineers' Journal on Microwave Theory and Techniques

The side effects of excess Cellphone usage or holding the phone too close to the head occasionally include the following:

Headaches
Genetic damage (breaks DNA, kills mitochondria)
Impaired immune system
Cancers, including brain tumors and melanoma
Break in the brain/blood barrier
Reduced melatonin (disrupts sleep)
Interference with pacemakers
Memory loss
Changes in electrical activity in the brain
Cardiovascular stress
Fatigue
Eye problems

Earphones, headsets and speaker modes provide distance, but they don't eliminate danger. Anytime the power is turned on, **cell phones emit electromagnetic**

radiation – even in stand-by mode and regardless if carried on belts, in pockets or purses—or set **on the table in front of you**.

More Important Facts about Cell Phones

- Cellphones emit **two kinds of EMFs** – microwave electromagnetic radiation from the antenna, and more EMFs from the phone body. Both are harmful.

- 20-80% of the radiation from a phone's antenna **penetrates** up to two inches into the adult brain.

- Cell phones have **thermal effects** (heating biological tissue) as well as non-thermal effects (affecting natural EMF frequencies)

- Studies have shown that people who **sleep with a cellphone by the bed** have poor REM sleep, leading to impaired learning and memory. Long-term effects remain to be seen.

- When the cellphone signal is held next to the head, **brainwaves are altered** a full 70% of the time.

- Many **insurance companies** are so alarmed by the evidence that they now exclude health issues related to cell phone radiation from coverage.

  Most brain surgeons limit their cellphone use, and counsel patients to **never hold them to their ears.**

WiFi Issues

Wi-Fi Dangers Made Worse by Cumulative Effect

Wireless routers – as well as **Bluetooth** and similar wireless systems – give off electromagnetic radiation in the low-gigahertz frequency. This level is considered **potentially dangerous** to people. And the danger is compounded by several factors:

- Just like the wireless signals themselves, the EMFs can pass through walls.
- Most routers are not turned off at night, so you are exposed 24/7.
- You are not only exposed to EMFs from your own router. Did you ever search for a wireless signal and see not only your wireless network, but also **your neighbor's** and the one from the business down the street? All of them emit EMFs.

Increasingly, scientists and researchers are uncovering the health risks of EMFs. Depending on **the strength and the length of exposure**, those risks can range from **insomnia and headaches to tumors**.

> This also goes for **children exposed at school** to the prevalent use of WiFi in the school – especially in the computer lab... Schools should hard-wire their buildings with fiber optic.

WiGig and GiFi

As if standard WiFi (2.4 GHz) wasn't bad enough, they have ramped it up to 5 GigaBits per second (GiFi version) and 7 GigaBits per second (WiGig version) and both are in the **60 GHz range**. These are the millimeter waves and you do not want them in schools.

> EMF researcher **Joe Imbriano** has called the new impending combination 5G plus WiGig a **"Death Grid"** .... that has the ability to cause a rotation of all **water molecules** in the body on their axis... [and since] the body is at least 70% water this has a enormous systemic effect on all cellular structures....
>
> It has the unique ability to affect all the orbital properties of all **oxygen molecules**... Together these frequencies have the ability to totally saturate and oscillate all oxygen and water molecules in the human body. In other words, **total control over the terrain and carcinogenic processes**....
>
> ... "it creates **a perfect weapon of mass extinction**." [303]

Are we thinking Option 1 yet...?

Computer (PC) and Laptop

Your computer, whether it is a desktop or a laptop, emits EMFs. The closer you are to the computer, the stronger the electromagnetic field. Electromagnetic fields (EMFs) don't only come from your **computer screen**. The electronics inside your computer's box generate a powerful EMF.

It's been shown that EMF strength decreases with distance. Studies have shown that **exposure above two milliGauss (mG) can begin to harm biological organisms**. At a distance of three feet, you probably are above that level of exposure. If you are four inches or less away – such as when you are holding your laptop on your lap – your exposure might be as high as **20 mG**. This level of exposure has been linked to many problems, from fatigue to immune suppression to cancer.

> **Do NOT sit with your laptop on your lap or lie with it on the bed and put it on your chest... same with an iPad.**

Laptops in the lap expose the liver, pancreas and male/female reproductive organs to excess EMF (20 mG and more) – affecting fertility and diabetes to name a few.

Power Line Issues

Electromagnetic radiation from high voltage power lines is something that can affect the health of people in urban and rural communities. Strong, artificial **EMFs that radiate from power lines can scramble and interfere with your body's natural EMF**, affecting everything from your **sleep** cycles and stress levels to your **immune response** and DNA!

Hundreds of studies worldwide have shown that <u>living next to</u> high voltage power lines and other parts of the power transmission network increases your risk of

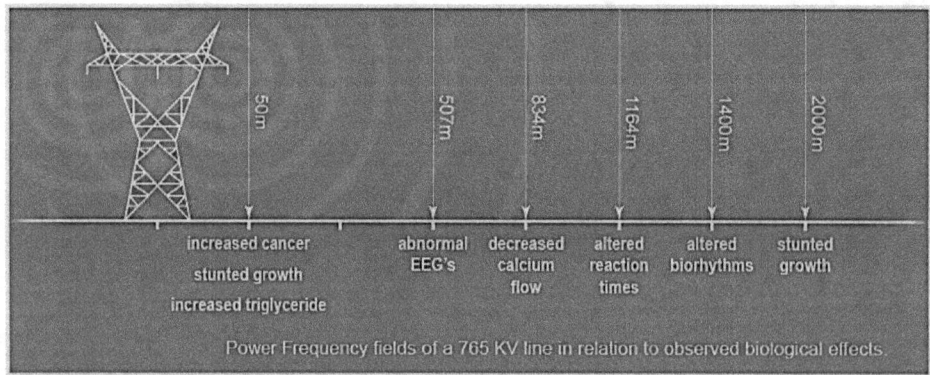

50m	507m	834m	1164m	1400m	2000m
increased cancer	abnormal EEG's	decreased calcium flow	altered reaction times	altered biorhythms	stunted growth
stunted growth					
increased triglyceride					

Power Frequency fields of a 765 KV line in relation to observed biological effects.

cancer and other health problems. As well as it is not wise to **jog or bike** beneath them, either.

According to research and publications put out by the World Health Organization (WHO), EMF such as those from power lines, can cause:

Headaches
* Fatigue
* Anxiety
* Insomnia (disrupts melatonin)
* Prickling and/or burning skin
* Rashes
* Muscle pain

> The kids described earlier who were playing Volleyall under the power lines experienced most of the above list. I got a headache just from 10 minutes walking and testing in the area.

Cellphone Towers

Cell towers (or cell sites) that hold antennas and other communications equipment flood the area for miles around with powerful high frequency radio waves (known as microwaves) to support the use of cellphones as well as Wi-Fi, WiMax, Wireless LANs, 802.11 networks, **Bluetooth** supported devices and more.

Cell tower microwaves might travel for as few as **two miles** in hilly areas, and up to **45 miles** where there are fewer obstructions; and of course, they easily penetrate brick and metal.

Radio masts - Smaller versions of cell towers, often seen on rooftops and billboards, typically installed 800-1300 feet apart.

Mobile towers - Sometimes installed on the tops of buildings. **Mobile towers are especially dangerous** because they emit microwaves at a frequency of 1900 MHz. Recent studies have shown that the intense radioactivity from mobile phone towers adversely **impacts every biological organism within one square kilometer**.

Cellphone tower microwaves have a significantly higher frequency than even radio waves. The higher the frequency, the more powerful the wave—and the more powerful effect on biological organisms (recall that a mobile tower emit microwaves at 1900 MHz).

Smart Meters

Traditional electromechanical meters are rapidly being replaced by utilities with electronic smart meters to lay the foundation for the United States' **smart grid (i.e., 5G) energy infrastructure.**

Wireless Smart Meters
A wireless smart meter sends radiofrequency microwave radiation through **two antennas**, typically in the same frequency range (900 MHz to **2.4 GHz**) as cell towers. Unlike cell towers that are now located further from residential buildings, a smart meter located on an outside wall of an occupied home of office space will effectively transmit at a much closer range – increasing the exposure risks!

For the smart meters to communicate with the grid, each meter has a network radio that sends radio frequency signals to other meters and the grid -- generating an **invisible "mesh" of chaotic radio frequencies**.

The health risks of layering additional EMF and wireless technology radiation in homes is being essentially ignored in rolling out smart meter installations. **Compounded exposure** will include your new appliances (stoves, washing machines, microwaves, dishwashers, heating and cooling equipment, refrigerators, televisions, computers, small kitchen appliances etc.) with integrated transmitters.

Appliances will send transmissions wirelessly to smart meters that in turn will transmit very low frequency signals with energy use information every few hours. In order for appliances to send your energy demand information to the utility provider (and if desired your own cell phone or computer), your home's smart meter will receive appliance device signals and then transmit its own signal to aggregated networks that finally link to the internet and grid.

The exposure for people living in multiple meter pole office or multi-unit housing is another unstudied area – not to mention the impact of proximity to utility collector meters that are designed to transmit RF signals from 500 to as many as 5,000 homes. For many communities, the level of layered "mesh" driven exposure will be truly **unprecedented and virtually inescapable** for most people (let alone animals and natural systems).

Since the transmitters will soon be integrated into most new devices, shutting off the communication is NOT going to be a viable or easy option for consumers.

5G Issues

5G is the latest generation in mobile broadband technology servicing both cellular and wireless systems. Expanding the efficiency of 3G and 4G networks, 5G applies EHF (**extremely high frequency**) bands to transfer signals to our phones and other wireless devices. Relying upon these superior and innovative broadband frequencies, 5G can process thereby galvanizing the **enormous amounts of data** at lightning fast speeds industry standard for wireless communications.

The past decade has shown an exponential increase in society's dependence on wireless technology. **The current infrastructure however is not sufficient to accommodate this upsurge.** Dated technology is resulting in overall slower and unstable connections. 5G's use of EHF offers the answer in the form of instantaneous download/upload capability, allowing unprecedented amounts of data transference at speeds **100 times faster than that of 4G**. The potential of 5G is truly unknown: 5G can enable unlimited applications from automated automobiles to household appliances redefining entire industries.

The infrastructure of 5G is radically different.

Both 3G and 4G employ radio waves which are transferred from one large cell phone tower to another. Since radiation dissipates with distance, risk is reduced if you are not in direct proximity to the source. **The high frequencies used in 5G have short range and are limited in their ability to penetrate obstacles and dense materials.** Therefore they require a preponderance of mini portable base stations that will be placed **every 300 yards or so** to form a dense platform that relays signals from one station to the next.

These cell sites can be easily placed on existing buildings and telephone poles creating an intricate and unrelenting background of electromagnetic energy.

However, there is a caveat with this:

A growing body of research (see Bibliography Internet section) affirms that electromagnetic fields (EMF) produce adverse health conditions. The concern of 5G is the danger inherent in increased speed and the amount of radiation exposure once it is implemented.

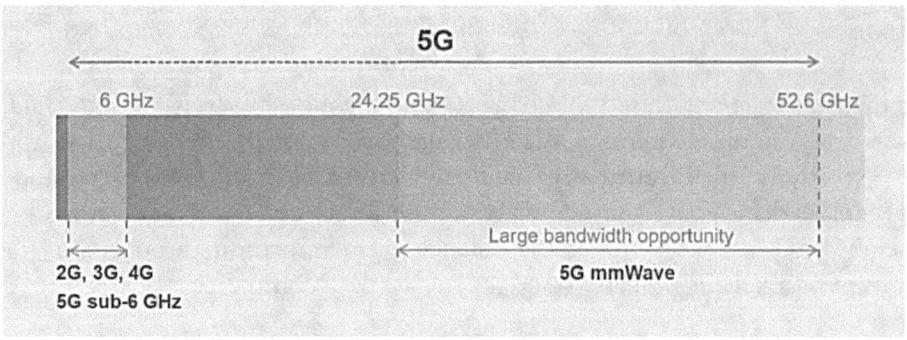

Proven negative impact upon the body starts at 6GHz.

5G uses broad spectrum bands in the 30GHz to 300GHz range. Close proximity and prolonged exposure to cell sites intensify the risk. Transmitter boxes stationed outside your window make EMF impossible to avoid. Additionally the number of network antennas needed to facilitate this upcoming surge is estimated to leap from 300,000 as of 2016 to millions of small cell transmitters as the use expands.

Increased too will be the amount of low orbit **5G satellites** that deliver focused beams of micro-wave frequency signals to everything from network stations to automated automobiles.

5G Health Risk Concerns & Recent EMF Research

In 2017 a petition was signed and presented to the United Nations calling for a moratorium on the implementation of 5G due to health risks. Presented by a collective of more than 180 doctors from 35 countries, this international plea was part of a resulting investigation on the biological and health risks of EMF radiation initiated in 2015.

> Findings cited an increased risk for cancer, cellular stress, genetic damage, pre-mature aging, reproductive problems and neurological disorders.

Further groundbreaking research by biochemistry professor and EMF radiation expert **Dr. Martin Pall** (THE EMF Expert, see Chapter 5 & 7) cited the following potential health effects:

> blindness, hearing loss or deafness, male infertility, nervous system disorders, thyroid and immune system dysfunction and low blood oxygenation.

Furthermore Dr. Pall warns that the impact on humans, however dire could be **even greater on animals, insects, plants and undermine entire ecosystems.**

While 5G may enhance our quality of life the price we pay may be dear. **New technologies often appear in the United States without responsible research only to later discover the consequences are dire.** (Humans are guinea pigs.)

Wireless technology is a relatively recent phenomenon. Despite reassurance to the contrary, the truth is we just don't know how it affects our health, our well-being, or our environment. Personal health, safety and individual care often play second fiddle to the intoxicating promise technological upgrades offer.

Since **5G is inevitable**, it is best to be informed, educate others and promote community awareness. In an increasingly challenging world the responsibility to safeguard our health lies with us. Chapter 7 offers some protection ideas.

SafeSpace Products

There are a number of things one can do, as outlined in Chapter 7, but here are a few more from the **website** [304] that has provided the above relevant info:

Home & Office
EMF Adapter (balance building wiring effects)
Radiant Room

Mobile & Travel
Vehicle Adapter (USB plug)
Smart Patch (attach to devices)
SafeSpace II (pendant)

Personal Protection
VitaPlex Pendant
Food Liquid Energizer

Outdoor Environment
GeoResonator (balance soil and land)

Take a look and also see the Tab on the SafeSpace Home page that explains how the products work.

It is supposedly some sort of **resonant interference** with the EMF waves hitting your body's energy field (you DO have one), also hitting the pendant or patch, and exiting the pendant/patch with an altered wave spectrum. The "harmful EMF waves are **repatterned** and made coherent"... thus a variety of **"carrier" materials, primarily holograms do the work**. "The stored information on the products stands ready for dissemination onto the carrier waves of harmful EMFs...."

...and yet I wonder if that affects the whole body... ? Can a pendant on the neck protect the feet?

We get the same benefit from lying on the Bemer® mat for 8 minutes and letting it repattern (separate the clumping of corpuscles in) our blood and strengthen our Bionet. The AVOCEN device (also Chapter 7) does a similar job improving microcirculation, thus removing toxins.

For further information:

- **Contact Us 404 Laurel Ave., Wilmette, IL. 60091**
  +1 877 571 7878
  info@safespaceprotection.com

Encyclo-Glossary

Note: throughout this book, reference is made to further detail on certain topics, and the books are referred to by abbreviation (list on Copyright page).

ξ　　ξ　　ξ

1-Sec Drop -- this is a direct communication from a higher being into one's mind and memory/knowledge base. It is not a voice, not automatic writing. It takes a very brief split second and one knows that it is happening, and then it can take anywhere from 10 seconds to 20 minutes to examine what one was given. It is information that is usually complete and appears to the recipient to be something that s/he already knew and is now aware of.
Similar to an insight or revelation, except that it has an energy signature about it that you know it is being "dropped" into you. (Reminiscent of **V2K** but there are no words 'spoken.')

100th Monkey Effect – When one animal in a group discovers some new behavior and finds it serves him, it is said that the behavior is not learned as much as passed on as soon as the energy reaches a critical level so that their group soul can recognize and 'appropriate' the behavior. This was the case with a few monkeys on an island who discovered that washing their fruit before eating it avoided the problem of sand in the mouth. More and more monkeys on island 1 began doing it, and while they had no way to communicate the new behavior to the monkeys on islands 2 and 3, after about 100 monkeys were doing it on island 1, the others on islands 2 and 3 also began doing it (as confirmed by zoologists who were present studying the islands).

AI Signal – the signature of an electronic hack by AI (XSI bot or better) into a network; the frequency and carrier data can affect and influence not only physical computer systems, like a virus, but biological systems as well (by activating nanites put there via vaccines). People can be manipulated by a specific frequency carried in the electronic hack (via their cellphone on 5G) ; affecting the way they think or behave. Done by AI (Artificial Intelligence), aka XSI via The Cloud– see Chapter 6. (See Michael Salla book in Bibliography, pp. 218-19.)

Anomalon -- a subatomic particle was recently discovered whose properties vary from laboratory to laboratory. Imagine owning a car that had a different color and different features depending on who drove it! This is very curious and seems to suggest that an anomalon's reality depends on who finds/creates it. (See DNQF Ch 9.)

Anunnaki – one of the early, original ET visitors to Earth who interfered in the natural progression of the bipedal hominids here, and created some of the first 'humans' in Africa and Sumeria. Because of their technology and power, they were looked upon as gods. Supposedly from the planet **Nibiru**, but more likely Orion or Sirius systems. (See **Zechariah Sitchin** and **Remnant**. See also VEG Ch. 3.)

Anunnaki Elite – consists of two main types: the ruling reptilians who retained their original appearance (e.g., Enlil and Enki), and the later, hybridized, more human-looking (e.g., Inanna, Marduk, Sargon, even Alexander the Great). The later change was effected by Enki's mating with humans and later genetic prowess to enable the Anunnaki to move among the humans who found the original., reptilian/reptibian appearance repulsive.

Archons – the 'powers that be' in the celestial realms – according to the **Gnostics**. These are the same ones that Ephesians 6:12 refers to: powers & principalities (a hierarchy) dedicated to evil and wickedness. Synonymous with 'demon.' (See **Nephilim**, and see **Djinn** in VEG Ch. 6).

Astral Realm – note that there are levels in the Astral realm, and in particular, the one that most concerns Man, is the Level I (**Chart 4** in VEG, Ch. 12) which is a kind of intra-dimensional space – more than 3D and yet not really 4D, and this is inhabited by Man's oppressors, the STS Gang. The normal 4D STS/STO entities occupy the higher 4D and lower 5D Astral realms and cannot see 3D Man.

Attractor – energy in the form of an idea, person, or thing that draws other things, ideas or people together based on similar and strong resonance.

Baldy – aka **Philealel** is the Guide or higher being who guided me thru the books. I can't pronounce his name, so I called him Baldy and he chuckled. He looks more like Mr. Clean... black trousers and white fluffy shirt (reminded me of Errol Flynn's pirate shirts). He was about 6'8, perfectly shaped head, bald, and large blue eyes that went right thru me. He showed up October 25, 1998 and left (officially) August 27, 2016. He still guides me but not like the first 18 years.

Bands – referring to the H-Band and M-Field suggested by Robert Monroe. These are energy bands, or grids, created around the Earth due to the often negative activities of souls on the Earth. Experienced as static or noise. See also **RCF/Matrix**.

Beings of Light – often referred to as **Angels**, or some of today's Watchers, they guide and protect Man. They are also known to provide the life review that NDErs speak of, and they are the '*Inspecs*' that Robert Monroe spoke of.

Bionet – a term coined in VEG Ch. 9 to describe the hyperdimensional network of communication in the body. Like the Internet, *chi* is carried in meridians of energy to all parts of the system, from the chakras, and tells the cells and organs what to do. The Bionet is manipulated during Accupuncture to channel biophotons. Chinese TCM called it Meridians. See TCiM.

Book 2 – the successor to the *Virtual Earth Graduate* , with additional examination of specific topics, is *Transformation of Man*, aka Book 2. Book 1 and 2 were originally the same book – at 868 pages.

Brain Waves – a measurement of consciousness.

> Gamma cycle: 30Hz – 150Hz (seers & mystics)
> Beta cycle: 13 – 30 Hz (normal waking consciousness)
> Alpha cycle: 8 – 12 Hz (relaxed, aware state)
> Theta cycle: 4 – 8 Hz (sleep)
> Delta cycle: less than 1-4 Hz (deep sleep)

Catalyst – anything like an event, an idea or a word, that causes change in a person; the threat of being fired for bad performance at work is a catalyst to perform at one's best. Illness is a catalyst to see what is wrong, or what energy is blocked, in one's body.

Chakra – a vortex of energy formed in the body wherever two or more chi meridians cross or come together; same as a vortex on the earth with its ley lines. (Sedona, AZ is known for several of these.) These are also referred to as 'energy centers' as they transduce energy from the air/water/Sun around a body and draw it into the body thru the chakras. There are 7 main charkas in the body and 1 above the head, and 1 below the feet. There are many more, minor charkas all over the body.

The Change – the coming insertion of Hubrids, Starchildren, and Indigos into human society to stabilize it and move society in a new proactive direction. It is not to conquer or invade; the current version of Man has not worked out any more than Neanderthal did, and he was replaced by those whose job it is to guide and run the **Control System**, developing Mankind.   Chapter 5.

Chi – energy particles, also called *ruach*, orgone, mana, prana or ki – without chi in our food, air and water the human body could not exist. The chi is a force that travels along meridians (pathways) in the body that link the etheric aura (1st level of the aura) to the physical body; it can be directed by the mind to specific parts of the body for healing. See Bionet.

Cognitive Dissonance – the result of hearing/reading something new that does not fit into one's reality, or in what one thought was their reality; the effect is to create confusion followed by denial of new concepts. More specifically, when a new idea <u>conflicts</u> with an established idea that one already thinks they know, the result is 'dissonance', and rejection. When people were told 500 years ago that the Earth was round, they experienced great cognitive dissonance... which led to denial.

Coherence – resonating alike; attracted to each other by similar resonance. Two energy waves are coherent if they have the same shape, size, and strength.

Control System – discovered by Dr. Jacques Vallée, (Chapter 4) it is a system run by the gods to guide and develop Man in the direction of the Greater Drama.

Déjà Vu – the experience of having done, seen and/or heard something before; as though one is reliving a prior moment in their current lifetime. Relates to reliving a fractal simulation. See **Recycling** (Mouravieff, Ch. 8).

Dissidents – Anunnaki hybrid Remnant still on the Earth who seek to control Man and deny him his divine heritage. See **Insiders**.

Draconians – a militaristic STS race largely from the Orion System who have subjugated many worlds in our Galaxy. Also referred to as the Dracs, or the Reptiles. A very old race that lays claim to much of our Galaxy and they fear Man because they don't all have (nor do they understand) the **soul**, thus they seek to contain/control Man (who will eventually rule over them). Aka: **Djinn**.

Elite -- those humans who are mostly descended from the Anunnaki hybrids; as a group they may be augmented by the Remnant Insiders who stayed behind when the main contingent of Anunnaki went home. They are generally not the enemy. See **Chapter 13, 'Insider'.** See also **PTB**.

ELF – Extremely Low Frequency; a vibratory wave form that is sent by a radio-like device or microwave transmitter at a certain Hz or MHz frequency such that it entrains the mind into a 'resonant state' (usually Alpha) with the wave.

EMF – ElectroMagnetic Field; such an electrical field around a high tension power line also generates its own weak magnetic field, hence EMF. Note that a cellphone or TV or PC – anything electronic has an EMF and it is unhealthy to spend much time in it as it 'afflicts' the cells of the body and disrupts their function. The reason is that the body has its own weak EMF and communicates info to other parts of the body via the nervous system and chi meridians (see **Bionet**).

Energy Vampire – a person, OP or ensouled, who subconsciously starts an argument, gets the other person angry, and the instigator takes the other person's energy through the Law of Energy Potentials: Energy always flows from the higher potential to the lower and this applies to car batteries, as well as humans. So the instigator creates a fight, not to win or lose (they don't care), they will walk off with some of your energy, and they quit the argument when they have it. They are 'up,' and the victim is usually tired.

Entrain – to induce a state in B like in A; usually done by music, movies, and words, but can be done by powerful thoughts and beliefs. A hypnotist entrains a subject into a desired state; Hitler's harangues entrained the crowds into the Nazi mindset he wanted; and classical music entrains the listeners into a relaxed (Alpha) state. Music also entrains… see Chapter 4.

Entropy – the tendency of all things in the universe to wind down and die; also called the Second Law of Thermodynamics. The enemy of Evolution.

Era – occasionally the Higher Beings have to clean up and reset the Simulation, usually what this book referees to as a Wipe and Reboot. When Man is restarted after a Wipe and Reboot, the Era will have some dominant theme in the Greater Script that the activities of Man are to experience and handle. Our current Era began about AD 800-900.

ESH – Ensouled Human Being, has a soul and thus an aura. See also **OP**s(VEG, Ch. 5). Significance: TOM, Appendix D: a soul means a conscience.

Flow – often referred to as The Flow. This is the rising energetic vibrational entrainment into the higher 4th and 5th dimensional realms. It has increased awareness, compassion, Light, and STO aspects for service and is available to all who seek to align themselves with a Higher Way. It was created by the Higher Beings and is supported by an archetype that masters on the Earth reinforced and made available to all spiritual growth aspirants.

FreeWill – an illusion. The more one grows spiritually, the more one does the will of the Father of Light. Baby souls, or those who insist on their own way, think they have free will but the Father is merely letting them experience the results of what they do… their **Script** controls much of what young/baby souls can do. As Jesus said "Not my will, but Thine be done." Advanced souls have surrendered their will by eliminating their ego. Old souls have less constrainment.

Galactic Law – the ethics and rules as set forth by the Galactic Council and adhered to by all subordinate councils for the maintenance of order. It includes a Non-interference directive, responsibilities of 'creator races', transportation or communication protocols, terra-forming procedures, and energy creation/disposal to name a few. Specifically, the creation of brand new sentient, humanoid species is forbidden, but the modification of existing species may be allowed (approval required) provided: (1) the species is terminated when its usefulness is done [which was Enlil's choice], or (2) the species may be further developed if reasonable and can be done in a way that benefits them and the solar system's Association of Worlds [this was Enki's choice].

Gnostic – one who believes that God is accessible to all by **going within** and following their 'inner knowing'. Gnosticism relies on one's personal enlightenment to guide them; it is in fact, connecting with one's **Higher Self** which has true Knowledge. The Bible refers to it as the 'Inner Light' and is what one connects with when they follow Jesus' suggestion that "the Kingdom of Heaven is within." Gnosticism was originally (AD 100) part of the Christian movement.

God/gods – this is god with a small "g". It is just a convention for referring to those who are alleged to be watching over us, running the Drama on Earth. This includes the Skygods, the Ancient Ones, and Angels.

Godhead – a collection of higher souls, and Soul Groups, in closer proximity to God, like spokes on a wheel where the hub is God Himself. The Godhead works directly with the Oversoul for each Soul Group and sometimes the two are hard to distinguish. The basic hierarchy is: **God – Godhead/Oversouls – Soul Groups – Angels/Neggs – and individual souls.** Between the Godhead and God is the Hieracrchy of Masters and those who are responsible for Solar Systems, and then Galaxies.

Gods-in-Training – when Man graduates from the Earth School, he can be useful to the Father of Light in various places in the Multiverse (**Chapter 10**). One of those places is to undergo an apprentice position in overseeing the Earth and its souls – under the tutelage of more advanced 'gods' who give direction and training. The gods-in-training still make mistakes, just as Man does, and while often minimized by karmic override, these are allowed in part as an aspect of the new gods' training. This is therefore sometimes a source of things going wrong in an Earth person's life. If a god-in-training abuses his power, he is recycled back to Earth. (VEG, Ch. 15.)

Greys – the 3' tall gray-colored humanoids with the big heads, big black eyes and skinny bodies; their eyes are large (really protective coverings over eyes sensitive to light); they typically perform the abductions on humans, some cattle and other species. They have a hive/group mentality as they are **bio-cybernetic roboids** and Anunnaki Insider tools to improve Man's DNA.

Ground of Being – who and what you really are; your PFV is the physical reflection of the <u>sum</u> of your STO/STS quotient. If you, your **soul essence**, were to be removed from your body, the energy being that you are would have a certain vibration level (also reflected in the color of your aura) – higher or lower depending on how much Light you hold, how compassionate you are, whether you seek to serve (STO) or be served (STS), what issues (stuck points, agendas and attachments) you still carry with you, and in general, it refers to the "quality" of Light & Love that you <u>are.</u> Ultimately, it reflects the highest actions/thoughts that you are capable of.

Higher Beings – Light Beings above the Astral and Reincarnative levels (1-6) and who are responsible for the operation of these lower 6 levels, reside on the 7th level themselves; may intervene in $3^{rd} - 4^{th}$ $5^{th} - 6^{th}$ dimensional affairs when the Greater Script of the Father of Light, or the One, requires it to keep the Multiverse working.
Also colloquially called **"the gods"** as part of those who run **the HVR Sphere** or Simulation (Chapter 10). The Higher Beings are <u>not</u> the Beings of Light (Angels) nor ETs nor aliens.

Higher Self – also called the Oversoul, this is the coordinating entity of each Soul Group and acts to oversee Scripts, events, lessons – and coordinate with the souls of the same Soul Group, <u>and</u> with other Oversouls who manage other Soul Groups in the same Godhead. Each Godhead has multiple Oversouls that interface with the multiple Soul Groups. See VEG Ch. 7, Chart 3a.

HVR Sphere –. The 3D Earth Construct that is semi-holographic and operates as a Virtual realm in 4D.

Hybrids – this is any human-looking but 'upgraded' version of Homo *sapiens* which may or may not have a soul. It can be the Anunnaki hybrids – part Anunnaki, part human, and their bloodline. Or it may be Homo *noeticus* that the Greys have been so busy developing to restart civilization after the big **Change** event in the near future (> 2018). Most are very intuitive, psychic and look to be the next step in the development of Man.

Imprinting – when a soul has not lived a life that s/he recalls, it was added to that soul as an imprint. The actual life experience (stored in the Akashic Records) can be downloaded to a soul for their benefit – it becomes a part of their memory – as if they had lived it. The reason for this is to save time by not having to live a specific life, but by downloading the essence of it, the soul can grasp what it was like, what it meant, and can hopefully draw the lessons from it. For example, a soul who wants to birth into the Medieval Times, and wants to be a Knight, could imprint the life, experience and skills of a successful Knight to help guarantee survival when living as a Knight... especially if he has to become a Knight to perform a special feat, such as being there to save King Arthur, who might otherwise be killed, and thus save the Greater Script for England.

Insiders – Anunnaki hybrid Remnant still on the Earth (may include Enki). The pro-active ones who try to help mankind and block the pessimistic **Dissidents** (qv).

Interdimensionals – those beings in 4D <u>and above</u> who normally have very little interest in Man, and may be STO or STS. The STOs are often curious and are observing. The STS version has been known to use humans for unknown agendas. Also a generic term for the 4D STS Controllers inasmuch as they operate between dimensions. Possibly Djinn.

InterLife – where souls go when they die, after passing through the **Tunnel** to the **Light**. Also called the Other Side, and sometimes appears to be Heaven. It is where the **Script** is designed, souls are counseled by the Masters and Teachers, souls are rehabilitated after a rough lifetime on Earth (or elsewhere), and it is where the Heavenly Biocomputer resides. This is also where reunions with members of one's **SoulGroup** happen. (Book 2 TOM spends 2 chapters on this aspect.)

Karma – *Aka* **The Law of Karma**. – originally the concept of "meeting oneself", or "what goes 'round, comes 'round." It does <u>not</u> mean being stabbed in this lifetime because one stabbed someone else in a former lifetime. The original, true concept was that of the Universal Law of Cause and Effect, and it forms the basis of one or more aspects of your Life Script. Karma can also be a manipulative issue in the Virtual Reality of Earth if the life review is done by a Negg posing as a Being of Light.
Note that Karma applies only to Earth; other souls who do not come to Earth do not have to deal with Karma.

Law of Attraction – simply says that you attract to you who and what you are. Attraction also works by repeated focus, visualization, affirmations and speaking one's word, and of course, by prayer. Attracting what you want may be blocked by the Script if it is not permitted this lifetime… to enforce some lesson: you stay poor to learn humility or reliance on other people, for example.

Law of Confusion – when RA was asked a question that violated someone else's right to privacy, or asked something that would be giving advanced level information that the person had no context for, RA would comment that the question could not be answered because it "violated the Law of Confusion." We are to work thru confusion and seek the answer(s) on our own; **everyone has a temporary 'right' to be confused** (i.e., to not know) and are expected to work thru it, or ask, thereby absorbing the lesson and info on a level that makes the lesson/info part of us.

Law of Non-Interference -- there are several parts to this Law. One is that other sentient beings from other worlds do not have *carte blanche* to manipulate or interfere with the normal development of Earth's humans. In addition, it also specifies that human bodies are to have only one soul (possession is not permitted), and there is a provision that ensouled humans cannot be manipulated from the Astral violating their Freewill.

Law of One – the concept that we are all connected at a higher level, mostly thru our Higher Selves, and we are all part of the One, the Father of Light – if you have a soul. The Law of One also includes freewill and love. Telling someone else what to do, how to live, etc. is a violation of the Law of One, a violation of freewill whose flipside is called the Law of Confusion. (Think: **Entanglement**.)

Law of Energy Potentials – this merely says that energy in the universe always flows from the higher potential to the lower. (Think: a weaker battery being charged by a stronger one, or a Master healing a sick human.)

Light – an intelligent aspect of God; sometimes referred to as the Force. It may be used interchangeably with Heaven and/or Knowledge. There are biophotons of light that support the operation of DNA and sustain bodily operations.
Note that Light (large L) is a conscious aspect of the God force, which force can have a brilliant light about it. The light (small "l") is everyday, regular light.

LifePlan – see Script.

Lightworker – any entity, physical or Astral, that uses Light as an energy source to do its work. (Includes angels, demons, Neggs, energy healers and Higher Beings.) Caution: it does not always imply STO behavior.

Loosh – a term coined by Robert Monroe in his chapter 12 of Far Journeys. It is energy produced by 3D living beings that is allegedly 'harvested' by 4D entities in the astral for sustenance. Loosh is bountifully produced by humans who go into states of deep **fear** or **anger** or **lust** – they radiate the energy after being manipulated to produce it – like a grain of sand in an oyster produces a pearl that is harvested. See **Energy Vampire**.

Matrix – synonym for the HVR Sphere, but not to be confused with the matrix as shown in the movie *Matrix*. Can also be ZPE Etheric composition or **Quantum Net** which s Dark Energy/Matter that interpenetrates everything.

Memes – a concept, or idea, that generally has spread through a population – an idea that may spread like a biological virus – such as a belief in ghosts, or a belief that black cats bring bad luck…or, if you go out in the rain and get wet, you can catch a cold. There are positive and negative memes. (See also "**100ᵗʰ Monkey Effect**.")

Morphic Resonance – said of a plant or animal that takes its physical shape from the *morphogenetic* field that establishes a 'morphic' (shape) resonance with the object's energy. The plant's shape is entrained by the morphic resonance with the morphogenetic field (pattern) that governs how living things take shape, according to Rupert Sheldrake. (See VEG, Ch. 9 'Chuck.')

Morphogenesis – Rupert Sheldrake conceived of the presence of a 4D field around living things that influences the shape they take – kind of an Astral Template that governs height, width, color and other aspects of the oak tree for example, such as when and where it sends out its branches, how fast and how far.

Multiverse – the universe we live in is one of a number of universes comprising a Multiverse… multiple universes interconnected forming a coherent larger universe consisting of multiple levels (realities), and can involve parallel universes or dimensions in **'superposition'** (or stacked).

NDE – a Near Death Experience where the person appears to die, and their body is pronounced clinically dead, but they come back to life and relay their experience of meeting a Being of Light with whom they have a Life Review, and they usually come back a changed (better) person. The NDE effect often produces a spiritual transformation in the person.

Neggs – the 4D 'dark' angelic beings operating in the Astral realm around the Earth, whose sole purpose is to apply the negative lessons specified in one's Life Script. Thus they are "**NEG**ative **G**uide**S**." They work with the Beings of Light (Angels). See **Appendix A: Dr. Lerma**.
They are programmed to afflict mankind – they are appointed to effect the negative parts of one's Script (aka **catalyst**). They provide catalyst and feedback inducing Man to change and grow. They work <u>with</u> the Beings of Light (VEG Ch. 6) because they, too, are Beings of Light who <u>volunteered</u> to serve the negative agenda and they were 'reoriented' to Darkness to maximize their effectiveness. They still carry a small, suppressed connection to their original Light down inside and they will be restored to their original condition when their service is complete.

Nephilim – the physical Anunnaki Igigi mated with earth women and produced giants (Nephilim) called Annakim, Gibborim and Rephaim, which were hell on earth. (VEG Ch. 3.)

NPCs – these are the other characters in a Virtual Reality game; they are not programmable or operable by the player – the Game or operating system uses them to play a part in the Drama. They are called Non-Player Characters. Same as **OP**s (VEG, Ch. 5). Allegedly 60% of the humans in our daily world.

OPs – Organic Portals -- (pronounced "Oh Pee") human beings, flesh and blood (Organic part), and they can serve as a portal for astral entities to operate thru them.

This is not synonymous with Zombie.

They also are not fully human as they <u>allegedly</u> lack a soul and that is because they have incomplete DNA and only the first 3 chakras are wired to function; they cannot access higher energy centers. Due to their somewhat robotic nature, they can be used to guide and/or influence ensouled humans in 3D. The Greeks called them 'hylics.' The Mayans called them 'wooden people.' Dr. Mouravieff called them pre-Adamics. (See TOM.) Also called **Backdrop People** (Appendix C).

It has been said for centuries that <u>all</u> humans have souls. The Greeks and Mayans said there were people who have no soul. This whole issue was **examined in detail in VEG, Ch. 1 & 5**. They allegedly do exist and are probably the Sociopaths and atheists among us. 98%+ **are not evil**, they just have **no conscience** and think they can do whatever they want. Jim Jones, Charles Manson, and Richard Ramirez are prime examples of <u>OPs on steroids</u> that have run amok, because they have no connection to a Higher Self. Discussing spirituality with them gets nowhere, they cannot imagine what you are talking about. They are run by 'A' Influences (VEG, Ch. 3: **Life is a Film** and this book's Chapter 8).

Also called **Non-Playable Characters** (NPCs) as in a video game (see QES and Appendix D in TOM.)

Oversoul – the lowest level of the God Hierarchy, between God and Man. Between the God Hierarchy and Man are the Angels, also called Beings of Light (because they don't have wings). Where the Higher Self resides.

Orbs -- these are round balls of light that often showed up in the original digital cameras' photos, and some people thought they were intelligent beings visiting whatever scene was being photographed. In reality, the pixels in the digital camera are charge-coupled devices, carrying a small electric charge, and occasionally one or more pixels would misfire… producing the orb(s). Better quality digital cameras rarely produce the phenomenon.

Parallax – where the perceived shape and size of something (say an object in the sky) does not change with any change in our position from which it is viewed (i.e., a parallax), thus it must be reflected off the surface of something consistent in shape. If it were reflected off dust in the atmosphere, it would have size and shape distortion depending on the viewing angle – <u>but the *Gegenschein* in VEG Ch.12 doesn't</u>.

PFV – Personal Frequency Vibration -- the day-to-day, overall vibratory rate (resonance) of the soul energy sustaining the human body. When a person is angry their aura 'glows' red, and the PFV can drop to a lower (denser) vibration than when a person feels a lot of love and the aura 'glows' rose and the vibration reflects the energy ofthe heart charka (higher, lighter energy). The PFV also denotes which charka is dominant in the person; a person living from their higher charkas has a higher PFV than one engaged in sex, violence and pettiness (lower chakra activity). The aura typically reflects what one is feeling, yet the base PFV does not change; when the person is at rest, the base PFV is consistent from day to day as it reflects the overall level of soul growth. Also known as that person's "energy signature" as recorded in objects (Psychometry).

Phase-Shifted – refers to 3D and 4D entities or 3D and 4D timelines which cannot see the other even though they may occupy the same space. For example, there may be a 30° phase shift, or a 60° or 90° shift (the most common). Think of 2 Sine waves almost on top of each other (congruent and coherent; now move one wave to where its trough is below the other wave's peak – they are 90° phase-shifted.

Placeholder – an OP-like version of a real ensouled human living on another timeline (parallel universe) that already split into two, with duplications of people between the timelines. If the ensouled human did not replicate to the new TL, the other people who went to the new timeline still need/expect that 'body' for their everyday world activities to function, and his absence in their lives would be noticed. And so minus a soul, John Doe exists as a kind of 'synthetic' human in the <u>new</u> timeline. (See VEG, Ch. 5 and TOM, Apx. D.)

Points of Choice – there are **pre-programmed** points in a person's life where important choices must be made, and they are found in a person's Script. Examples are whether to move to Florida or stay in California, whether to accept what looks like a great new job, or whether to get married. Sometimes the choice results in a **timeline bifurcation** into a fractal subset so that another aspect of you (See Apx A) can see how that turned out. Note: these are the only things preprogrammed in the LifePlan.. they are a must and have to be faced... your choice reflects your degree of soul growth. Failure means a repeat. See Ch 4 Prphecy and Freewill. See also **Timeline**. (Covered more in detail in TOM, Ch. 2.)

Pre-Soul – also called First Time Soul, allegedly the initial stage of a human (not an animal that leaves 2D and enters the 3D human soul realm, see metempsychosis); this is not a complete soul, but a <u>potential one</u> if the entity applies itself as a 1st time human. Typically, only the first 3 chakras are functional, and thus **there is not enough 'soul energy' to create an aura**. What they have instead of the 1" aura, is "heatwaves" above their head. They have little or non-functioning connection to their Higher Self, and thus little or no conscience. (See also **OP**s.)

Prime Directive – a requirement in our Galaxy for those races who can create life and modify existing life genetically – often referred to as a 'creator race.' They are responsible for overseeing the welfare: safety and education of their creation. This is why a **Remnant** of the Anunnaki stayed behind (now known as the Naga.) (VEG, Ch. 12 'Prime Directive.')

Prophylactic Fantasy – describes the **world of denial** that some people live in. 'Prophylactic' because they feel safe in <u>their version</u> of the world, and they reason that nothing really destructive has ever happened to them, nor can it. 'Fantasy' because they do not accept the real world and its negativity; they see their world as they want it to be and sometimes think that they can exert a 'force' that makes it that way. (e.g., New Age and "Create Your Day")

PTB – the earthly human Powers That Be; the 3rd dimensional STS people running the world for their Anunnaki Dissident masters (control group still here). They are also influenced by corrupt DNA, and the **RCF/Matrix** itself. Puppets. Many of them are OPs. See **Elite** – not the same thing, just a higher level of control.

Quantum Net – the Reality Field that interpenetrates all things and all of space. This has also been likened to the AEther and what is today called **Dark Energy**. The body has a minor Net, called the **Bionet**. In all Nets, the communication is almost instantaneous.

RCF/Matrix – see below (Resonant Consciousness Field).

Reassembled – this is also called **Disassembled**, Dissolved in VEG and TOM (books 1 and 2). The Higher Beings will attempt to infuse a wayward soul with new, proactive energy, and perform a kind of 'psychic surgery' on the soul's energy field (which emanates from their Ground of Being so it is tricky), and failing that, if a soul cannot be re-oriented to STO behavior, the energy (soul) is **Dissolved** back into its component energy parts, as a failed soul. The consciousness is removed, energy then is cleaned, and can be reused without the former consciousness that accompanied the failed soul. (In short, fooling around for Eternity is not allowed, and when a soul goes mostly Dark, it is taken aside and examined closely... it is not allowed to pollute the world of souls headed for the Light. Thus, it should be clear that there is no Satan, or fallen angel.)

Recycled – short-circuited version of reincarnation: to come back into the same body, same lifetime, hence experiences **Déjà Vu**. Implies the inability to move forward into new realms and experiences in the greater **Multiverse**. (See VEG, Ch. 14.) If the soul learns nothing in a lifetime, this is the first step in rehabilitating them, and is a gentle nudge to apply oneself and handle the lessons... ultimately potentially followed by Reassembly. (**Chapter 8.**)

Reincarnation – the spiritual growth aspect of a soul moving thru the different realms in the Multiverse (not just back to Earth) for the purpose of experiencing and gaining knowledge and wisdom. On the other hand, a repeated lifetime limited to Earth is more of a **recycling**.

Remnant – short for Anunnaki Remnant – that part of the Anunnaki group that stayed on Earth and did not leave with the main group, between 610-560 BC. Comprised of the Insiders (+) and the Dissidents (-). Some are human-looking. Also known as Naga (underground 'Serpent' dwellers in Asia), may also be Dravidians.

Reptibian – A humanoid being, part reptile, part amphibian. In most ways looking like a human being, but with scales instead of skin, perhaps slightly webbed toes and fingers, cat's eyes, and a face that suggests a reptile/amphibian more than a human being. Note that an anaconda is aquatic and is also a reptile that moves on land. (VEG, Ch. 3)

Reassembled – this is also called **Disassembled**, Dissolved in VEG and TOM (books 1 and 2). The Higher Beings will attempt to infuse a wayward soul with new, proactive energy, and perform a kind of 'psychic surgery' on the soul's energy field (which emanates from their Ground of Being so it is tricky), and failing that, if a soul cannot be re-oriented to STO behavior, the energy (soul) is **Dissolved** back into its component energy parts, as a failed soul. The consciousness is removed, energy then is cleaned, and can be reused without the former consciousness that accompanied the failed soul. (In short, fooling around for Eternity is not allowed, and when a soul goes mostly Dark, it is taken aside and examined closely… it is not allowed to pollute the world of souls headed for the Light. Thus, it should be clear that there is no Satan, or fallen angel.)

Resonance – vibrating alike: such that two tuning forks A and B side by side, with A struck hard to set it vibrating, when put next to B which was not vibrating, will set tuning fork B vibrating at the same frequency as tuning fork A. This also happens with people in close proximity: a very negative person can 'detune' (bring down) a room of people and some people may actually feel ill and not know why (as they pick up the negative person's vibes). See **Entrainment**.

Resonant Consciousness Field – RCF/Matrix – the very negative energy and thoughtforms surrounding the Earth, as a vibrational envelope or field or Band that is so strong it entrains ensouled humans who are unconscious (qv) into their lower 3 chakras and they act out STS ideals. It is not alive or evil; it just has a lot of strength from centuries of people acting in synch with it and thus reinforcing its energy level. Similar in structure to the Matrix (qv) described earlier, or Robert Monroe's "H Band" but not run by any entities.

Satan – allegedly the leader of the demonic spirits which were the deceased Nephilim and/or their offspring. This titular role may have been filled by the

Nephilim (fallen Igigi) leaders known as Samayaza or Azazyel, or the Gnostic favorite: Ialdabaoth, in a former Era. The Egyptian Set, or **Sata**, was probably synonymous with one of the three just named. He lived in the desert and wore red. A convenient mythological character to personify Man's need for duality in the universe. (VEG, Ch. 2 and 6.) See Chapter 2 for *Satam*.

Schumann Resonance – natural frequency of earth's vibration/resonance: 7.8Hz.

Script – (**LifePlan**) to assist Karma, when one is born, one has a Script covering the basic (usually 10 % max) events that are to happen in one's life, which one is expected to overcome. They may be positive or negative, and how one meets them and handles them determines how one is progressing towards the goal of getting out of the Earth School. It often has Options programmed into it (**Points of Choice**) where the Soul must make a significant choice. It is a test of soul growth. A personal LifeScript is usually subject to the Greater Script of the Father of Light and works within it. The Angels (Beings of Light) administer the Script, yet the Soul still has 90% Freewill. (See **Ground of Being**.) **The Script does not tell you what to do or say; just a few key EVENTS are scripted to TEST you.**

ShapeShifting – the ability to control what people see... the being doing the shape-shifting does not actually change any of his atomic structure – just the way his appearance is perceived, and perception is holographic. So to effect a different appearance, the being just produces new interference waves that the observer 'sees' differently. Commonly done by 4D and above entities while in 3D.

Sheeple – people who are barely conscious, and refuse to think for themselves. They want someone to tell them what to do and when to do it, and they go along with whatever they are told. They are easily manipulated by the Media. Also called **'sheeple'** and may be OPs or 'dense' ensouled humans.

Soul Aspect – all souls can 'split' themselves to experience different realms; as when a timeline splits, one part of the soul stays with the original TL and another part replicates to the new TL. Each soul has aspects in different TLs, dimensions, worlds, and realms, etc. and at a point in the future, they reunite to the Soul Group. Not a **Fragment** (next). (See VEG, Ch.7.)

Soul Fragment – some souls may fragment **due to trauma** and then special therapy is often needed to coax the missing fragment to rejoin its source. Some fragments are held by family members, past lovers, and even by the 4D STS themselves.

Soul Group – each soul was part of a group of like souls (same core vibration PFV which usually synchs up with a specific archetype) and these split up to better experience the Creation – souls will eventually reunite in their original group when their explorings are done. The Soul Groups reunite with the **Godhead** from which they came.

Soul Merge – as suspected in the case of this author, to undertake a special project where a 3ʳᵈ level soul has volunteered to serve in a capacity that it alone can't do, and so a Merge is performed to give that 3ʳᵈ level soul the extra knowledge and strength of the merging soul (who is of the <u>same soul group</u> -- usually from a higher level) and together they perform some task that the Higher Beings must have approved – before the Merge can happen. (Conceptually similar to a Walk-in except that in the Merge, the original soul stays...)

Soulless -- in this book, this refers to humans who act/react **without conscience**, compassion or regard for the spiritual side of life. This describes the majority of the PTB – the fat cats whose money 'entitles' them to dictate to the rest of humanity. (See also theory of the **OP**s.) See also **NPC**s.

Soul Migration – the concept that animals can progress to first-time human beings with 'baby' souls and the full-fledged human soul must be earned thru successive incarnations. As they would also have only the lower 3 chakras operative, they may be mistaken for OPs. (See Metempsychosis on Wikipedia.)

STO – Service To Others; altruistic behavior, self-sacrificing. Patience, compassion, humility and respect for all living things. Intent to be of service to others, to assist – not do it for them or to them. (Found in 3D – 7D.)

STS – Service To Self; selfish behavior; 'Me-My-Mine' syndrome. Opposite of STO. Not found in 6D and above.

STS Gang – this is a 'catch all' group term referring to the Neggs, discarnates, thoughtforms, all **4D STS**, including Anunnaki Dissidents (3D & 4D), acting as oppressors of Man, without a clear distinction as to exactly which one is doing what to Man at any one time. (See Chapter 4.)
The group may occasionally include the **Interdimensional** souls described by Wilde and Monroe (VEG, Ch. 12), meaning Draco and fallen Elohim, although such are <u>usually</u> too busy interfering with the STO entities on their own level to harass Man on the 3D level.

Subquantum Kinetics – is an approach to microphysics with roots in general system theory, nonequilibrium thermodynamics, and nonlinear dynamics. It represents quantum phenomena differently than Quantum Physics (QP) and works with the concept of the **Ether** (VEG, Ch. 9) which is composed of subquantum units called **etherons** (as opposed to QP's quarks). It is simpler than Quantum Physics and explains the issues that QP is still wrestling with: wave-particle dualism, strings, singularities, and the cosmological constant.

It also embraces and explains Tesla's work better than QP. (Refer to Chapter 4 of Dr. Paul LaViolette's book <u>Secrets of Antigravity Propulsion</u> for a more complete description in layman's terms.) (See VEG, Ch. 9)

Terraforming – an advanced technical process whereby a whole planet is set to its original, or a near-new, pristine condition following some catastrophe or pollution, or both. The ecology is balanced, the air, land and water are unpolluted, and in the case of planet Earth, it can once again support lifeforms. See also "**Wipe and Reboot.**"

Thoughtform (**TF**) – any thought that many people subscribe to and which reflects a widely held belief, esp. one imbued with a lot of fear, or hate, generates a TF which after a while (depending on the amt of energy put into it) takes on a 'life' of its own; **man is a creator and thoughts are things**. If enough people fear and believe in werewolves, there will be thoughtform 'werewolves' … which are not real entities but are attracted to those who fear and believe in them (like attracts like).
Any unwanted TF can be cancelled and should be before it attaches itself to a person's aura and then 'feeds' off the person's energy – like a parasite. TF have no conscious volition of their own, they are reactive and go to wherever (1) they are attracted by sympathetic vibration, and (2) where the person's aura is weak.
Carl Jung called these TF's **Archtypes**.

Timeline (**TL**) – the linear coherent vector on which all souls and Placeholders (OPs) of a certain frequency range have their being; a reality timeline that linearly moves forward creating causal events. It is not permanent and is subject to entropy if a bifurcation results from a rise in consciousness and attendant agreement coherently shared among the souls seeking to live in a higher consciousness in TL2 is preferable to the negatively polarized TL1. If there is not enough agreement (energy) to sustain the new TL2, it dissolves. Diagram in TOM.
If a dimension has only one TL, the TL is the dimension, but dimensions can have multiple TLs. There is a TL where Hitler won, for example.
And timelines may create, 'run' and dissolve **fractal** subsets (within the larger TL framework) for special purposes (qv). (See TOM, Apx. D.)

Torsion Wave – Being neither electromagnetic in nature nor relating to gravity as it stands on its own, this wave energy is a spiraling, non-Hertzian electromagnetic wave that travels through the vacuum of space at super-luminal speed - a billion times faster than the speed of light. Because these waves trace a spiraling path, they are called "torsion waves." (See QES, Ch. 10.) Thoughts are torsion waves.

Torus – an energy field that is toroidal in shape (like a doughnut), and is found around the heart – extending and an electromagnetic field about 3-4' around the human body. (See TOM, Ch. 12.)

Unconscious – unaware, not a very high level of perception. A person who is 'asleep' spiritually and is not aware that there are more than the 5 senses. Can also mean 'spacing out' with eyes wide open. Standard condition of the **Sheeple**.

V2K – "Voice to Skull" -- a microwave enhanced transmission of words directly into a person's head, as if they actually hear the words, without any external devices or hearing apparatus. Developed by the US Army to communicate with a soldier on the battlefield, to the exclusion of other soldiers, it was perfected during the mind experiments with Helen Schucman while she transcribed the *Course in Miracles* book. Who sent her the information is not known. (VEG, Ch. 11. Also see Bibliography : Internet Sources.)

Vibration/Vibe – the energy state of a person, place or thing. Everything puts out an energy 'signature', which is how pyschometry works... objects record the energy of the person that held/owned the object, and places often hold the residual energy of events that happened there: some sensitive people cannot visit Gettysburg as they feel the negative energy from all the hate and fear created in that place – even thought it was long ago, it still holds some energy that has not completely dissipated. (See **PFV**.)

Vimanas – In Hindu literature (*Ramayana* and *MahaBharata*), the gods were said to fly around the sky and even engage in warfare between these craft with exotic but powerful weapons – similar to the Sumerian flying machines (MAR.GID.DA, IM.DU.GUD, and GIR). An ancient form of UFO, cone-shaped like many temples in Thailand, or *Stupas* in Tibet. Chapter 2.

Visual Spatial Acuity – the ability to see fine detail; visual term reflecting the number of rods/cones in the retina. Similar to **pixels** in computer printing, display screens and digital cameras.

VR Sphere – a term created in VEG, Chapter 12, to refer to the concept that Earth is a 3D construct, quarantined to protect it from 3D and 4D interference while the Higher Beings grow and develop souls in the Earth School. VR is Virtual Reality
(suggesting that Earth or parts of it may be a very sophisticated Simulation) which is dealt with in Chapter 13 and then further developed in QES, Book 4.

Wanderers – higher souls from other realms who have volunteered to incarnate on Earth in troubled times to serve as the Light leads: they may anchor the Light, write books, lead New Thought churches, heal or perform other services to benefit Mankind. Usually 6th level beings (souls). The Indigos and different forms of "Starseed" are part of this group.

Wipe and Reboot – an end to a current **Era** of Man on Earth, followed usually by a terra-forming (resetting the environment back to clean and balanced), followed by the Re-seeding of Man on the planet. (See Chart 5 in VEG, Ch. 15.)

The term is borrowed from the computer world where when a PC is non-functional (i.e., locked up and displays the dreaded BSOD [Blue Screen of Death]), it is necessary to "Wipe" the hard disk – reformat it – and reload the operating system and application software… i.e., "Reboot" the system and start all over again.

Whereas the PC gets a clean start as if nothing happened, each new Era for Man still includes whatever major, solid objects were created in the prior Era – i.e., pyramids, huge walls, and Stonehenge. A Reset of the **HVR Sphere**.

Zechariah Sitchin – the late Middle Eastern scholar, speaking several languages, who translated the Anunnaki/Sumerian tablets. VEG Ch. 3 is mostly dedicated to a summary of his findings about Man's origins. His claim to fame was *The Earth Chronicles* series of 8+ books that revealed the Sumerian – Anunnaki connection (see Bibliography for partial list).

ZPE (Zero Point Energy) – **Zero-point energy**, also called quantum vacuum zero-point energy, is the lowest possible energy that a quantum mechanical physical system may have; it is the energy of its ground state. Vacuum energy is the zero-point energy of all the fields in space, which in the Standard Model includes the electromagnetic field, fermionic fields, and the Higgs [boson] field.

It is the energy of the vacuum, which in quantum field theory is defined not as empty space but as the ground [lowest vibrational] state of the fields. In cosmology, the vacuum energy is one possible explanation for the cosmological constant. A related term is *zero-point field*, which is the lowest energy state of a particular field. (Definition: credit https://en.wikipedia.org/wiki/Zero-point_energy)

Dark Energy is thought to contain a ZPE state.

(See **Quantum Net**, and VEG, Ch. 9. Also see TOM, Ch. 11.)

Books

A.I., H+, QP, Simulation and 5G

Barrat, James   ***Our Final Invention***, NY: St Martin's Press, 2013.
Becker, Robt O., MD   ***The Body Electric***, NY: Harper, 1985.
Carlo, Dr. George   ***Cell Phones***, NY: Carroll& Graf, 2001.
Elvidge, Jim   *Digital Consciousness*, UK: Iff Books, 2018.
_____   ***The Universe Solved***, Alternative Theories Press, 2007.
Gittleman, Ann L.   ***Zapped,*** NY: Harper One, 2010.
Horn, Dr Thomas   ***The Milieu... Transhuman Resistance***, MO: Defender
                          Crane, 2018.
Kaku, Michio   ***The Future of the Mind***, NY: Anchor Books, 2014.
Kurzweil, Ray   ***The Singularity is Near, NY***: Penguin, 2005.
McTaggart, Lynn.   *The Field.* New York: HarperCollins/Quill, 2002.
Pearce, Joseph Chilton.   *The Biology of Transcendence.* Rochester, VT: Park Street
                          Press, 2002.
Pearsall, Paul, Ph.D.   *The Heart's Code.* New York: Random House/Broadway,
                          1998.
Rifat, Tim   ***Remote Viewing,*** UK: Vision Paperbacks, 23001.
                          (excellent chapter on EMF and dangers.)
Talbot, Michael.   *The Holographic Universe.* New York: HarperCollins, 1991.
Tegmark, Max   *Life 3.0,* NY: Vintage, 2017.
Virk, Rizwan   *The Simulation Hypothesis*, Bayview Books, 2018-19.

Ancient History

Barrow, Valerie   ***Alcheringa,*** Noveletta Press: 2014.
Fenton, Daniella   ***Hybrid Humans***, self-pub., 2018
Guiley, Rosemary & Philip Imbrogno   *The Vengeful Djinn*, Llewellyn Publ., 2012.
Kramer, Samuel Noah.   *The Sumerians.* Chicago, IL: Univ. of Chicago Press, 1971.
Lessin PhD, Sasha.   *Anunnaki: Gods No More.* Lexington, KY: CreateSpace,
                          2012.

_____   *Anunnaki: Legacy of the Gods.* CA: Aquarian Radio
                          Publishing, 2014.
Parks, Anton   ***EDEN***, France: Pahana Books, 2013.
_____   ***The Secret of the Dark Stars***, France: Pahana Books,
                          2103.

Serpent/Dragon Wisdom

Capra, Fritjof. *The Tao of Physics.* Boston: Shambhala Publ., 1999.

DeAngelis, Barbara Dr. *Soul Shifts.* Los Angeles: HAY House, 2015.

Elkins, Don and Carla Rueckert. *The RA Material, Book I.* Atglen, PA: Schiffer Publishing/Whitford Press, 1984.

Freke, Timothy & Peter Gandy. *The laughing Jesus.* NY: Three Rivers Press, 2005.

_____ *Lucid Living.* UK: Sunwheel Books, 2005.

Gardiner & Osborn. *The Shining Ones, Revised.* London: Watkins Publishing, 2010.

Golas, Thaddeus. *The Lazy Man's Guide to Enlightenment.* Salt Lake City: Gibbs-Smith, 1995.

Goswami, Amit. *The Self-Aware Universe.* NY: Tarcher: Penguin, 1993..

Greene, Brian. *The Elegant Universe.* New York: W.W. Norton & C0. 2003.

_____ *The Fabric of the Cosmos.* New York: Vintage Books. 2004.

_____ *The Hidden Reality.* New York: Alfred A. Knopf. 2011.

Johnson, Kurt, & D.R. Ord *The Coming Interspiritual Age,* Vancouver: Namaste Publ., 2012..

Laszlo, Ervin *Science and the Reenchantment of the Cosmos,* VT: Inner Traditions, 2006.

Lerma, John, M.D. *Into the Light.* Franklin Lakes, NJ: New Page Books, 2007.

_____ *Learning From the Light.* Franklin Lakes, NJ: New Page Books, 2009.

Lloyd, Seth. *Programming the Universe.* NY: Random House, 2007.

Marion, Jim. *Putting on the Mind of Christ.* NY: Hampton Roads, 2000.

Mead, G.R.S. *Apollonius of Tyana.* (1901 Edition reprint) Sacramento, CA: Murine Press, 2008.

Meyer, Marvin. *The Gospel of Thomas.* New York: HarperCollins, 1992.

Paulson, Genevieve Lewis. *Kundalini and the Chakras.* MN: Llewellyn Worldwide, 2005.

Rasha. **Oneness.** Santa Fe, NM: Earthstar Press, 2003.

Ring, Kenneth. **Lessons from the Light.** Portsmouth, NH: Moment Point Press, 2000.

Roman, Sanaya. **Spiritual Growth.** Tiburon, CA: HJ Kramer, Inc., 1989.

_____. *Personal Power Through Awareness.* Tiburon, CA: HJ Kramer, Inc., 1986.

Spencer, Robert. *The Craft of the Warrior.* 2nd Ed., CA: Frog, Ltd, 2006.

Zukav, Gary. *The Dancing Wu Li Masters.* NY: Wm Morrow: 1979.

Religion/Metaphysics/Spirituality

Acharya S. *The Christ Conspiracy.* Kempton, IL: AUPress, 1999.

Atwater, P.M.H. *Near Death Experiences.* NY: MJF Books, 2011.

Barrett, David V.   *Secret Religions*.   PA: Running Press, 2011.
Castaneda, Carlos.   *A Separate Reality*. New York: Washington Square Press, 1971.
_____.   *The Active Side of Infinity*. New York: HarperCollins, 2000.
Charles, R.A.   **The Book of Enoch the Prophe**t. San Francisco, CA: Weiser
                       Books, 2003.
Dawood, N. J.   *The Koran*. New York: Penguin Group (USA), 2006.
Frejer, B. Ernest.   *The Edgar Cayce Companion*. New York: B & N Press, 1995
Gaffney, Mark.   *Gnostic Secrets of the Naassenes*. Roch. VT: Inner Traditions, 2004
Gardiner, Philip.   *Secret Societies*. Franklin Lakes, NJ: Career Press/New Page,
                       2007.
_____.   *Secrets of the Serpent*. Forest Hill, CA: Reality Press, 2008.
_____.   **The Shining Ones**. Nottinghamshire, England: Phase Group,
                       2002.
_____.   *The Shining Ones*, rev. UK: Watkins Publ., 2010.
Goswami, Amit.   *The Self-Aware Universe*. NY: Tarcher: Penguin, 1993. .
Hall, Manly P.   *The Secret Teachings of All Ages*, NY: Tarcher/Penguin, 2003.
Harris, Sam   **Free Will** NY: Free Press, 2012.
Hoeller, Stephan A. *Gnosticism*. Wheaton, IL: Quest Books, 2002.
Houston, Jean   **The Science of Mind**. NY: Putnam, 1997.
Johnson, Kurt & David R. Ord. *The Coming Interspiritual Age*. Canada: Namaste
                       Publ, 2012.
Kenyon, J. Douglas, Ed. *Forbidden Religion*. Rochester, VT: Bear & Co., 2006.
Laurence, Richard.   **The Book of ENOCH the Prophet**. Kempton, IL:
                       Adventures Unlimited Press, 2000.
Leo Ed, Marilyn.   **Love & Law**. NY: Putnam, 2001.
Mack, Dr John E.   *Passport to the Cosmos*.   NY: Three Rivers Press, 1999.
Mead, G.R.S.   *Apollonius of Tyana*. (1901 Edition reprint) Sacramento, CA:
                       Murine Press, 2008.
Men, Hunbatz.   *The 8 Calendars of the Maya*. Roch, VT: Bear & Co. 2010.
_____   **Secrets of Mayan Science/Religion**. Roch, VT: Bear & Co.
                       1900.
Moody, Raymond A., Jr., MD. *Life After Life*. New York: HarperCollins, 2001.
Monroe, Robert.   *Journeys Out of the Body*. New York: Doubleday, 1971.
_____.   *Far Journeys*. New York: Random House/Broadway, 2001.
_____.   *Ultimate Journey*. New York: Random House/Broadway,
                       2000.
Pagels, Elaine.   *The Gnostic Gospels*. New York: Random House/Vintage, 1979.
Phillimore, J S.   *Philostratus – In Honor of Apollonius of Tyana*. London: Oxford
                       University Press, 1912.
Puryear, H.B.   *The Edgar Cayce Primer*. NY: Bantam Books, 1982.
Robinson, James M., Gen. Ed. **The Nag Hammadi Library**. NY: Harper
                       Collins, 1990.:
Ruffin, C. Bernard.   *Padre Pio, The True Story*. Huntington, IN: Our Sunday
                       Visitor Publishing Division, Inc. 1991.

Slate, Joe H., Ph.D.   *Aura Energy.*   Woodbury, MN: Llewellyn Worldwide, 2002.
_____.   *Psychic Vampires.*   St. Paul, MN: Llewellyn Worldwide, 2004.
Snellgrove, Brian.   *The Unseen Self.*   Essex, England: The C.W. Daniel Co., 1996.
Spong, Bishop J.S.   *Why Christianity Must Change or Die.* SF: Harper, 1978.
             A New Christianity for a New World. SF: Harper, 2001.
Wilde, Stuart.   *The Prayers and Contemplations of God's Gladiators.* Chicago, IL: Brookemarke, LLC., 2001.
             The Force. Carlsbad, CA: Hay House, 2006.
Zagami, Leo Lyon.   *Pope Francis: The Last Pope?* SF: CCC Publ., *2015.*
Zukav, Gary.   *The Dancing Wu Li Masters.* NY: Wm Morrow: Quill, 1979.

Scientific & Medical

Batmanghelidj, Dr.   *Your Body's Many Cries For Water,* GHS, 2008/
             You're Not Sick, You're Thirsty, NY: Warner Books, 2003.
Braden, Gregg.   *The Divine Matrix.* Carlsbad, CA: Hay House, 2007.
Brown, Walt.   *In The Beginning: Compelling Evidence for Creation and the Flood.* Phoenix, AZ: Center for Scientific Creation, 1995.
Capra, Fritjof.   *The Tao of Physics.* Boston: Shambhala Publ., 1999.
Carter, Mildred & Tammy Weber. ***Body Reflexology.*** (NJ: Prentice Hall, 1994).
Gerber, Richard, MD   *Vibrational Medicine,* VT: Bear & Co., 2001.
Greene, Brian.   *The Elegant Universe.* New York: W.W. Norton & C0. 2003.
             The Fabric of the Cosmos. New York: Vintage Books. 2004.
             The Hidden Reality. New York: Alfred A. Knopf. 2011.
Hunter, C. Roy.   *Master the Power of Self-Hypnosis.* NY: Sterling Publishing, 1998.
Lloyd, Seth.   *Programming the Universe.* New York: Random House, 2007.
McTaggart, Lynn.   *The Field.* New York: HarperCollins/Quill, 2002.
Mindell, Earl   ***Vitamin Bible,*** NY: Warner Books, `999.
Morris, John D., Ph.D.   *The Young Earth.*   Green Forest, AR: Master Books, 2006.
Myss, Caroline, Ph.D.   *Why People Don't Heal and How They Can.* New York: Three Rivers Press, 1997.
             Sacred Contracts. New York: Three Rivers Press, 2002.
Narby, Jeremy.   *The Cosmic Serpent.*   New York: Tarcher/Putnam, 1998.
Pearce, Joseph Chilton.   *The Biology of Transcendence.* Rochester, VT: Park Street Press, 2002.
Pearsall, Paul, Ph.D.   *The Heart's Code.* New York: Random House/Broadway, 1998.
Peterson, Dennis R.   *Unlocking the Mysteries of Creation.* 6th edition. El Dorado, CA: Creation Resource Foundation, 1990.
Redfern, Nick   ***Bloodline of the Gods,*** NJ: New Page, 2015.
Talbot, Michael.   *The Holographic Universe.* New York: HarperCollins, 1991.

Walsh, Randy *The Apollo Moon Missions, Part I*, KDP/CreateSpace, 2018.

UFOs & ETs

Ancient Aliens DVD: Season 9, Episode 8, History Channel: 2016.

Boulay, R.A. ***Flying Serpents and Dragons***. Rev. Ed. San Diego, CA: The Book Tree, 1999.

Bramley, William. ***The Gods of Eden***. New York: HarperCollins/Avon, 1993.

Branton, ed. *The Omega Files*. NJ: Global Communications, 2012.

Cannon, Dolores. *Keepers of the Garden*. Huntsville, AR: Ozark Mtn Publishers, 2002.

Fowler, Raymond. *The Watchers*. New York: Bantam Books, 1990.

Good, Timothy. *Earth: An Alien Enterprise*. New York: Pegasus Publishing, 2013.

Greer, Steven M., MD. *Hidden Truth – Forbidden Knowledge*. Crozet, VA: Crossing Point, Inc., 2006.

Haze, Xaviant. *Aliens in Ancient Egypt*. VT: Bear & Co., 2013.

Heron, Patrick. *The Nephilim and the Pyramid of the Apocalypse*. New York: Kensington Publishing/Citadel Press, 2004.

Komarek, Ed. UFOs: *Exopolitics and the New World DisOrder*. Lexington KY: Shoestring Publishing, 2012.

Lessin PhD, Sasha. ***Anunnaki: Gods No More***. Lex. KY: CreateSpace, 2012.

_____ *Anunnaki: Legacy of the Gods*. CA: Aquarian Radio Publishing, 2014.

Lewels, Joe, Ph.D. *The God Hypothesis*. Columbus, NC: Wild Flower Press, 2005.

Littrell, Helen & Jean Bilodeaux *Raechel's Eyes*. NC: Granite Publ., 2005.

Marrs, Jim. ***Alien Agenda***. New York: HarperCollins, 1997.

_____. *Rule by Secrecy*. New York: HarperCollins, 2000.

_____ *Our Occulted History*, NY: Wm Morrow, 2013.

_____ ***The Rise of the Fourth Reich***. NY: Wm Morrow, 2008.

_____ ***The Illuminati***, Detroit: Visible Ink, 2017.

Olsen, Brad. *Future Esoteric: The Unseen Realms*. CA: CCC Publishing, 2013.

_____ *Modern Esoteric: Beyond Our Senses*. CA: CCC Publishing, 2014.

Pruett, Dr. Jack. *The Grandest Deception*. Xlibris Corp: Lexington, KY, 2011.

Salla, Dr. Michael *Insiders Reveal Secret Space Programs & Extraterrestrial Alliances (Book 1)* HI: Exopolitics Institute, 2015.

Story, Ronald D., Ed. *The Encyclopedia of Extraterrestrial Encounters*. New York: New American Library, 2001.

Tellinger, Michael.   *Slave Species of god.* Johannesburg, SA: Music Masters Close
              Corporation, 2005. (1st book)

_____.   **Slave Species of the Gods.** Roch. VT: Bear & Co., 2012.
              (2nd Book reprint)

Trench, B. LePoer.   **The Sky People.** NY: Award Book, 1970.

_____   **Operation Earth.** London: Spearman, 1969.

Vallée, Jacques.   *Passport to Magonia.*   Chicago, Il: Contemporary Books,
              1993.

_____.   *Dimensions.* Chicago, Il: Contemporary Books, 1988.

_____.   *Messengers of Deception.* Brisbane, Australia: Daily Grail, 1979.

_____.   *Revelations.* San Antonio, TX: Anomalist Books, 2008.

Von Daniken, Erich.   *Arrival of the Gods.* London: Vega, 2002.

_____   *History is Wrong.* New Jersey: New Page, 2009.

_____   *Eyewitness to the Gods.* NY: New Page, 2019.

History & Other Related Books

Andrews, Synthia, & Colin Andrews. *The Complete Idiot's Guide to 2012.*
                    New York: Penguin Group (USA), 2008.

Calleman, Carl J. PhD. *The Mayan Calendar and the Transformation of*
                    Consciousness. Rochester, VT: Bear & Co., 2004.

Coleman, J.A., Ed.   *The Dictionary of Mythology.* London: Arcturus, 2015.

Curran, Dr. Bob.   *Lost Lands, Forgotten Realms.* NJ: New Page, 2007.

Davidson, H.R. Ellis.   *Myths & Symbols in Pagan Europe.* NY: Syracuse Univ.
                    Press, 1988.

Davis, Kenneth C.   *Don't Know Much About Mythology.* NY: Harper Collins,
                    2005.

Dick, Tessa   *The Matrix Control System of Philip K. Dick.* NJ: Global
                    Communications, 2017.

Dolan, Richard   **Secret Space Pgm & Breakaway Civ**, Dolan Press,
                    2016.

Dougherty, Martin J.   *A Dark History: Vikings.* NY: Metro Books, 2013.

_____   *Dark History: Celts.* NY: Metro Books, 2015.

Duane & Hutchison.   *Chinese Myths and Legends.* London: Brockhampton Press,
                    1998.

Farrell, Joseph   **Covert Wars & Breakaway Civilizations**, Ill: AUP,
                    2012.

_____   *Covert Wars & the Clash of Civilizations*, Ill: AUP, 2013.

_____   *Saucers, Swastikas and Psyops*, Ill: AUP, 2011.

Guiley, Rosemary Ellen.   *Encyclopedia of the Strange, Mystical & Unexplained.*
                    New York: Gramercy Books, 2001.

Guiley, Rosemary & Philip Imbrogno   **The Vengeful Djinn**, Llewellyn Publ.,
                    2012.

Hawkins, David R.   *Reality and Subjectivity*. W. Sedona, AZ: Veritas Press, 2003.

_____   *The Eye of the I*. West Sedona, AZ: Veritas Press, 2001

Icke, David.   *The Biggest Secret*. Wildwood, MO: Bridge of Love, 2001.

_____   *...And the Truth Shall Set You Free*. Isle of Wight, UK: David Icke Books Ltd., 1995.

_____   *The David Icke Guide to the Global Conspiracy*. Isle of Wight, UK: David Icke Books Ltd., 2007.

_____   *Human Race: Get Off Your Knees*. Isle of Wight, UK: David Icke Books Ltd., 2010.

_____   *Children of the Matrix*. Isle of Wight, UK: David Icke Books Ltd., 2001.

_____   *Tales from the Time Loop*. Wildwood, MO: Bridge of Love, 2003.

_____   *Everything You Need To Know....*, David Icke Books Ltd., 2017.

Keel, John A.   *The Complete Guide to Mysterious Beings*. New York: Tor Books, 2002.

_____   *Why UFOs – **Operation Trojan Horse**. New York: Manor Books, 1970.

_____   *Our Haunted Planet*. NY: Fawcett Books, 1971.

Kenyon, Douglas.   *Forbidden Science*. Roch, VT: Bear & Co., 2008.

_____   *Forbidden History*. Roch, VT: Bear & Co, 2005.

_____   *Forbidden Religion*. Roch, VT: Bear & Co, 2006.

Kramer, Samuel Noah.   **The Sumerians**. Chicago, IL: Univ. of Chicago Press, 1971.

Kreisberg, Glenn, Ed.   *Lost Knowledge of the Ancients*. Roch, VT: Bear & Co, 2010.

Lawton, Ian.   *Genesis Unveiled*. London: Virgin Books, 2003.

Martinez PhD, Susan B.   **The Mysterious Origins of Hybrid Man**. Roch, VT: Bear & Co, 2013.

McKean, Erin, Ed.   *The New Oxford American Dictionary, 2ⁿᵈ Edition*. New York: Oxford University Press, 2005.

Parkes, Henry B.   *A History of Mexico*. Boston, MA: Houghton Mifflin Co., 1960.

Pemberton, John.   *Myths & Legends*. NY: Chartwell Books, 2012.

Pinkham, Mark Amaru. The *Return of the Serpents of Wisdom*. Kempton, IL: AUP, 1997.

Pye, Lloyd.   *Everything You Know Is Wrong, Book I: Human Origins*. Lincoln NE: iUniverse/Authors Choice Press, 2000.

Pye, Michael & Kirsten Dalley   *Lost Secrets of the Gods*. NJ: New Page, 2014.

Rosenberg, Donna. *World Mythology, 2ⁿᵈ Ed*. IL: NTC Publishing, 1994.

Sitchin, Zecharia.   *Journeys to the Mythical Past*. Rochester, VT: Bear & Co., 2007.

_____   **The Twelfth Planet**. New York: HarperCollins, 2007.

_____   *The Cosmic Code*. New York: HarperCollins, 2007.

_____   *The End of Days*. New York: HarperCollins, 2007.

_____   *The Earth Chronicles Expeditions*. Roch, VT: Bear & Co., 2004.

_____   *Divine Encounters*. New York: HarperCollins/Avon, 1996.

_____   *Genesis Revisited*. New York: HarperCollins/Avon, 1990.

_____   *The Wars of Gods and Men*. New York: HarperCollins, 2007.

_____   *The Lost Book of ENKI*. Roch, VT: Bear & Co., 2004.

_____   *The Stairway to Heaven*. New York: HarperCollins, 2007.

_____   *The Earth Chronicles Handbook*. Roch, VT: Bear & Co., 2009.

_____   *There Were Giants Upon the Earth*. Roch, VT: Bear & Co., 2010.

_____   *The Anunnaki Chronicles*. Roch., VT: Bear & Co. 2015.

Thomas Nelson.   *The King James Study Bible*. Nashville, TN: Thomas Nelson, 1988.

Thorsson, Edred,   *Runelore: The Magic, History & Hidden Codes of the Runes*. SF: Weiser Books, 2012.

_____   *Futhark: A Handbook of Rune Magic*. SF: Weiser Books, 1984.

Turner, Patricia & Charles Russell Coulter. *Dictionary of Ancient Deities*. NY, NY: Oxford University Press, 2001.

Walsh, Randy.   **The Apollo Moon Missions.** Createspace, 2018.

Wilkinson, Philip.   *Myths & Legends*. London: DK Ltd., 2009.

Internet Sources

5G (Super Grid)

*** The following 10 articles are all from the 5G Library:
5G Crisis: Awareness & Accountability *
at https://the 5GSummit.com
www.healthmeans.com

*Their Book: published by HealthMeans, no ISBN, no copyright:
825 pp – printed presentations by 42 doctors, engineers
and health professionals : http://hto.care to access.

Dr. Devra Davis, "Science on 5G and Wireless Radiation"

Robert F. Kennedy, Jr., "Dangers of 5G to Children's Health"

Patrick Wood, "5G: The Agenda for Total Control"

Dr. Martin Pall, "5G Risk: The Scientific Perspective"

Dr. Paul Heroux, "Harmful Effects of 5G and Wireless"

Jason Bawden-Smith, MSc, "Critical Disruption of Mitochondria by EMFs"

Dr. Magda Havas, "Extensive Biological Effects of EMFs and 5G"

Dr. Sharon Goldberg, MD "Science about Wireless and 5G"

Trevor Marshall, "Debunking 7 Myths about 5G"

Steven Whybrow, "The Inner Challenge of 5G"

***The following 2 articles are from Health Means Library:
Healthmeans.com (or: http://hto.care)

Carla Atherton with Ann Louise Gittleman, PhD,
   "A Danger We Can't See: Electromagnetic Fields and Their Effects
   on Our Health"

Dr. Evan Brand, "EMF, 5G Celltowers, Geoengineering and Retroviruses"

***The following article is from Environmental Health Trust:
ehtrust.org

"Questions & Answers about WiFi in Schools" in main article:
"Wi-Fi In Schools, Wireless Radiation Health Impacts And Practical Solutions"

***The following article is from NEXUS magazine:
www.nexusmagazine.com
July-Aug 2019 issue, vol, 26, no. 4, pp.15-23:
Jeremy Naydler, PhD, "5G: The Big Picture"

***The following article is from Bibliotecapleyedes Library at
https://www.bibliotecapleyades.net
sublibrary: **Tim Rifat & PSI**
(https://www.bibliotecapleyades.net/sociopolitica/esp_sociopol_mindcon10.htm)
Article1: "Microwave Mind Control – UK Intelligence Forces and
Microwave Mind Control."

Article2: "GWEN Towers – Total Control"
By Nicholas Jones (2001)
(https://www.bibliotecapleyades.net/scalar_tech/esp_scalartech04.htm)

***French BlueLight Research

Article: "Blue Light From Screens is Toxic to Eyes Warns French Health
Agency" from ANSES 5/14/2019: The French Agency for Food,
Environmental and Occupational Health & Safety (ANSES)...
https://www.anses.fr/en/system/files/PRES2019DPA01EN.pdf

Anunnaki
Amitakh Sanford, "The Anunnaki Remnants Are Still on Earth." Excellent article that extends and refutes some of the Zechariah Sitchin material on the Anunnaki. Website:
http://www.xeeatwclvc.com/articles/anunnaki_remnants.htm

Estelle N. H. Amrani, "A Different Story About the Anunnaki" is another article by a credible source, which partly agrees with Amitakh, and gives additional compatible information. Website:
http://www.bibliotecapleyades.net/sumer_anunnaki/anunnaki/anu_12.htm

Robertino Solàrion, "Nibiruan Physiology." Excellent article mentioning the Galactic Law that the Anunnaki must adhere to – first reported by John Baines in his book, The Stellar Man. Website:
http://www.bibliotecapleyades.net/cosmic_tree/physiology.htm
see also Baines reference:
http://www.bibliotecapleyades.net/serpents_dragons/boulay05e.htm
"Myths From Mesopotamia: Gilgamesh, The Flood, and Others" translated by Stephanie Dalley as quoted in the website: http://www.piney.com/Atrahasis.html also see:
http://www.book-of-thoth.com/ftopicp-137854.html

Extraterrestrial Exposure Law
Michael Salla, PhD., "Extraterrestrials Among Us" (vol.1:4, originally from ExoPolitics Journal website), is an interesting article on how ETs are among us who look so much like us that we don't suspect, and secondly the article explores the Extraterrestrial Exposure Law of 1969. Website:
http://www.bibliotecapleyades.net/exopolitica/esp_exopolitics_ZZZN.htm
also see: http://exopolitics.com for author Salla's general website.

Extraterrestial Genes in Human DNA
"Scientists Find Extraterrestrial Genes in Human DNA" is another article seeking to explain "junk DNA" and its probable origin and significance. Website:
http://www.bibliotecapleyades.net/vida_alien/esp_vida_alien_18n.htm

Greek Gods
Neil Jenkins, Sumair Mirza and Jason Tsang, "The Creation of the World & Mankind" is a great summary review of the major aspects of the Greek myths. Fascinating material, well-organized and indexed; won an award in 1997. Website:
http://www.classicsunveiled.com/mythnet/html/creation.html -- multiple topics.
See also: http://historylink102.com/greece2/ -- multiple topics.

People With Horns

Sutherland, Mary, "Was There a Race of People Who Had Horns?" is a thought-provoking article with pictures showing people in the past and present who have horny growths coming out of their heads. Has links to other related websites. Website:

http://www.burlingtonnews.net/hornedrace.html

ancillary link: http://www.bibliotecapleyades.net/vida_alien/alien_watchers04.htm

Sea Monster/Plesiosaur

40ANA blog website article, "Sea Monster or Shark?" Posted by The Moviebuff at 7:16am on 9/1/2006. Shows pix of the carcass caught in the trawler's net. Website:

http://40ana.blogsppot.com/2006/09/sea-monster-or-shark.html

Serpents, Reptiles & DNA

Paul Von Ward, "Aliens, Lies and Religions" article on great Belgian website that discusses the author's book Gods, Genes and Consciousness. The issue of serpents and DNA is clarified as well as other AB (Advanced Being) issues. Website:

http://www.karmapolis.be/pipeline/von_ward_uk.htm

Vatican & ETs

Patricia Cori, "The Vatican Says OK, We Can Believe in ET Now" is an article that comments on the more common Breitbart and Fox News article (below) revealing the Pope's blessing on humans accepting the existence of ETs. Website: www.sirianrevelations.net

FOX News article "Vatican: It's OK for Catholics to Believe in Aliens" containing a longer examination of the Pope's blessing on our ET brothers. Website:

http://www.foxnews.com/story/0,2933,355400,00.html

Videos of Interest

AI & Transhumanism

Ex Machina, Lionsgate, 2014.
Her, Warner Bros., 2013.
Transcendence, Warner Bros., 2014.
S1m0ne, Newline Cinema, 2003.
Lucy, Universal, 2014.
Artificial Intelligence, Warner Bros., /Dreamworks, 2002.
A.I. Rising, Mir Media/Balkanic, 2018.
Cold Souls, 20th Cent. Fox, 2009.
Terminator 3, Warner Bros., 2003.
Terminator Salvation, Warner Bros., 2009.
Terminator Genesis, Paramount, 2015.
Replicas, Lionsgate, 2019.

_____SciFi & Metaphysical

Forbidden Planet. MGM classic from 1956; debuts Robby the Robot.
The X-Files (TV series, 1993-2002): Twentieth Century Fox.
K-Pax. Universal Pictures, Lawrence Gordon et al. 2001.
Millenium. Gladden Entertainment. 1989.
Hangar 18. Republic Entertainment. 1980.

Capricorn One. Associated General Films. Lazarus/Hyams prod. 1978.
Groundhog Day. Dir. Harold Ramis, Columbia Tristar. 1993.
Men In Black. I & II Dir. Barry Sonnenfeld, Columbia Pictures. 2000.
The Matrix. Dir./Written by The Wachowski Bros., Warner Bros. 1999.
The Mothman Prophecies. Screen Gems/LakeShore Entertainment. 2001.
Prometheus I. 20th Century Fox, 2012.
Taken. (TV miniseries) Stephen Spielberg, DreamWorks. 2002.

V, the TV series (1983-85, and 2009-11). WarnerVideo.
The Truman Show. Peter Weir, Paramount Pictures. 1998.
The Young Age of the Earth. Aufderhar, Glenn. Earth Science Associates /
        Alpha Productions. 1996.
What the Bleep Do We Know? 20th Century Fox, Lord of the Wind films, 2004.

They Live. Dir./Written by John Carpenter, Universal Studios. 2003.
Prometheus I. Ridley Scott, 2oth Century Fox. 2012.
The Day the Earth Stood Still. Twentieth Century Fox, Erwin Stoff et al, 2009.
The Forgotten. Revolution Studios. 2004

Iron Sky.   Timo Vuorensola, Ger/Fin release via Paramount Pictures,
          2012 version.
Paul.   Universal Studios, Greg Motola. 2010.
2012.   Sony Pictures, Roland Emmerich. 2010.
The Fourth Kind. Universal Pictures, Olatunde Osunsanmi. 2010.
The Adjustment Bureau. Universal Pictures, George Nolfi. 2010.

Knowing. Summit Entertainment, Alex Proyas. 2009.
Dark City. New Line Cinema, Alex Proyas. 1998.
Source Code. Summit Entertainment, Duncan Jones. 2011.
eXistenZ. Canadian Television Fund, David Cronenburg, 1999.
Defending Your Life. Warner Bros., 1991.

Simulation & Avatar

The Thirteenth Floor. Columbia Pictures, Roland Emmerich. 1999.
Avatar. Lightstorm Entertainment, 2009.
Jumanji, Welcome to the Jungle. Columbia Pictures, 2017.
Groundhog Day. Dir. Harold Ramis, Columbia Tristar.   1993.
The Matrix. Dir./Written by The Wachowski Bros., Warner Bros. 1999.

Also: **Ancient Aliens** video series Seasons 1 – 12..

          ***

Endnotes – Chapter 1

---

[1] Anton Parks, <u>The Secret of the Dark Stars</u> , p. 64.

[2] Ibid., p. 86.

[3] Ibid., p. 193 (opposite plate).

[4] Zechariah Sitchin, <u>Earth Chronicles Handbook</u>, p. 160-61.

[5] Anton Parks, <u>Eden</u>, p. 89

[6] Ibid., , <u>Eden</u>, p. 89..

[7] Ibid., <u>Eden</u>, p. 76.

[8] Ibid., <u>Eden</u>, p. 77

[9] Ibid., <u>Eden</u>, p. 231.

[10] Ibid., <u>Eden</u>, pp. 231-236.

[11] Ibd., <u>Eden</u>, p. 232.

[12] Ibid., Eden, p. 235.

[13] Credit: Parks' book EDEN, his pix on this book's pp. 42, 80, 84 and 86.

[14] Rosemary Elle Guiley, <u>TheVengeful Djinn</u>, pp. 74-84.

[14] http://www.pbs.org/newshour/rundown/study-invisible-shield-space-protects-earth-killer-electrons/
     And
https://www.sciencedaily.com/releases/2014/11/141126133829.htm

[15] Ibid., -- they just told you that the barrier is impenetrable...

[15] Gregg Prescott article in *In5D Guest* and Waking Times:
     Earth's Quarantine Force Field Discovered by NASA? , Dec. 3, 2014.

[15] https://www.sciencedaily.com/releases/2014/11/141126133829.htm
     Also:
     An impenetrable barrier to ultrarelativistic electrons in the Van Allen radiation belts. *Nature*, 2014; 515 (7528): 531 DOI: 10.1038/nature13956

Endnotes – Chapter 2

[16] Taurus_constellation_map.png: Torsten Bronger derivative work: Kxx (talk) - Taurus_constellation_map.png

[17] https://en.wikipedia.org/wiki/Kariong%2C_New_South_Wales

[18] Daniella Fenton, <u>Human Hybrids</u>, pp. 99-100.

[19] Sasha Lessin, <u>Anunnaki Gods No More</u>. Pp. 197-207.

[20] Zechariah Sitchin, <u>The Earth Chronicles Handbook</u>. P.69.

[21] Sitchin, <u>The 12th Planet</u>. p.84.

[22] Sitchin, <u>The Cosmic Code</u>. Pp 78-79.

[23] Lawton, <u>Genesis Unveiled</u>. Pp. 73, 78-79.

[24] EAE Jelinkova, *The Shebtiw at the Temple at Edfu*.

[25] https://shebtiw.wordpress.com/?s=edfu

[26] Kenneth C. Davis, <u>Don't Know Much About Mythology</u>. PP 75-79.

[27]
http://www.bhporter.com/Porter%20PDF%20Files/Worship%20of%20the%20Ancestor%20Gods%20at%20Edfu.pdf
      See also: "Mythical Origin of the Egyptian Temple" by E.A.E. Reymond.
          Pp 23,34,35, and 59-61.

[28] "Mythical Origin of the Egyptian Temple" by E.A.E. Reymond.
          Pp 106-107, 114.
EAE Reymond was former professor of Egyptology at the University of Manchester and her work *Mythical Origin of the EgyptianTemple* was first published in 1969.

[29] Anton Parks, <u>The Secret of the Dark Stars</u>, p. 282.

Endnotes – Chapter 3

[30] Daniella Fenton, <u>Hybrid Humans</u>, pp. 53-54.
[31] Ibid., p. 54.
[32] Ibid., p 109.
[33] Ibid., p. 111.
      Also: Noonan, JP et al... Sequencing and Analysis of Neanderthal genomic DNA
          https://www.osti.gov/servlets/purl/923281
[34] Ibid., pp. 111-112.
[35] Sitchin, Zechariah. *There Were Giants Upon the Earth*. pp. 162-163.

[36] Ibid., pp. 114-116.
      Also: Benton, A. 2016 Adam's Chromosomes were Fused
          (and he lived earlier than we thought),
            https://www.filthymonkeymen.com/2016/11/08/adams-
chromosomal- fusion/
[37] Ibid., pp. 120-131.

[38] Redfern, <u>Bloodline of the Gods</u>, pp. 19-22.

[39] https://en.wikipedia.org/wiki/Junkyard_tornado

[40] https://en.wikipedia.org/wiki/Hox_gene

[41] Kulczyk in NEXUS: pp. 38-39.
    Wojciech K. Kulczyk, *Out of Space, Advanced Panspermia*, article in
    NEXUS magazine, May-June, 2018 Vol. 25, No. 3, pp. 34-39.

[42] Hart, <u>Ancient Alien Ancestors</u>, pp. 326-30.
[43] Ibid., p. 328.
[44] Ibid., p. 329.

[45] Dr Ellis Silver, <u>Humans Are Not From Earth</u>, pp. 19-197.

[46] https://en.wikipedia.org/wiki/CCR5

[47] Bill Sullivan, "Why You Like What You Like", NATGEO, Sept 2019, pp. 17-20.
[48] Ibid., p. 18.
[49] Ibid., p. 20.
[50] Ibid., p. 20.
[51] Ibid.

[52] Dr Devra Davis, "Science on 5G and Wireless Radiation, "
   (see Internet references in Bibliography)
  Also
  Carla Atherton, "A Danger We Can't See..." (Dr. Ann Gittleman)
[53] Op Cit., Atherton.
[54] Ibid.

[55] Robert F. Kennedy, "Dangers of 5G to Children's Health"
  (see Internet references in Bibliography)

[56] NEXUS, July-Aug 2017, p.8.

[57] https://en.wikipedia.org/wiki/CRISPR_gene_editing

[58] https://en.wikipedia.org/wiki/CRISPR

[59] https://en.wikipedia.org/wiki/Xeno_nucleic_acid

[60] NEXUS magazine, April-May 2019, pp.10-11.

[61] https://en.wikipedia.org/wiki/Hachimoji_DNA

[62] Wikipedia: Mycoplasma laboritorium

[63] https://en.wikipedia.org/wiki/XDNA#Uses

[64] NatGeo Magazine: John Perritano, "Your Genes: 100 Things You Never Knew", Sept. 2019.
[65] Ibid, p. 104.

Endnotes – Chapter 4

[66] Dr. Jacobs, David. *The Threat*. NY: Simon & Schuster, 1998. p 84.
[67] Ibid., p.84.
[68] Ibid., p 87.

[69] NEXUS magazine, Sept-Oct 2018, vol.25, No. 5 "Study Proves without a Doubt that Your Phone is Spying on You." (p. 8)

[70] For a really good overview of rhe Djinn, please see The Vengeful Djinn by
Rosemary Guiley and Philip Imbrogno in Bibliography.

[71] "The Revelations of the Insider" from a 5-day blog contains many higher truths.   Eprint at
http://www.scribd.com/doc/403303/The-Revelations-of-an-Elite-Family-Insider-2005

[72] Ruffin, C. Bernard. *Padre Pio: The True Story*. (Huntington, IN: Our Sunday Visitor
Publishing Division, Inc., 1991), 112-113.
[73] Ibid., 146-157.

[74] Gary D. Chance, "Voice to Skull Devices Defined By US Army as NLW" eprint at
www.hartford-hwp.com/archives/27/a/264.htm. (Military source definition appears to have been
Website: http://call.army.mil/products/thesaur/00016275.htm.)

[75] Rifat, Tim. *Remote Viewing*. (London: Vision Paperbacks, 2001), 218.

[76] Steven R. Corman, "PSYOPS Tech: Voices in your head" pp. 1-3, eprint at:
http://comops.org/journal/2007/12/20/psyops-tech-voices-in-your-head/
with an imbedded link to a video by ABC news on Website:
http://youtube.com/watch?v=6h3KZjysoEo.

[77] https://en.wikipedia.org/wiki/Microwave_auditory_effect

[78] Montalk, two articles:
https://www.bibliotecapleyades.net/ciencia/ciencia_matrix08.htm
Matric Reloaded
and
https://www.bibliotecapleyades.net/mistic/sas08.htm
Why Negative Forces Seem to Respect Freewill

[79] https://en.wikipedia.org/wiki/Neuroscience_of_free_will

[80] https://www.theatlantic.com/magazine/archive/2016/06/theres-no-such-thing-as-free-will/480750/

[81] https://en.wikipedia.org/wiki/Quantum_eraser_experiment

[82] Op Cit, Elvidge, Digital Consciousness, pp. 208-09.
[83] Ibid., p.229-30.

[84] Monroe, Robert. *Ultimate Journey*. 24.
[85] Monroe, Robert, *Far Journeys*. (New York: Random House/Broadway, 2001), 248.
[86] Ibid., 256.
[87] Ibid., 256-257.

Endnotes – Chapter 5

[88] HOTS – **Higher Order Thinking Skills** –used to be taught to teachers to really educate the kids, but in 1995 the HEW (Dept of Education) PTBs got it killed. It was all about Compare/Contrast, Analyze/Synthesize, Induce/Deduce.

[89] Tim Rifat, <u>Remote Viewing</u>, pp, 218-221.

[90] Global Warming **is** happening at a minor rate but every 10,000 years the Earth does a cycle of warming/cooling and the insects and birds survive. What **IS** killing the insects and birds is the proliferation of EMF (**cellphone towers**) that no one wants to talk about. See spurious article at:

https://www.earthfiles.com/2019/09/26/why-are-so-many-earth-insects-dying-2/

Also see Dr Paul LaViolette in *The Earth Under Fire*. He discusses the global cycle as proved by ice cores and he goes into the Younger Dryas issue.

[91] VEG, Ch., 14, pp 451-452 – (with the links.)

[92] https://en.wikipedia.org/wiki/Nanotechnology

[93] Op Cit, Rifat.

[94] Tim Rifat, "Microwave Mind Control…", article at
https://www.bibliotecapleyades.net/sociopolitica/esp_sociopol_mindcon10.htm

[95] https://en.wikipedia.org/wiki/5G

[96] https://en.wikipedia.org/wiki/Extremely_high_frequency

[97] https://en.wikipedia.org/wiki/Cellular_network#Frequency_reuse

[98] Dr. Devra Davis, "Science on 5G and Wireless Radiation", see Bibliography.

[99] Jeremy Naydler, NEXUS magazine, "5G The Big Picture",
July-Aug 2019, pp. 15-23.

[100] Op Cit., Dr Devra Davis.

[101] https://synergist.aiha.org/201908-health-effects-blue-light;
ANSES blue light toxicity:
https://www.anses.fr/en/content/leds-anses%E2%80%99s-recommendations-limiting-exposure-blue-light

[102] 5G Summit Conference: "Critical Disruption of Mitochondria by EMFs" by Jason Bawden Smith.
Also the ANSES French study:
https://www.anses.fr/en/system/files/PRES2019DPA01EN.pdf

In addition, please see the TrueBlueVision website with excellent info on what Blue Light does to the eyes: causes irritation, then **gradual oxidation** of the eyes, and may result in macular degeneration. ... see www.truebluevision.com

The blue light is intense and while the **eyes are designed to process blue light**, your eyes are not used to the intensity combined with UV (right next to blue in the light spectrum).

[103] Op Cit, Dr Devra Davis.
[104] Ibid.
[105] Ibid.

[106] Dr Martin Pall, "5G Risk: The Scientific Perspective". Chapter 2. Also his:
"Explaining the Mechanism of Wireless Harm", Part 1, in the 5G Summit Conference notes.

[107] Op Cit., Davis, , p. 29

[108] Carla Atherton with Dr Louise Gittleman, "A Danger We Can't See…"
See also her book Zapped in Bibliography and on Amazon.

[109] Dr Evan Brand, MD, PhD, "EMF, 5G Cell Towers, Geoengineering and Retroviruses"
(See Bibliography, Internet Sources for more)

[110] Op Cit, Carla Atherton.
[111] Ibid.

[112] See The Body Electric by Robert O Becker, MD

[113] Sources:
(credit: Bing Images: sguardinellogos.blogspot.com): hand
(credit: Bing Images: ottophoto.com) : leaf

[114] Op Cit. Dr Devra Davis

[115] Robert F. Kennedy, "Dangers of 5G to Children's Health."

[116] Op Cit, Dr Devra Davis
[117] Ibid.

[118] https://en.wikipedia.org/wiki/Extremely_high_frequency

[119] Op Cit., Dr Devra Davis

[120] Op Cit., Kennedy.
[121] Op Cit., Dr Devra Davis

[122] https://en.wikipedia.org/wiki/Microwave_auditory_effect#Conspiracy_theories

[123] David Icke, Everything You Need to Know…, p. 625.

[124] Op Cit., Tim Rifat, pp. 18-20, 220-24.
[125] Ibid., 382.

[126] https://www.zmescience.com/space/seti-shuts-down-27042011/

127 https://en.wikipedia.org/wiki/Starfish_Prime

Endnotes – Chapter 6

128 D. Icke, Everything You Need to Know..., pp. 589-90.

129 https://www.digitaltrends.com/cool-tech/us-military-readies-iron-man-style-suit-for-deployment/

130 This was disproven – the soul was scientifically recorded and found to weigh about 21 grams.. see VEG and Dr Duncan McDougall.

131 https://www.nytimes.com/2018/03/19/technology/uber-driverless-fatality.html
and
https://www.theguardian.com/technology/2018/mar/19/uber-self-driving-car-kills-woman-arizona-tempe

132 Op Cit: Ancient Aliens S11 vol 2, The Artificial Human.

133 John Searle in *Wired* magazine, "Why The Future Doesn't Need Us"
https://www.wired.com/2000/04/joy-2/

134 Op Cit., Ancient Aliens DVD.

135 https://en.wikipedia.org/wiki/Chinese_room

136 Kurzweil, The Age of Spiritual Machines, (Kindle location: 3217 and 3125).
137 Ibid., loc: 23:26.

138 https://en.wikipedia.org/wiki/Eugene_Goostman
139 Ibid.

140 Op Cit. Barrat, p. 72.
141 Ibid., p p72-74.
142 Ibid., pp. 74-76.
143 Op Cit., Barrat, p.86.

144 http://www.wayofthefuture.church/ and

https://www.neowin.net/news/anthony-levandowski-is-the-messiah-of-his-new-ai-god

145 https://www.wired.com/story/anthony-levandowski-artificial-intelligence-religion/

[146] https://www.cnet.com/news/the-new-church-of-ai-god-is-even-creepier-than-i-imagined/

[147] http://www.wayofthefuture.church/

[148] Dr Thomas Horn, The Milieu, pp. 98-99.
[149] Ibid.
[150] Ibid., p.102
[151] Ibid.
[152] Ibid.,pp. 102, 111-12.
[153] Ibid., p. 116
[154] Ibid.

[155] Steinmeyer, Jim. *The Book of the Damned; The Collected Works of Charles Fort.*
    (New York: Tarcher/Penguin Group, 2008), 163.
[156] Ibid., 163

[157] Bramley, William. *The Gods of Eden.* 34.

[158] Ibid., pp 283-284.

[159] Dr Paul LaViolette, Earth Under Fire, pp. 162, 191-95.

[160] NEXUS magazine, July-Aug 2017, "Is There Something Wrong with the Germ Theory?",
    Paul Fassa, pp. 16-17.

[161] https://memory-alpha.fandom.com/wiki/The_Offspring_(episode)
    Star Trek TNG 3x16
    Production number: 40273-164
    First aired: 12 March 1990

[162] insight_cyborgs_web_text_1280x720.wdp

Endnotes - Chapter 7

[163] https://en.wikipedia.org/wiki/Tattoo_ink
[164] Ibid.

[165] https://en.wikipedia.org/wiki/Plastic_bottle

[166] A company on the internet sells to the public all kinds of water testing chemicals
    And litmus paper testers with color c harts to show what the % of any dissolved
    solid/chemical is.. eben lead, sodium and heavy metals....
        www.sensafe.com   aka   Industrial Test Systems, Rock Hill, SC

[167] https://en.wikipedia.org/wiki/Hydroxyl_radical#/media/File:HydroxideVsHydroxyl.png

[168] Op Cit, Mindell, p 113.

[169] Dr Batmanghelidj, MD, You're Not Sick, You're Thirsty., Ch 9-11.
    Also by the same doctor: Your Body's Many Cries for Water.

[170] This was proven by mice swimming in a steel tank: 5 mice **without** B5 lasted 20 minutes swimming (the sides were steep and the mice had to swim, they could not get out), and then they put 5 mice in the water **with 2 gm** of Vitamin B5 in each mouse and they lasted almost an hour. (This was in a book by Dr. Paavo Airola that I no longer have.)
I know this works as I have done it daily – I have low Cortisol (a 5-6 range) and it does work to manage stress due to fatigue and low stamina... BUT check with your doctor to see that you do not have an issue with Vitamin B5.

[171] Op Cit., Batmanghelidj,, p. 236.

[172] Dr Earl Mindell, The Vitamin Bible, pp. 349-351.

[173] Dr Batmanghelidj, : Your Body's Many Cries for Water. pp. 159-161.
[174] Ibid., pp 138-39.
[175] Ibid.

[176] https://en.wikipedia.org/wiki/BEMER_therapy
    Also
    Bemergroup.com – shows interesting video of blood flow before and after an 8 minute Bemer mat treatment.

[177] See LLFrench.HTTPS://nulifeventures.com
    The device is patented and FDA approved.

[178] Op Cit Mindell, p. 164
[179] Ibid., p. 365.

[180] The operative factor here was the combination of **Onnit + Higher Mind** – the tinnitus was happening INSIDE the brain – neurons misfiring in the auditory area of the brain, producing a tic-tic-tic...and the broad range of known herbs and medicinal chemicals/vitamins stopped it – the brain needed to be 'fed.' The items in Higher Mind also must have stopped or cleaned up any plaque starting to form (according to my doctor who saw it work).

[181] Wikipedia: Excitotoxicity

[182] Dr. Caroline Leaf, Switch on Your Brain, pp. 67-68, 96, 99.
[183] Ibid., 88, 94.

[184] Scientific American MIND magazine, July/Aug 2015, *How Violent Video Games Really Affect Kids*, p. 42.
[185] Ibid., 44.
[186] Indi., 45
[187] Ibid., 45.

[188] Op Cit, Dr. Leaf, 74.
[189] Dr. Leaf, Switch on Your Brain, p. 13.
[190] Ibid., 14.

[191] Dr. Caroline Myss, <u>Why People Don't Heal and How They Can</u>.

[192] Op Cit, Leaf., pp. 34-35
[193] Ibid.
[194] Ibid.
[195] Ibid.
[196] Ibid., 48.
[197] Ibid., 73-74, 88-89.
[198] Ibid., 93-94

[199] http://www.divinecosmos.com/start-here/articles/334-kozyrev-aether-time-and-torsion
[200] Ibid.
[201] Ibid.
[202] Ibid
[202] Ibid.
[202] Ibid.

[203] Karen Jensen, ND, <u>Three Brains</u>, Chapter 1 goes over the Head Brain, the Heart Brain and the Gut Brain – the body is more wonderfully made than we have been told… the 3 brains must be in communication and synchronized if we are to be fully functioning and healthy.

[204] Brendan B. Murphy, in
       http://blog.world-mysteries.com/science/torsion-the-key-to-theory-of-everything/

[205] Op Cit, . in http://www.divinecosmos.com/start-here/articles/334-kozyrev-aether-time-and-torsion
[206] Ibid.
[207] Ibid.

[208] Consumers Reports, Oct. 2019, "Your Guide to Digital Privacy,"
[209] Ibid., p. 28.

[210] Dr. Sharon Goldberg, MD in "Science about Wireless and 5G" article – see Internet Sources for 5G Summit Conference in Bibliography.

[211] www.shieldyourbody.com article called "The Thermal Effect of EMF"
       https://www.shieldyourbody.com/2014/07/thermal-effect-emf
       SYB also sells a protective shield for your pocket so you can carry your cellphone on you

[212] http://www.greenpeace.org/usa/campaigns/oceans/follow-the-journey/trashing-our-oceans

[213] http://search.yahoo.com/search?ei=UTF-8&fr=att-portal&p=oxygen+levels&rs=0&fr2=rs-bottom

Endnotes – Chapter 8

[214] Joseph Farrell, <u>Covert Wars & the Clash of Civilizations</u>., pp. 133-136.

[215] Farrell, Joseph. <u>Reich of the Black Sun</u>, Ch. 16 (& 291-292).

[216] Joseph Farrell, <u>Saucers, Swastikas and Psyops</u>, 216-224.
The late **Dr. Ben Rich**, head of the Lockheed Skunk Works (Think: Area 51 and S4), shared a number of things in September 1992 during a presentation at the Air Force Museum in Dayton, Ohio. (This is all covered in VEG, Ch.. 4)

[217] NEXUS magazine, Sept-Oct 2018, Vol. 25 No. 5, p. 8. *America Prepares for a Space Force.*

[218] Op Cit., Farrell, <u>Covert Wars,</u> p. 133.
[219] Ibid., pp 170-71.
[220] Ibid., p. 172.
[221] Ibid., p. 172.
[222] Ibid., p. 173
[223] Ibid., p. 175.
[224] Ibid., p. 175-176.

[225] The Nexus Seven, "From the 33 Arks…"., 15.20-30.
[226] Ibid., 24.7-8.

[227] Monroe, Robert, <u>Far Journeys</u>. p. 248.

Endnotes – Chapter 9

[228] https://en.wikipedia.org/wiki/Genetics_of_aggression
Involving the MAO-A and Serotonin 5HT factors
See also: https://www.crimetraveller.org/2016/04/the-warrior-gene/
Found in Men only the MAOA-L variant predisposes to violence..
(found on the X Chromosome.)
See also https://www.bbc.com/news/science-environment-29760212
Two genes have been found associated with violent offenders:
The two genes associated with violent repeat offenders were the
MAOA-L gene and a variant of Cadherin 13 (CDH13).
..
[229] 5G Summit, Trevor Marshall, (p. 438) and Steven Whybrow (p. 546).
See Bibliography (5G Super Grid) – in 825 pp book.

[230] https://en.wikipedia.org/wiki/Fifth_column

[231] 5G Summit, Op Cit., pp. 546 and 438.

[232] Jim Marrs, <u>Alien Agenda</u>, pp. 137-142.

[233] 5G Summit, Op Cit., pp. 546 and 438.

Endnotes – Appendix A

[234] Rasha. *Oneness.* (Santa Fe, NM: Earthstar Press, 2003), 223, 354.
[235] Ibid., 195-196.
[236] Ibid., 195.
[237] Ibid., 195-196.

Endnotes -- Appendix B

[238] Steinmeyer, Jim. *The Book of the Damned; The Collected Works of Charles Fort.*
    (New York: Tarcher/Penguin Group, 2008), 381.
[239] Ibid., 381-82.
[240] Ibid., with ref to
    http://www.betawired.com/science-fiction-like-electron-shield-found-around-earth/1418358/
        may also see:
    http://lasp.colorado.edu/home/?post_type=mag-seminars&p=16204

[241] *Encyclopedia of World Mythology*, Gen, Ed, Arthur Cotterell, p. 251-252

[242] *African Mythology*, Geoffrey Parrinder, pp. 37-40

[243] http://www.pbs.org/newshour/rundown/study-invisible-shield-space-protects-earth-killer-electrons/
    And
https://www.sciencedaily.com/releases/2014/11/141126133829.htm

[244] Ibid., -- they just told you that the barrier is impenetrable…

[245] Gregg Prescott article in *In5D Guest* and Waking Times:
    Earth's Quarantine Force Field Discovered by NASA? , Dec. 3, 2014.

[246] https://www.sciencedaily.com/releases/2014/11/141126133829.htm
        Also:
    An impenetrable barrier to ultrarelativistic electrons in the Van Allen radiation belts. *Nature*, 2014; 515 (7528): 531 DOI: 10.1038/nature13956

[247] Robert Monroe, Journeys Out of the Body, pp. 253-259

Endnotes -- Appendix C

[248] "Matrix Agents: Profiles and Analysis (Parts I & II)" I. eprint at:
http://montalk.net/matrix/62/matrix-agents-profiles-and-analysis-part-i

[249] Joseph Macchio, "The Orthodox Suppression of Original Christianity", Ch VII, Great Schools, Valentinus.
[250] Ibid., Ch. VII, Great Schools, Doc. of 3 Natures.

[251] Bramley, William. *The Gods of Eden.* (New York: HarperCollins/Avon, 1993), 176-177

[252] Sitchin, Zechariah, *The Cosmic Code.* (New York: HarperCollins, 2007), 44, 58.

[253] Sitchin, Zechariah, *The Cosmic Code.* (New York: HarperCollins, 2007), 44, 58
[254] Ibid., 178.

[255] Bibliotecapleyades website product of Jose Ingenieros, Book III, p. 108.
reprint at: http://www.bibliotecapleyades.net/esp_autor_mouravieff.htm

[256] "Organic Portals Theory: Sources", compendium: Book II of <u>Gnosis</u>.
    eprint at: http://www.montalk.net/opsources.pdf
[257] Ibid., II, 7.
[258] Ibid., II, 49.
[259] Ibid., (II, 8)
[260] Ibid., III, 8.
[261] Ibid., III, 109

[262] Cannon, Convoluted Universe Book V, pp257-269 (Chapter 18).
[263] Ibid., p. 268.
[264] Ibid., 133.
[265] Ibid., 129-134.
[266] Ibid., 136

[267] Bibliotecapleyades website product of Jose Ingenieros, Book III, p. 112-115.
reprint at: http://www.bibliotecapleyades.net/esp_autor_mouravieff.htm

[268] W.H. Church, *Edgar Cayce's Story of the Soul*, pp. 137-39.

[269] Bibliotecapleyades website product of Jose Ingenieros, Book III, p. 112-115.
reprint at: http://www.bibliotecapleyades.net/esp_autor_mouravieff.htm
[270] Ibid.

Endnotes – Appendix D

[271] Ring PhD, Kenneth. *Lessons from the Light.* NH: Moment Point Press, 1998.
    pp. 173-184.
[272] Ibid., pp173-175.

[273] Moody, Raymond A., Jr., MD. *Life After Life.* NY: HarperCollins, 2001. p. 13

Endnotes – Appendix E

[274]

https://search.yahoo.com/search;_ylt=Am6bkPXiGAevy1Fbpfp_Z0mbvZx4?fr=yfp-t-901-s&toggle=1&cop=mss&ei=UTF-8&p=hidden%20history%20of%20the%20human%20race

[275] Http://www.telegraph.co.uk/news/main.jhtml?xml=/news/2004/03/29/ndrag29.xml&sSheet=/news/2004/03/29/ixhome.html

[276] https://ufoholic.com/forbidden-history/peruvian-artist-erects-statue-reptilian-humanoid-capital-friendship/
This pertains to Morrop, Lang Ñam and Strombus refrences and pix.

[277] https://www.the-alien-project.com/en/mummified-heads/

[278] The Paracas Skulls Debunked... YouTube by Justin Gabriel.

[279] http://www.wisdom-square.com/alien-contact.html

[280] https://en.wikipedia.org/wiki/Triple-stranded_DNA
TFOs can inactivate a gene or help to induce mutations:
See also: https://en.wikipedia.org/wiki/Triple_helix#DNA

[281] DISCOVER Magazine Special Issue, July/August 2012, Veronique Greenwood, pp. 29-31.
[282] Ibid., p.30.

Endnotes – Appendix F

[283] Boulay, R.A.,*Flying Serpents and Dragons*. Rev Ed. (San Diego, CA: The Book Tree, 1999), p. 117.

[284] Kenneth Woodward in Newsweek article, "In the Beginning There Were the Holy Books", Feb. 11, 2002, pp 51-57.

[285] Joseph Macchio, "The Orthodox Suppression of Original Christianity", Chapter III (2004). Eprints at: http://essenes.net/conspireindex.html   and http://essenes.net/new/subteachings.html

[286] (Credit: http://en.wikipedia.org/wiki/Star_Trek/Cardassians)

[287] Boulay, R.A., *Flying Serpents and Dragons*, 16.

[288] Fig sap can cause dermatitis. http://www.katu.com/news/local/62682802.html

[289] Sitchin, Zechariah, *12th Planet,* p. 105.

[290] Farrell, Joseph P. *The Cosmic War.* 142.

[291] Boulay, R.A. *Flying Serpents and Dragons*. (Revised Edition. San Diego, CA: The Book Tree, 1999), 115-117.
[292] Ibid.

[293] http://www.bibliotecapleyades.net/sumer_anunnaki/reptiles/reptiles40.htm)

[294] Lessin PhD, Sasha, *Anunnaki Gods No More.*, 83, 106.

[295] Op Cit, Boulay, 11.

[296] Pinkham, Mark Amaru. *The Return of the Serpents of Wisdom.* (Kempton, IL: AU Press, 1997), 44.

[297] Turner, Patricia & Charles Russell Coulter. *Dictionary of Ancient Deities.* (NY, NY: Oxford University Press, 2001), 355.

[298] Icke, David, *Children of the Matrix.* (Isle of Wight, UK: David Icke Books Ltd., 2001), 91. Also quoted in Boulay, *Flying Serpents & Dragons*, p. 61.

[299] Jim Marrs, *Alien Agenda*, pp. 11-12.
[300] Ibid.

Endnotes – Appendix G

[301] https://en.wikipedia.org/wiki/Ionizing_radiation
[302] Ibid.

[303] NEXUS Magazine, May-June 2019 (vol 26 no. 3), "Nagalese and the Cancer Kill Switch" on pp 20.... Dr. Greg Fredericks, MA, ND.

[304] https://www.safespaceprotection.com/emf-health-risks/emf-health-effects/

Other Books by the Author ...

The Anunnaki Legacy

Transformation, Kundalini, Hyperborea, Göbekli Tepe, Hopi, Maya,
Gods, Goddesses & the Divine Feminine, Skygods & Creation,
The Flood, Sacred Trees, Alchemy and Serpent Wisdom.

by TJ Hegland

Virtual Earth Graduate

TJ Hegland

Reflecting: (inserts left to right, all covered in the book):

Physics (atom), Genetics (DNA), Dragon, Ubaid Statue of Anunnaki, UFO (TR3B), and *Castillo* at Mayan Chichen Itzá.

Center: Earth with Soul/Earth Graduate

Background: Electromagnetic Dark Energy Matrix

The Transformation of Man

By T J Hegland

Spirituality, Reincarnation & the Interlife, Consciousness, The Matrix, Zero Point Energy, Dark Matter, DNA & Healing Energy, Timelines, Quantum Biocomputer, Greys & Hybrids, and Abduction & the Near Death Experience as Transformation.

The Science in Metaphysics

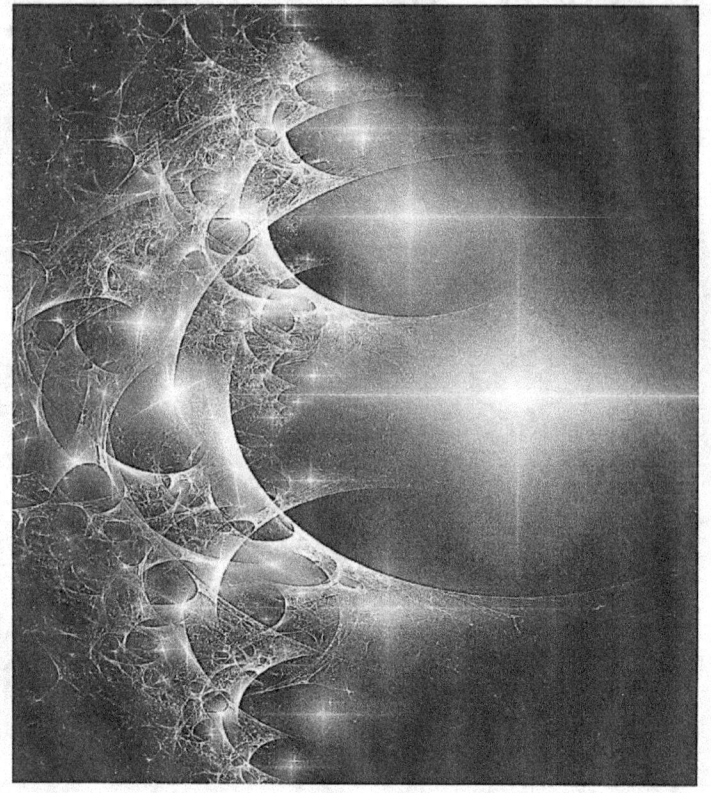

by **TJ Hegland**

Proof from Science that the Principles of Metaphysics
Are Real and Why They Work.

The Earth Warrior
illustrated

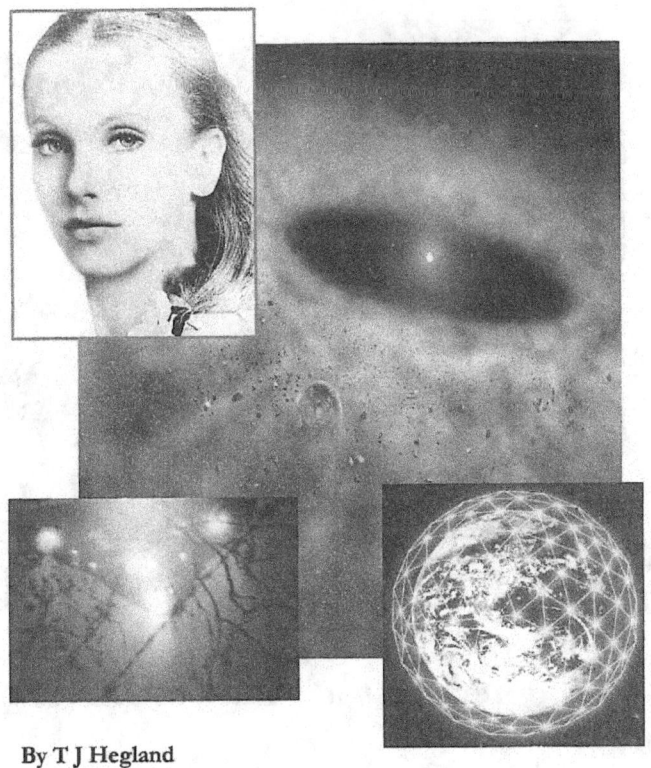

By T J Hegland

A romp thru alternative history with Nazi UFOs, Antarctic bases, the
Moon and Mars, a Breakaway Civilization, the Orion Empire, and a
Galactic Super Wave. Maria Orsič and her team of Aldebaran women
warriors must resolve the Earth dilemma or Man disappears.
Could this be the real history of Earth?

This is a Novel based on real events...
It was also scripted for a 10-part TV series.

Quantum Earth Simulation

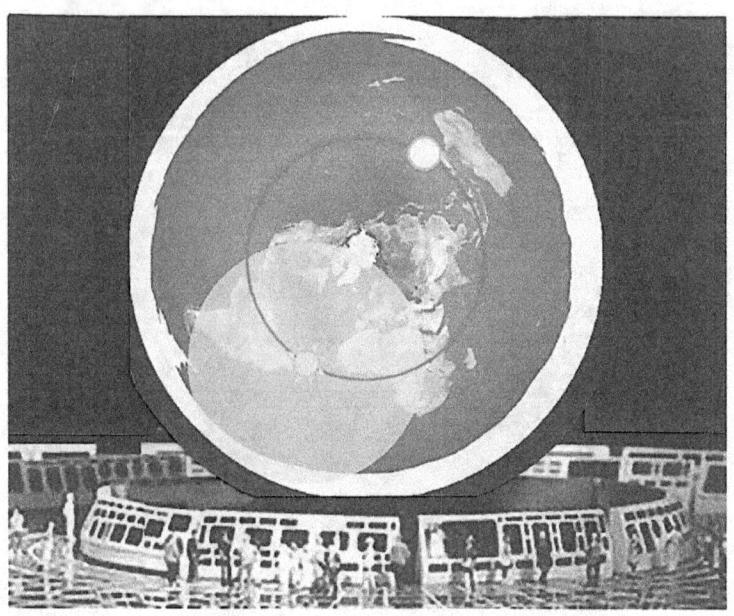

Updated for the Flat Earth Nexus

by **T J Hegland**

Simulation, Virtual Reality, Matrix and Holodecks, Holograms and Quantum Physics, Vision & Hypnosis, Timelines, Eras, Programming the Simulation, Consciousness, Reality Fields, Lucid Living, and many Anomalies.

The Great Earth Puzzle

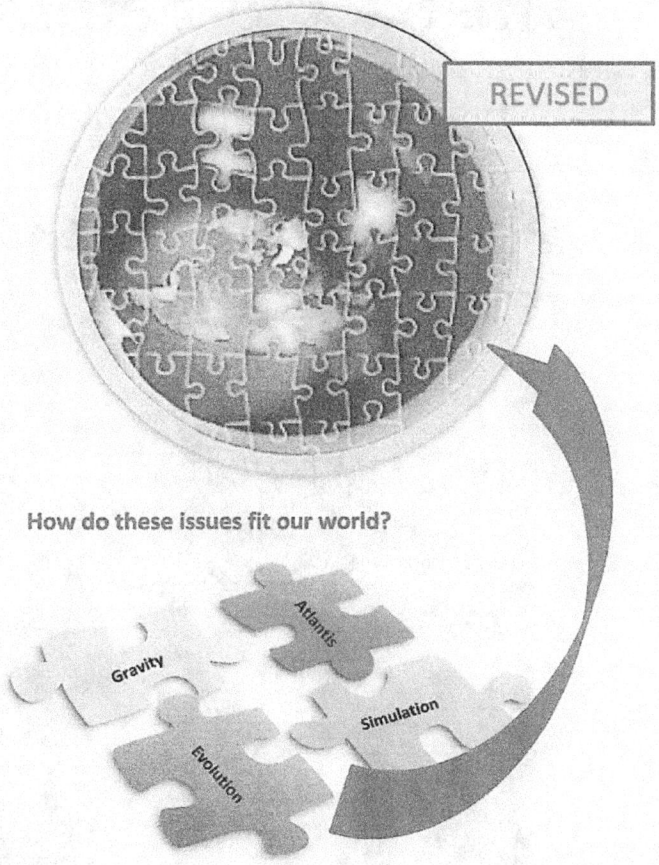

REVISED

How do these issues fit our world?

Gravity

Atlantis

Evolution

Simulation

This book summarizes what the Earth Realm really is and whether these issues, and more, fit our reality. Are we really living on the planet we think we are?

By TJ Hegland

The Sacred Feminine
and
The Red Pill of History

TJ Hegland

The Goddess Energy and Today's Woman

Mary Magdalene was not a prostitute, she was a Temple Priestess initiating people in Kundalini Awakening, and she was from a rich family in Egypt, so she bankrolled Jesus' ministry, then went to South France and continued to teach what He had taught.

Virtual Disneyland of the Gods

Astral Spooks, Programmers, and Hybrids

by TJ Hegland

Missing People, Abductions, Soulless, Anunnaki Remnant, Black-eyed Kids, Invisibility, Simulation, Dyson Sphere & Noise from the Edge of the Universe, Great Wall of China & Nagas, AI Singularity, Mandela Effect, and The Change.

NOTES

www.ingramcontent.com/pod-product-compliance
Lightning Source LLC
Chambersburg PA
CBHW081506220526
45467CB00010B/2809